STORM WATER

STORM WATER

Editor

Clemens Vonnegut

scitus
academics

Storm Water

Edited by **Clemens Vonnegut**

This Book is Printed on Acid-Free Paper. ♾

ISBN: 9781681179711

Library of Congress Control Number: 2017949109

SCITUS Academics LLC,

616, Corporate Way, Suite 2, 4766,

Valley Cottage, NY 10989

www.scitusacademics.com

Printed in United States of America.

Preface

Storm runoff is considered one of important water resources for urban areas. It can be applied in rural areas (e.g. to harvest precipitation water), but is essential in urban areas where run-off cannot infiltrate because the surfaces are impermeable. Traditional storm water management was mainly to drain high peak flows away. Unfortunately, this only dislocates high water loads. Modern approaches aim to rebuild the natural water cycle, i.e. to store runoff water for a certain time, to recharge ground water and to use the collected water for irrigation or household supply. Storm water runoff from urban catchments, including runoff from buildings, roads, and paved areas, is considered a major source of environmental pollution. Best management practices are widely applied to reduce their quantity and to improve the quality of storm water runoff entering receiving waters. A variety of centralized and decentralized facilities have been developed over the past two decades to treat urban storm water runoff and there is now a great deal of focus on the technical performance and economic viability of these systems. New technologies for storm water development in the sense of Integrated Urban Water Management (IUWM) have been developed since the 1970s and include detention and retention ponds, permeable surfaces, infiltration trenches, surface and surbsurface groundwater recharge, and other sources control measures.

This volume aims approaches and strategies to protect, maintain and enhance the public health, safety, and general welfare by establishing minimum requirements and procedures to reduce the adverse impacts associated with increased storm water runoff. It gives readers an overview of the environmental effects of urban storm water pollution and how both centralized and decentralized facilities systems are being implemented to effectively manage and treat polluted storm water runoff to protect the water environment. With the latest observations in climate change, the uncertainty in the design flood has imposed unprecedented challenges in flood mitigation designs. The flexibility in freeboards and easements need to be refined in order to accommodate the changes in extreme rainfall events. This book presents a summary of the Green approach in storm water management and LID designs as the engineering measures to preserve the watershed regime.

Table of Contents

CHAPTER 1

Stormwater Management: Calculation of Traffic Area Runoff Loads and Traffic Related Emissions

Maximilian Huber and Brigitte Helmreich*

Chair of Urban Water Systems Engineering, Technical University of Munich, Am Coulombwall 3, 85748 Garching, Germany

ABSTRACT

Metals such as antimony, cadmium, chromium, copper, lead, nickel, and zinc can be highly relevant pollutants in stormwater runoff from traffic areas because of their occurrence, toxicity, and non-degradability. Long-term measurements of their concentrations, the corresponding water volumes, the catchment areas, and the traffic volumes can be used to calculate specific emission loads and annual runoff loads that are necessary for mass balances. In the literature, the annual runoff loads are often specified by a distinct catchment area (e.g., g/ha). These loads were summarized and discussed in this paper for all seven metals and three types of traffic areas (highways, parking lots, and roads; 45 sites). For example, the calculated median annual runoff loads of all sites are 355 g/ha for copper, 110 g/ha for lead (only data of the 21st century), and 1960 g/ha for zinc. In addition, historical trends, annual variations, and site-specific factors were evaluated for the runoff loads. For Germany, mass balances of traffic related emissions and annual heavy metal runoff loads from highways and total traffic areas were calculated. The influences on the mass fluxes of the heavy metal emissions and the runoff pollution were discussed. However, a statistical analysis of the annual traffic related metal fluxes, in particular for different traffic area categories and land uses, is currently not possible because of a lack of monitoring data.

Keywords: de-icing salt; Germany; heavy metal source; highway; mass balance; parking lot; pollution; road; site-specific factors; urban

1. INTRODUCTION

In most cases, four types of traffic related loads are used for calculating mass balances and fluxes [1]. Specific substance emission loads are presented in the literature for each vehicle per kilometer traveled (mg/(vehicle·km)) or for all vehicle kilometers traveled in a distinct catchment area during a specific period (mg/(ha·km)). Annual loads can be specified by the road length (g/km) or by the traffic area (g/ha). For the description of the pollution of a traffic area, a differentiation is necessary between total traffic related emission loads, atmospheric deposition loads (wet and/or dry), and runoff loads from traffic areas. For the calculation of runoff loads, long-term measurements of runoff concentrations and water volumes of a monitoring site with a distinct catchment area must be recorded, analyzed, and published. Although several researchers have evaluated the runoff concentrations of different traffic areas [2,3], only a part of these monitoring programs included the calculation of annual loads. This is based on the fact that during most monitoring programs, only a limited number of samples is collected and most objectives do not require the calculation of loads (e.g., manual (grab) samples for stormwater permit applications). Nevertheless, annual heavy metal loads normalized per hectare of impervious catchment area are necessary for mass balances and their determination is subsequently important for several purposes (e.g., modeling of stormwater quality and transportation into receiving water, calculation of metal masses removed by treatment systems, determination of pollution loads for laboratory test methods, calculation of regional or national mass balances, or the assessment of the environmental impact and benefit of different techniques to remove toxic substances in relation to their costs).

For stormwater management, determining traffic area runoff loads for different traffic area categories is necessary to design effective stormwater management practices [4]. Therefore, detailed information on differentiated traffic related sources (emissions because of leaded gasoline, tire wear, brake lining wear, roadway abrasion, weights for tire balance, guardrails, lampposts/signs, and de-icing salts) and areas (traffic area categories) is needed to determine their contributions to the total pollution of environmental compartments and to develop cost efficient mitigation strategies. Previously, urban area pollutant categories were often confined to broader land use categories such as commercial, industrial, or residential [4]. By this approach, the different pollution loads of traffic areas cannot be considered although all road surfaces represent approximately 10%–15% of the total urban area [5,6] and in commercial and industrial areas, parking lots can constitute up to 46% of the total area [5]. Thus, it is essential to consider the runoff loads more differentiated because of the large percentages of different traffic areas to the total urban area. However, most values for specific traffic area categories were available and summarized for highways and less were published for parking lots and other roads [7-9]. Because most values and ranges of pollutant loads published in books, reports, regulations, and standards are only based on a few studies, no reliable values are currently available for the main traffic area categories and a relation of the loads to climatic factors was also not evaluated previously. In addition, most published data summaries of runoff loads are based on the last century and they do subsequently not represent the latest trends.

In traffic area runoff, the occurrence of substances depends on several processes. Dustfalls and dry deposition during periods without rain entrain contaminants and remove them from the atmosphere [6,10]. These mechanisms are especially relevant for urban areas [11]. The processes during rain events are wet deposition (removal of additional substances from the atmosphere) and wash-off of previously deposited pollutants. Concerning the wash-off, Racin et al. [12] concluded that splashing and washing of pollutants from vehicles is more important than the wash-off of pollutants accumulated on road surfaces. Both the wash-off and the substances in the atmosphere can be linked to traffic related sources. For most heavy metals, the wear of brakes and tires are relevant sources. Moreover, braking, acceleration, and steering activities lead to increased abrasion of tires, higher use of brake linings, and increased automotive exhaust gas emissions [13,14]. The corrosion and subsequent dissolution of zinc (Zn) from the surface of galvanized elements during rain events is also a relevant source [15]. Further traffic related sources are the use of catalytic converters in vehicles [16], road maintenance with de-icing salts [17], road wear, and drip losses. All of these traffic related sources emit heavy metals, hydrocarbons, and further substances and subsequently lead to a pollution of runoff from traffic areas. Thus, the substances analyzed in traffic area runoff waters include solids, organic parameters, heavy metals, and compounds of de-icing salts [18,19]. Metals such as cadmium (Cd), chromium (Cr), copper (Cu), nickel (Ni), lead (Pb), antimony (Sb), and Zn are crucial pollutants because of their toxicity, non-degradability, and their increase in all environmental compartments in consequence of their widespread industrial use [20,21]. Among these parameters, the three heavy metals Cu, Pb, and Zn were mostly measured in the highest concentrations in runoff from traffic areas [2]. Because of the relevant heavy metal concentrations, a treatment of the runoff water is often necessary to prevent negative effects on surface water, groundwater, and aquatic biota [22-24]. These stormwater treatment systems can be designed as a part of existing urban infrastructure for stormwater management [25].

The focus of this paper was set on the metals in traffic area runoff and their traffic related emissions. The hypothesis of this study was that a distinct calculation of mass balances for traffic related metals (emissions and runoff loads) of all types of traffic areas (i.e., highways, roads, and parking lots) is possible as a result of a literature study. The objectives of this paper were as follows: to summarize the distributions of annual metal loads of runoff from different types of traffic areas, to identify relevant trends and factors, to update and expand existing mass balances (heavy metal runoff loads and emissions) for Germany, and to determine the mass fluxes of traffic related metals.

2. MATERIALS AND METHODS

The annual loads summarized in this paper were selected from a database of German and international studies with descriptions of the runoff from traffic areas. The database includes information on the monitored traffic area (categories: bridges, highways, parking lots, and roads), the substance concentrations and loads, and all influencing factors reported by the authors (e.g., land use characteristics, vegetation, topography, road design, operational characteristics, climatic factors, type of sampling, sampling strategy, sample preparation, sample analysis, and calculation methods). In addition to peer-reviewed journal papers, reports, books, and non-reviewed journal articles that presented data of metals in traffic area runoff were considered after validating the published results (i.e., to fulfill requirements such as a detailed description of the applied methods, the performance of quality control measures, and the reliability of the monitoring setup). The metal runoff concentrations and site-specific and method-specific (i.e., sample collection, preparation, and analysis) factors influencing the results were presented in another review [2]. From this database with more than 300 monitoring sites, this review includes all studies that presented traffic area runoff loads for the metals Cd, Cr, Cu, Ni, Pb, Sb, and Zn. The values were always reported as total (recoverable) metals that were measured after an appropriate digestion (i.e., strong acid or aqua regia digestion; sometimes combined with an autoclave or microwave). The sample preparation and analysis conducted by the selected studies had a low influence on the results because of quality assurance/quality control procedures [2]. All studies used automatic composite samplers to obtain a large number of samples for each rain event (active sampling) or a mixed sample of a part of the complete runoff (passive sampling). The highest uncertainties regarding all selected long-term monitoring programs are related to the handling of the raw data (e.g., time interval of the data logging and the link between flow-/rainfall-measurements and measured runoff concentrations) and the calculation methods for the annual loads (e.g., values below the detection limit or missing values because of a malfunction of a device). All monitored traffic areas were categorized in one of the three categories highway (H), parking lot (P), and road (R). Sites with reported runoff loads that are not completely linked to impervious traffic areas were categorized as special (S). Each monitoring site got an identification number (ID) that consists of one capital letter according to the category (H, P, R, or S) and two consecutive numbers, e.g., H01. These IDs were used in the Tables and Section 3.

The characteristics of the sites used for this review are summarized in Table 1. Table 1 includes data about the location (country), the period of sampling, the land use, special fixed site-specific factors, the road maintenance (i.e., winter services and sweeping), the average annual daily traffic (AADT), the discharge area of the monitoring site (i.e., the catchment area connected to the sampler), and the annual rainfall depth. The runoff loads of the metals are presented in Table 2. This novel evaluation of different types of traffic areas (H, R, and P) only used annual metal loads per hectare of catchment area (g/ha) that were calculated by the respective authors of the selected studies from runoff concentrations of impervious surfaces without a subsequent (pre-)treatment (i.e., some sites presented in [26–30] are summarized in the category S). Therefore, a conversion of units was sometimes performed (lb/acre, mg/m^2, or kg/ha in g/ha). In addition, if all required information was available, annual metal loads presented as g/km were recalculated to g/ha. A calculation of loads from runoff concentrations and the corresponding water volumes was not performed because of climatic site-specific factors (i.e., variable at each site) and several uncertainties (e.g., missing data about the percentage of analyzed runoff volumes to total volumes, the percentage of collected rain events to total rain events, the representativeness of the analyzed rain events in terms of intensities and durations, and the ambiguity between rainfall and runoff volumes).

The software package SPSS 22 (IBM) was used for statistical analysis and plotting. For box and whisker plots, the bottom and top of each box are the first and third quartiles and the band inside the box is the median. The whiskers represent 1.5 times the interquartile range (IQR). Outliers (>1.5 times IQR) are marked as small circles and extreme values (>3.0 times IQR) as stars. To analyze historical trends, the period of the monitoring program was used instead of the publication date. Because of the non-normal distribution of most parameters, the non-parametric Spearmen rank-order correlation was subsequently used for the correlation analysis following the example of Mosley and Peake [31]. For the correlation analysis, the sites H27, H28, and R02 were split up into two, four, and nine independent data sets, respectively, to consider the heavy metal loads and annual rainfall depths for each year. For H05, H06, and H08–H13, no values for the annual rainfall depths were used for the statistical analysis because of missing mean values.

For the calculations of median values, mean values, and standard deviations (SD) in Section 3.3, some data were excluded because of special site-specific factors (cf. Section 3.2.1). To avoid a bias because of the historical trends of the Pb data (cf. Section 3.2.2), the values of the 20th century were excluded for the discussion of recent loads. All of these exceptions are directly specified at the presentation and discussion of the particular results.

Table 1. Characteristics of traffic area sites with heavy metal loads reported in the literature.

ID	Literature	Location *	Period of Sampling	Study Site Characteristics **	AADT (veh./d)	Discharge Area (m^2)	Rainfall (mm/yr)
H01	[32]	GBR	1973–1974	urban, B, D, S	57,600	36,000	533
H02	[1]	CHE	1976–1977	non-urban, D	72,000	54,300	1132
H03	[33]	FRA	1978–1979	–	13,600	–	–
H04	[26,27]	DEU	1978	non-urban, D, nS	41,000	13,000	792
H05	[34]	USA	1978–1981	urban	53,000	4937	810–1140
H06	[34]	USA	1978–1981	urban	42,000	401	810–1140
H07	[26,27]	DEU	1979	non-urban, D, nS	47,000	25,200	737
H08	[34]	USA	1979–1981	non-urban	8600	1133	1020–2290
H09	[34]	USA	1979–1981	non-urban	7700	728	1520–2540
H10	[34]	USA	1979–1981	non-urban	7300	1133	1780–2540
H11	[34]	USA	1979–1981	non-urban	2000	5059	178–381
H12	[34]	USA	1979–1981	urban, B, Zn smelter	17,300	890	432–711
H13	[34]	USA	1979–1981	non-urban	2500	1012	254–483
H14	[35]	FRA	1980–1982	non-urban	5500	1470	801
H15	[35]	FRA	1980–1982	non-urban	7000	–	620
H16	[36]	FRA	1993–1994	non-urban	30,000	13,000	–
H17	[30]	USA	1993–1995	urban, B	58,150	5341	1160
H18	[30]	USA	1994–1995	urban, B	8780	526	1670
H19	[37]	USA	1995–1996	urban	41,000	1670	945
H20	[37]	USA	1995–1996	urban	51,000	4940	945
H21	[28]	USA	1995–1996	urban, B, D	25,000	1497	–
H22	[38]	FRA	1995–1996	urban, B, D, S	12,000	3200	656
H23	[37]	USA	1995–1997	urban	120,000	1880	945
H24	[39]	DEU	1998–2000	urban, D	>50,000	75,800	774
H25	[39]	DEU	1998–2000	urban, D	>50,000	51,300	774
H26	[40]	CHE	1999–2000	non-urban	60,000	75,000	1050
H27	[41]	CHE	2006–2007	urban, D, S	59,000	25,000	876–1184

H28	[41]	CHE	2006–2009	non-urban, D, S	74,000	20,000	477–726
H29	[42]	DEU	2011–2012	non-urban, D	85,600	27.5	538
H30	[42]	DEU	2011–2012	urban, D	66,200	28.0	471
H31	[42]	DEU	2011–2012	non-urban, D	45,000	27.3	454
H32	[43]	CHE	2012–2013	urban, D, S	39,000	42,000	1147
R01	[44]	DEU	1988–1989	urban	18,129	162.2	609
R02	[45]	DEU	1996–2005	urban, D	6800	17.3	334–863
R03	[14]	CHE	2002–2004	non-urban	17,000	1500	1000
R04	[14]	CHE	2002–2004	urban	17,000	14.1	1000
R05	[46]	BRA	2002–2004	urban	9000	1300	–
R06	[47]	DEU	2006–2007	urban, D, S	57,000	100	739
R07	[48]	DEU	2014–2015	non-urban, D	20,600	9870	1200
R08	[48]	DEU	2015	urban, nD	9500	1600	800
P01	[29]	USA	1998–1999	urban	–	1050	1000
P02	[29]	USA	1998–1999	urban	–	1050	1000
P03	[49]	DEU	2001	non-urban, D, S	–	10,000	635
P04	[49]	DEU	2001	non-urban, D, S	–	5000	635
P05	[49]	DEU	2001	non-urban, D, S	–	17,700	769

Notes: * BRA (Brazil), CHE (Switzerland), DEU (Germany), FRA (France), GBR (Great Britain), and USA (United States of America); ** B = Bridge, D = De-icing salt application, nD = no De-icing salt application, nS = no Sweeping, S = Sweeping reported in the literature.

3. RESULTS AND DISCUSSION

3.1. Annual Metal Runoff Loads for Different Traffic Areas

A summary of annual total metal loads of 45 sites is presented in Table 2 for all three traffic area categories (H, R, and P) and the six most relevant heavy metals (Cd, Cr, Cu, Ni, Pb, and Zn). Most of the values were published for highway sites (H, n = 32) with fewer values for other roads (R, n = 8) and parking lots (P, n = 5). The traffic area runoff loads of Cu, Pb, and Zn were mostly measured and published by other researchers. Less data are available for Cd, Cr, Ni, and Sb For Sb, the annual average runoff load was only published for one urban road (16.8 g/ha at R02; [45]) and three highway sites (11.2 g/ha at H19, 5.6 g/ha at H20, and 17.9 g/ha at H23; [37]).

The most concentrated metal was Zn, followed by Pb, Cu, Ni, Cr, Cd, and Sb. The SDs between all monitoring results are quite high for all heavy metals summarized in Table 2: 12.9 g/ha Cd, 291 g/ha Cr, 587 g/ha Cu, 540 g/ha Ni, 3180 g/ha Pb, and 3330 g/ha Zn. In addition, the runoff loads of one to four sites are calculated as extreme values for most metals (Figure 1). Compared with the other heavy metals, the variations of the Pb and Zn data (e.g., cf. the SDs above and the IQRs presented in Figure 1) are the highest. For Pb, it is linked to the historical trends (cf. Section 3.2.2). In contrast, the contributions of different Zn sources to the runoff loads are highly variable between each site (e.g., brake linings, galvanized bridge parts, galvanized car parts, guardrails, lamp-posts, motor oil, tires, safety fences, and signs; [38,50,51]). Thus, these high loads are based on different fixed site-specific factors (cf. Section 3.2.1), operational characteristics (cf. Section 3.2.2) but can also be linked to climatic (cf. Section 3.2.3) and method-specific influences [2].

Table 2. Annual total heavy metal loads in runoff from traffic areas. The characteristics of all traffic area sites are described in Table 1.

ID	Literature	Cd Load (g/ha)	Cr Load (g/ha)	Cu Load (g/ha)	Ni Load (g/ha)	Pb Load (g/ha)	Zn Load (g/ha)
H01	[32]	–	972	3780	1000	12,900	19,000
H02	[1]	25	–	350	–	2500	1900
H03	[33]	–	–	–	–	1200	2300
H04	[26,27]	37	62	621	–	1330	2330
H05	[34]	–	–	223	–	4420	2320
H06	[34]	–	–	732	–	15,920	4210
H07	[26,27]	29	100	544	–	1160	2890
H08	[34]	–	–	34	–	130	220
H09	[34]	–	–	361	–	1670	2010
H10	[34]	–	–	565	–	5480	2560
H11	[34]	–	–	30	–	80	470
H12	[34]	–	–	118	–	690	10,400
H13	[34]	–	–	65	–	320	390
H14	[35]	–	–	–	–	410	1170
H15	[35]	–	–	–	–	820	1410
H16	[36]	–	–	–	–	140	1390
H17	[30]	–	–	230	–	470	1300
H18	[30]	–	–	80	–	200	450
H19	[37]	5.6	50.4	235	–	112	1300
H20	[37]	4.5	40.4	269	–	112	874
H21	[28]	30	90	220	90	200	–
H22	[38]	8.1	–	249	–	567	2080
H23	[37]	19.1	381	628	–	425	3230
H24	[39]	–	–	440	–	217	1960
H25	[39]	–	–	480	–	250	2120
H26	[40]	30	–	1070	–	560	1990
H27	[41]	–	–	440	–	80	1190
H28	[41]	–	–	359	–	55	1270
H29	[42]	1.4	–	805	–	88	2440
H30	[42]	0.6	–	241	–	44	878
H31	[42]	0.8	–	350	–	72	834
H32	[43]	–	–	750	–	–	3930
R01	[44]	–	–	500	–	1800	10,600
R02	[45]	1.7	41	225	41	83	2510
R03	[14]	4.0	119	471	–	179	3230
R04	[14]	3.2	75.4	398	–	170	2100
R05	[46]	8.0	–	310	–	500	–
R06	[47]	–	–	970	–	210	4820
R07	[48]	–	–	600	–	–	1300
R08	[48]	–	–	184	–	–	1060
P01	[29]	–	–	42	–	18	174
P02	[29]	–	–	33	–	17	147
P03	[49]	–	–	400	–	150	2400
P04	[49]	–	–	250	–	110	1000
P05	[49]	–	–	500	–	200	2000

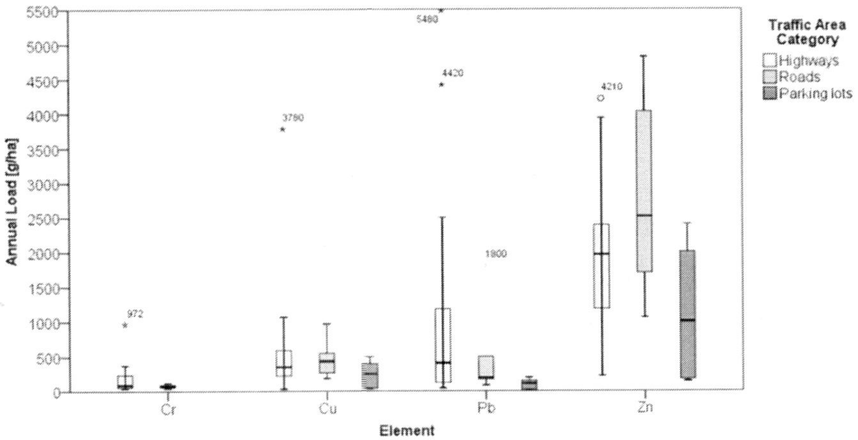

Figure 1. Annual total heavy metal loads of Cr (n = 10), Cu (n = 41), Pb (n = 42), and Zn (n = 43) in runoff from traffic areas for each of the three categories highways (H), roads (R), and parking lots (P).

For most values, the loads published for runoff from parking lots were lower than the loads of highway and road runoff (Table 2 and Figure 1). One further trend between the traffic area categories was found for the runoff loads from parking lots. The two parking lots P01 and P02 at the Florida Aquarium (USA) were less polluted compared with the three frequently used German highway parking lots for trucks (P03), cars (P04), and all types of vehicles (P05). Thus, the frequently used highway parking lots with high percentages of trucks have increased runoff loads and these values are comparable with the annual metal loads of the runoff from roads.

Strong and statistical significant correlations were determined for Cr with Pb, Cu, and Zn (Table 3) that represent the particulate fraction of these metals because Cr and Pb are mostly particulate in the runoff from traffic areas and Cu and Zn are also particulate but more dissolved [2]. Some significant correlations between the catchment area and the metal runoff loads were also calculated (Table 3). However, only an effect of the area on the first flush was previously reported [52].

3.2. Influences on the Metal Runoff Loads

3.2.1. Fixed Site-Specific Influences

One often-discussed factor is the impact of the AADT on the runoff pollution of traffic areas. However, AADT can only explain approximately

30% of the variations between different sites [53]. In this review, the AADT varies between 2000 vehicles per day and 120,000 vehicles per day for the highway and road sites (Table 1). Significant correlations between AADT and the heavy metals were found for Cd, Cr, and Cu (Table 3). Cr was the only metal with a strong correlation (δ = 0.66). Thus, further site-specific factors have an influence on the annual loads. A general description of these site-specific factors that influence the metal runoff concentrations and loads from traffic areas is given in [2]. In the following, only the fixed site-specific factors that were responsible for the extreme values calculated for all 45 sites were discussed with the exception of Pb (cf. Section 3.2.2).

Table 3. Spearman correlation coefficients of average annual daily traffic (AADT), catchment area, annual rainfall depths, and heavy metal runoff loads: correlation coefficient ≤0.39 = weak correlation, 0.40–0.59 = medium correlation, and ≥0.60 = strong correlation. Strong correlations are in bold type and no values are presented for Ni because of the small n (5 or 6 for all correlations).

Variable	Cd	Cr	Cu	Pb	Sb	Zn	AADT	Area	Rainfall
Cd	–	– –	– –	– –	– –	– –	– –	– –	– –
Cr	**0.76** ** $n = 17$	–	– –	– –	– –	– –	– –	– –	– –
Cu	0.49 * $n = 24$	**0.77** ** $n = 18$	–	– –	– –	– –	– –	– –	– –
Pb	**0.94** ** $n = 24$	**0.90** ** $n = 18$	0.43 ** $n = 50$	–	– –	– –	– –	– –	– –
Sb	−0.22 $n = 12$	0.08 $n = 12$	0.41 $n = 12$	−0.10 $n = 12$	–	– –	– –	– –	– –
Zn	0.21 $n = 22$	**0.69** ** $n = 17$	0.56 ** $n = 51$	0.50 ** $n = 52$	0.63 * $n = 12$	–	– –	– –	– –
AADT	0.44 * $n = 24$	**0.66** ** $n = 18$	0.47 ** $n = 48$	−0.01 $n = 49$	−0.32 $n = 12$	−0.03 $n = 50$	–	– –	– –
Area	**0.74** ** $n = 24$	0.54 * $n = 18$	0.41 ** $n = 53$	0.25 $n = 52$	−0.39 $n = 12$	−0.14 $n = 53$	0.43 ** $n = 50$	–	– –
Rainfall	**0.64** ** $n = 22$	0.29 $n = 17$	0.04 $n = 43$	0.22 $n = 42$	0.07 $n = 12$	−0.11 $n = 45$	0.02 $n = 40$	0.22 $n = 44$	-

Notes: * $p < 0.05$, statistical significance of the correlations; ** $p < 0.01$, statistical significance of the correlations.

Most of the largest annual loads were measured for the urban site H01, which has a high AADT of 57600 vehicles per day. Further factors that might have led to the high runoff concentrations were the extreme use of de-icing salts, which had increased levels of metal contents (cf. Section 3.4.2), by the winter services (ca. 2100 g/(m^2·yr)), the grade of congestion during rush hours, and the road design (e.g., a section of the monitoring site was a bridge with concrete pillars) [32]. In addition to these high concentrations, uncertainties regarding the discharge area because of the steep slope at the ends of the drainage area causing additional water flowing into the monitored area might have led to the increased calculated annual runoff loads [32]. At H12, the presence of a nearby Zn smelter was responsible for the high Zn runoff loads [34]. The road design of R01 resulted in increased runoff loads because of the complete drainage of the catchment area by curbs and gutters (runoff volumes were calculated from the rainfall depths and an assumed runoff coefficient of 0.9), the presence of guardrails (at the median and at both sites of the road), and the stop-and-go traffic.

For the special sites (S, n = 10), grass swales were often used instead of conventional curbs and gutters to drain the catchment areas and the samples were often taken at the drainage pipes. The category S consists of the highway sites S01–S04 and the parking lot sites S05–S10. Most metal runoff loads of category S (Table 4) are much lower than the ones reported for non-pre-treated runoff loads for comparable sites (Table 2). At S01, the surface was only sealed by 39.6% with asphalt and the rest of the drainage area were adjacent areas with surfaces vegetated by grass [26]. At S02, the catchment area consisted of 61% asphalt, which was drained to a catch basin, and the grassy shoulder that were both monitored [28]. A similar situation was present at S03 with 45% impervious asphalt and a grassy shoulder [28]. At S04, 37.6% of the drainage area was paved with asphalt and the highway runoff drained into a large grassy median that were both simultaneously monitored in mixed samples [30]. For the parking lots, several loads for pre-treated runoff were reported [29]: asphalt surfaces with subsequent vegetated swales (S05,S06), cement surfaces with subsequent vegetated swales (S07,S08), and pervious surfaces with subsequent vegetated swales (S09,S10). A comparison of the sites S05–S10 with the sites P01 and P02 shows that small alterations to parking lot designs can dramatically decrease runoff metal loads. The reduction of pollutants by grass swales was also measured for the adjacent sites H21/S02 and the data presented for S04 in [30]. These results concerning the runoff load reductions by pervious surfaces and vegetated swales can be used for designing stormwater management practices. For the design of vegetated swales, further site-specific factors such as the road design (e.g., crossings and roundabouts) and the grade of congestion (i.e., sites with frequent stop-and-go traffic) must be considered [54].

3.2.2. Historical Trends

The historical trends of traffic area runoff loads published in literature were analyzed for Cu, Pb, and Zn from the 1970s to date. The loads were aggregated into five decades and plotted as box and whisker plots in Figure 2.

Table 4. Annual total heavy metal loads in pre-treated runoff from traffic areas.

Site	Literature	Cd Load (g/ha)	Cr Load (g/ha)	Cu Load (g/ha)	Ni Load (g/ha)	Pb Load (g/ha)	Zn Load (g/ha)
S01	[26,27]	7.2	12	130	–	360	715
S02	[28]	10	20	70	20	70	–
S03	[28]	50	50	100	50	130	–
S04	[30]	–	–	15	–	10	60
S05	[29]	–	–	8	–	2	37
S06	[29]	–	–	25	–	7	79
S07	[29]	–	–	8	–	3	42
S08	[29]	–	–	9	–	4	56
S09	[29]	–	–	3	–	1	20
S10	[29]	–	–	6	–	3	36

For the Pb data, a significant decrease of the loads was detected as a function of time because of the phase-out and substitution of leaded gasoline. The decrease of Pb use as an anti-knocking agent started in the USA in the mid-1980s and was completed in 1996 [55]. In Europe, the situation was not as homogenous as in the USA. For example, Pb was almost completely phased-out in gasoline in Germany and several other countries of Western Europe since 1986 [56,57]. The phase-out in the European Union ended in January 2002, with the elimination of leaded gasoline in Italy [58]. At the same time, leaded gasoline was banned in China [15]. Further reductions of Pb usage are related to the substitution of Pb in tires and brake linings [38,59,60], lubricating oil and grease [50], and weights added to vehicles for tire balance [61,62]. These developments are responsible for the high Pb loads reported in older publications (cf. Table 2). However, tires still contain Pb [63,64] (cf. Section 3.4.2) and Pb is added to the gasoline used in classic cars. As a consequence, Pb is still found in runoff samples of the 21st century in low concentrations [2] and in younger vegetated infiltration swales in low contents [54]. Thus, only Pb data of the 21st century should be used for current mass balances (cf. Sections 3.3 and 3.4). For Cu and Zn, no historical trends were detected. The median values of the decades varied because of different characteristics of the sites monitored per decade. The evaluation of the Cu and Zn loads confirms the higher variability of Zn (e.g., cf. the IQRs presented in Figure 2) compared with Cu that is based on the different sources (e.g., presence of Zn in galvanized structures and crumbs of car tire rubber; cf. Section 3.4.2). In addition, Cu and Zn are simultaneously present in some sources such as brake linings [38,65] and, therefore, their runoff loads have a medium correlation (Table 3)

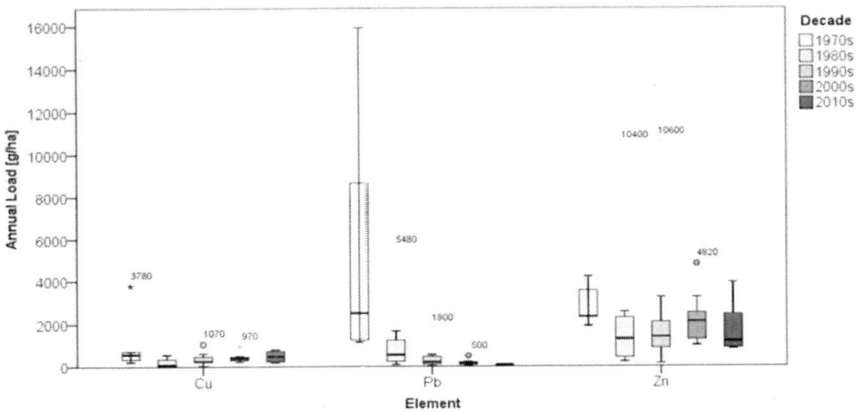

Figure 2. Historical trends of Cu (n = 41), Pb (n = 42), and Zn (n = 43) in traffic area runoff loads from the 1970s to date.

3.2.3. Annual Variations and Climatic Site-Specific Influences

Since most of the other sampling periods were only 1 to 2 years long (cf. Table 1), values of the annual runoff loads and annual rainfall depths are only available for each monitored year for the three traffic area sites R02, H27, and H28. Variations of the annual runoff loads (n = 9) were calculated for a nine-year monitoring program (R02) from the data presented by Nadler and Meißner [45]. The minima and maxima of the annual runoff loads calculated for each twelve months are 0.5–3.1 g/ha Cd, 17–76 g/ha Cr, 116–324 g/ha Cu, 20–60 g/ha Ni, 28–143 g/ha Pb, 10–23 g/ha Sb, and 1530–3550 g/ha Zn. The coefficients of variation (n = 9) for the annual runoff loads of R02 are 56% for Cd, 51% for Cr, 33% for Cu, 43% for Ni, 44% for Pb, 27% for Sb, and 30% for Zn. The catchment area, the sampling procedure, and most fixed site-specific factors (e.g., presence of a guardrail) were the same for the complete monitoring campaign. Only a small increase of the AADT was measured (6100 vehicles per day at the beginning and approximately 6800 vehicles per day in the second half of the monitoring program) and a different agricultural land use occurred in the east of the monitoring site. Since most annual metal loads of the second half of the monitoring program are lower than the loads of the first half, the influence of the AADT on the loads is not significant. Therefore, the variation of the heavy metals might be correlated with climatic factors such as the annual rainfall depth that varied largely between 334 mm and 863 mm (mean 667 mm; coefficient of variation 26%).

However, a strong correlation ($\delta \geq 0.60$) between the seven metals and the annual rainfall depth was only calculated for Sb ($\delta = 0.68$, $p < 0.05$). Thus, the annual rainfall depths also had no significant influence on the variability of the annual heavy metal runoff loads for this site although most of the lowest loads were calculated for the year with low rainfall depths and long dry weather periods in the spring and summer.

Smaller annual variations of the heavy metal runoff loads were published for H28 ($n = 4$) for each of the four years that were monitored [41]: 340–381 g/ha for Cu, 47–63 g/ha for Pb, and 1200–1370 g/ha for Zn. The coefficients of variation ($n = 4$) are 5% for Cu, 13% for Pb, and 6% for Zn. The annual rainfall depths varied between 477 mm and 726 mm (mean 591 mm; coefficient of variation 22%) at this highway site. The small variations can be partially explained by varying street sweeping intervals of the hard shoulders [41]. Because of the small n, no correlation analyses were performed.

The variations of the annual rainfall depths at both sites were comparable (coefficients of variation are 26% and 22%). However, the variations between the metal runoff loads were much higher for R02 than for H28. These differences between the monitored years and sites can depend on a multitude of climatic boundary conditions (not only the rainfall depths), which can generally induce different deposition and wash-off processes, and on fixed site-specific factors (i.e., different median pollution loads of both sites). These influences on the median loads can be related to the different AADT and speed limits (50 km/h at R02 and 120 km/h at H28) and the presence of noise barriers, a forest, and hard shoulders at H28, which can also reduce the influences of climatic factors (e.g., wind turbulences) on the runoff loads [41]. Thus, the fixed site-specific factors also affected the influences of the climatic factors on the annual variations of the runoff loads.

At H27, with a speed limit of 120 km/h, the published annual heavy metal runoff loads were 470/410 g/ha for Cu, 71/88 g/ha for Pb, and 1010/1370 g/ha for Zn for the years 2006/2007 [41]. The annual rainfall depths were 876 mm and 1184 mm for the years 2006 and 2007, respectively. Thus, the rainfall depth was higher for 2007 and the runoff loads increased for Pb and Zn but not for Cu. Consequently, median runoff concentrations from the literature multiplied by different annual rainfall depths of a specific site cannot be used to describe the variance of the annual runoff loads because of climatic variations.

For all monitoring sites, the annual rainfall depth varied between 254 mm and 2540 mm (median 792 mm; 90th percentile 1200 mm). A strong correlation between the annual rainfall depths and the heavy metals was only calculated for Cd ($\delta = 0.64$, $p < 0.01$) for all monitoring sites (Table 3). The influence of the seasons on the pollution of traffic area runoff was evaluated by Helmreich et al. [66] with a considerable seasonal increase of pollutants in the runoff during the cold season. A correlation of the seasons could not be performed in this study because all published annual loads did not distinguish between different seasons. Therefore, only the results of single events can be used for this purpose. For example, the first storm of each season in (semi)arid regions is a special case for the climatic influences that is characterized by higher runoff loads [67,68].

In addition, the atmospheric deposition (wet and dry) must be considered as a climatic factor that affects the runoff loads. The deposition fluxes on traffic areas are strongly influenced by site-specific factors [69]. For example, the absence of curbs, hard shoulders, and noise barriers are site-specific influences that have a positive effect on reducing runoff concentrations by decreasing the deposition rates and the subsequent wash-off [41,70]. The deposition is also influenced by the traffic volume and the surrounding land use. Less data are available for urban traffic area sites (in particular roads and parking lots) at which higher fractions of metal emissions can be deposited and washed-off compared with non-urban traffic areas [11]. These rates are also influenced by the antecedent dry period [71]. However, Racin et al. [12] concluded that splashing and washing of pollutants from vehicles is more important than the wash-off of pollutants accumulated on road surfaces by precipitation (wet deposition). An increased importance of precipitation for runoff pollution of traffic areas in industrial zones is proposed by Dannecker et al. [72] and it is highly variable for each metal. Dry deposition at industrial sites also has an important effect, whereas the influences of traffic related emissions become more important at non-industrial urban sites [72]. For the non-urban sites H04, H07, and S01, the influence of the wet and dry deposition rates on the corresponding annual runoff loads was determined. The ratios of deposition loads to runoff loads for the sites H05/H07/S01 were 35%/31%/42% for Cd, 65%/2%/125% for Cr, 36%/23%/46% for Cu, 24%/27%/50% for Pb, and 54%/31%/80% for Zn [27]. Thus, the deposition affects the runoff loads at these non-urban highway sites and the influences depend on both the site and the metal.

In summary, the deposition is highly variable for each site because of site-specific factors and a robust data set is currently not available to determine the deposition loads for all types of traffic areas and the most relevant land use categories. For a rough estimation, the average deposition loads summarized by Zessner [73] can be used, which consist of traffic related and non-traffic related metals. He determined annual deposition loads of 1.4 g/ha Cd, 50 g/ha Cu, and 300 g/ha Zn for rural areas and annual deposition loads of 6.0 g/ha Cd, 250 g/ha Cu, and 1000 g/ha Zn for urban areas. Ilyin et al. [74] modeled annual deposition loads of 0.25–0.65 g/ha Cd and 8–20 g/ha Pb for Germany on a large scale. However, these data might underestimate the real values of metals deposited on traffic areas because of the large-scale model. For comparison, Kocher et al. [75] determined annual deposition loads for three different German highways and the rates varied strongly between each year, season, metal, and monitoring point (at ground level or 1.5 m height and with distance from the highways). Thus, method-specific factors also have an influence on the results. At 1.5 m height, the deposition rates determined by Kocher et al. [75] are in the range of the ones summarized by Zessner [73]. However, the values at ground level can be higher by a factor of up to approximately five for the measurements directly at the highway site. Thus, traffic related emissions highly affect the deposition rates on traffic areas.

A general description of the influences of climatic factors such as deposition rates, the antecedent dry periods, rain characteristics (volume, intensity, and duration), seasonal effects, and wind turbulence on the runoff from traffic areas is given in [2]. For rain characteristics, mostly poor correlations were obtained for the metal runoff concentrations. This is in accordance with the findings of this review (low correlation between annual rainfall depths and runoff loads). Thus, the differences of the runoff loads cannot be based on the uncertainties of the measured values and the high variability of the metal runoff loads between different years must be based on the variability of climatic factors and some fixed-site specific factors such as road maintenance (e.g., the variability of the use of de-icing salts by winter services; cf. Section 3.4.2). However, further data are needed to calculate the correlations between runoff loads and rainfall intensities or antecedent dry periods.

3.3. Average Runoff Loads for Different Traffic Area Categories

The values of the runoff loads presented in Table 2 were used to calculate median values, mean values, and SDs that could be used for mass balances (no values from H01 and Zn from H12). The calculated median annual runoff loads for all traffic areas are presented in Table 5.

In a literature study, Hullmann and Kraft [76] presented annual average Cu and Zn loads for runoff from traffic areas that were summarized from several studies as approximately 800 g/ha for Cu and 2700 g/ha for Zn. These values are higher than the ones calculated in this study. For comparison, Driscoll et al. [7] reported the following ranges of annual heavy metal runoff loads measured in the USA (n = 24): 7.2–37 g/ha for Cd, 12–120 g/ha for Cr, 30–4670 g/ha for Cu, 70 g/ha for Ni, 80–21,200 g/ha for

Pb, and 220–10,400 g/ha for Zn. The surrounding land use characteristics comprised nineteen urban sites (commercial, residential, and suburban) and five non-urban sites (forest, agricultural, and desert). Burton and Pitt [8] also summarized the runoff loads for parking lots, highways, and different land use categories that were published in five reports of the 1970s and 1980s. For parking lots, the reported values of the annual runoff loads are 10 g/ha Cd, 60 g/ha Cu, 800 g/ha Pb, and 800 g/ha Zn [8]. The published annual runoff loads are higher for highway sites: 20 g/ha Cd, 90 g/ha Cr, 370 g/ha Cu, 4500 g/ha Pb, and 2100 g/ha Zn [8]. With the exception of Pb, these annual highway runoff loads are only slightly higher than the median values calculated for highways in this review. In contrast, these summarized parking lot runoff loads are highly different compared with the parking lot values of Table 2 because the use of parking lots differ widely. Subsequently, the runoff loads of parking lots should not be summed up but considered in accordance to their type of use (cf. Section 3.1). Another relevant factor for parking lots is the contamination of parked vehicles with dirt, rust, and further substances [77] that must be considered for planning stormwater management strategies.

Table 5. Median annual runoff loads (g/ha) for all traffic areas. For Pb, only data measured in the 21st century was used.

Parameter	*n*	Median	Mean	SD
Cd	16	6.8	13.0	12.9
Cr	9	75	107	106
Cu	40	355	384	254
Ni	2	66	66	35
Pb	13	110	149	119
Sb	4	14.0	12.9	5.7
Zn	41	1960	2018	1751

For the highway data, the calculated median runoff loads are presented in Table 6.

3.4. Heavy Metal Load Balances for Germany

3.4.1. Mass Balances of Heavy Metal Traffic Area Runoff Loads

In Germany, there are approximately 281.3 km^2 of highways and 339.5 km^2 of state roads [78]. The total amount of all types of traffic areas, which also include the areas for railroads, airports, and waterways, is 18,070 km2 in

Germany and the fraction of roads, places (e.g., parking lots and marketplaces), and sidewalks is 15,750 km2 [79].

Table 6. Median annual runoff loads (g/ha) for all highway sites. For Pb, only data measured in the 21st century was used.

Parameter	*n*	Median	Mean	SD
Cd	12	14	16	14
Cr	6	76	121	130
Cu	27	355	388	260
Ni	1	90	90	–
Pb	5	72	68	18
Sb	3	11	12	6
Zn	29	1900	1773	1004

Since median values are preferred for the analyses of such field data [80], the median annual loads for highway runoff and the area of highways in Germany (281.3 km2) were used to calculate the total amount of metal loads in runoff from German highways.. These total loads are presented as t/yr (1 t/yr = 1000 kg/yr) and were calculated as approximately 0.38 t/yr Cd, 2.1 t/yr Cr, 9.8 t/yr Cu, 2.5 t/yr Ni, 2.0 t/yr Pb, 0.31 t/yr Sb, and 53.4 t/yr Zn.

The calculated median annual heavy metal loads for all traffic areas, which are similar to the ones for the highway sites, and the total amount of all transportation areas (15,750 km^2) were used to estimate the total traffic area runoff loads for Germany. Thus, approximately 10.7 t/yr Cd, 119 t/yr Cr, 558 t/yr Cu, 103 t/yr Ni, 173 t/yr Pb, 22.1 t/yr Sb, and 3087 t/yr Zn are transported by runoff from traffic areas in Germany. Since less polluted traffic areas are also summarized in the total amount of transportation areas and no data are available for these categories regarding runoff loads, the calculated values tend to exaggerate the total runoff loads for Germany.

3.4.2. Mass Balances of Traffic Related Heavy Metal Emissions

Hillenbrand et al. [81] published heavy metal emission loads of different sources in a literature study for Germany. For the traffic related heavy metal emissions, they calculated the average loads for Cu, Pb, and Zn for the following sources: brake lining wear, tire wear, roadway abrasion, and Pb weights for tire balance. These data were adapted, updated, and extended by Cd and further sources (Zn weights for tire balance guardrails, lampposts/signs, and de-icing salts) for this review. The emission of Pb by leaded gasoline used by classic cars was not considered (only 0.6% of the cars in Germany are classic cars and their kilometric performances are low). A summary of the following results for the four metals Cd, Cu, Pb, and Zn is presented in Table 7. Both the metal contents of different sources (mg/kg) and the emission rates (mg/(vehicle_ km)) are highly variable and only average values for each source are published to indicate their importance in relation to other emission sources.presented in Section 3.4.2 that are based on different

Table 7. Traffic related heavy metal emissions in Germany (summary of mean values of the calculationsliterature data).

Source	Cd (t/yr)	Cu (t/yr)	Pb (t/yr)	Zn (t/yr)
Tire wear	0.19	0.79	2.0	1255
Brake lining wear	0.03	928	61.5	309
Roadway abrasion	–	4.3	7.1	149
Weights for tire balance	–	–	2.2	6.6
Guardrails	–	–	–	313
Lampposts and signs	–	–	–	60
De-icing salts	0.70	1.75	11.6	1.75
Sum of all seven sources	0.92	935	84.4	2094

The first four sources are directly related to vehicles and the loads are determined by multiplying the metal contents of the material by their total abrasion masses. The tire wear was determined by Kocher et al. [82] based on 62 different tire samples for 13 metals and further substances. Kocher et al. [82] measured the metal contents of each tire (n = 62) and the coefficients of variation for the metal contents are 70% for Cd, 38% for Cr, 56% for Cu, 74% for Ni, 95% for Pb, 330% for Sb, and 35% for Zn. This is based on the different ages, manufacturers, and types of tires. The complete mass of tire wear is based on the emission factors reported in seven studies, which determined the tire wear as (mg/(vehicle_ km)) for four vehicle categories (summarized in [81]), and the kilometric performances of the vehicles in Germany. By using the mean metal contents published by Kocher et al. [82] and an annual mass of tire wear of 111,420 t (coefficient of variation 29%) for Germany, the mean total emission loads for tire wear in Germany were calculated as 0.19 t/yr Cd, 0.51 t/yr Cr, 0.79 t/yr Cu, 0.97 t/yr Ni, 2.03 t/yr Pb, 0.42 t/yr Sb, and 1255 t/yr Zn. In contrast, Hillenbrand et al. [81] proposed annual emission loads of 2.7 t/yr Pb and 1620 t/yr Zn for tire wear. Thus, the calculated Pb and Zn loads of tire wear differ by approximately 20%–25% between these two studies. A summary of measured metal contents in tires and brakes was also presented by McKenzie et al. [64]. The contents also have a high variation for each metal (mostly one order of magnitude) and are in the order of the values summarized and measured by the German researchers [81,82]. The Cd emission load caused by brake wear was calculated with the metal content of 2.7 mg/kg presented by Legret and Pagotto [38], which is the median value of the contents summarized by McKenzie et al. [64] (coefficient of variation 120%), and the total brake lining abrasion mass of 12,500 t/yr published by Hillenbrand et al. [81]. The loads of brake lining wear for Cu, Pb, and Zn are taken from [81], which are based on the heavy metal contents of brake linings and their emissions published in eight studies separately for cars and trucks. In Germany, the brake linings of cars were responsible for 83%, 94%, and 80% of the total brake lining emission loads for Cu, Pb, and Zn, respectively. The emissions by weights for tire balances (Pb emissions reduced by 75% because of its substitution, mainly by Zn) and roadway abrasion were adapted from [81]. The roadway abrasion loads are based on the mean values of 18 road samples. The metal contents of the road surfaces varied between 0.0 and 31.5 mg/kg Cu, 0.9–29.0 mg/kg Pb, and 4–614 mg/kg Zn. Since only few data were available about the wear emission rates, in particular for different types of vehicles (e.g., cars, trucks, and buses), the uncertainties of the emission loads are high. Consequently, more data are needed to calculate reliable tire and break wear loads.

Guardrails, lampposts, and signs are parts of the road design and their heavy metal emissions must also be linked to road transportation [81]. The Zn emissions of lampposts and signs can only be a rough estimation because the number and surface area of traffic signs and further posts at traffic areas cannot be determined exactly. Most values are based on the assumptions proposed by Hillenbrand et al. [81].

The traffic related heavy metal emissions by de-icing salts for road maintenance were not previously evaluated for Germany. The novel calculated mean de-icing salt emissions (0.70 t/yr Cd, 1.75 t/yr Cu, 11.6 t/yr Pb, and 1.75 t/yr Zn) are based on an annual consumption of 3.5 million tons in Germany [78] and heavy metal contents of 0.2 mg/kg for Cd, 0.5 mg/kg for Cu and Zn, and 3.3 mg/kg for Pb in the de-icing salts [38]. However, these annual de-icing salt emissions are highly variable because of the different metal contents of the salts and the different annual consumptions that strongly depend on climatic factors. According to German regulation, the maximum permitted values for the soluble fraction of de-icing salts in Germany (sodium chloride, calcium chloride, and magnesium chloride) are 2 mg/kg Cd, 5 mg/kg Cu, 5 mg/kg Cr, 5 mg/kg Ni, 5 mg/kg Pb, and 20 mg/kg Zn [83]. In Great Britain, higher values were measured for a sodium chloride de-icing salt with 4.7 mg/kg Cr, 12 mg/kg Ni, 8.7 mg/kg Pb, and 6.0 mg/kg Zn [32]. Values of 1.2 mg/kg Cu and 1.6 mg/kg Pb were reported in [84] for sodium chloride (Cd and Zn not detected). For another sodium chloride de-icing salt, metal contents of 0.94 mg/kg Cd, 1.02 mg/kg Cr, 3.15 mg/kg Cu, 6.29 mg/kg Ni, 6.29 mg/kg Pb, and 1.57 mg/kg Zn were reported by Kobriger and Geinopolos [85]. Variations of the metal contents for highway de-icing salts were also published by Gupta et al. [86]: 0.003–0.01 mg/kg for Cr, 0–0.0004 mg/kg for Cu, 0.002–0.003 mg/kg for Ni, and 0.08–0.09 mg/kg for Pb. A comparison of the metal contents in de-icing salts with the ones measured in cooking salts [87] confirms that these small impurities can occur in all salts used in large quantities. Nevertheless, the total emission loads can be high because of the large amounts of salts used in Germany per year..

The total emissions are influenced by the highly variable annual consumption in Germany (0.46–5.26 million tons per year for the years 1990–2013). The metal contents of Cd, Cu, Pb, and Zn published in five references were multiplied by the annual consumption values in Germany for the years 1990–2013 (Table 8 and Figure 3). Each box and whisker represents the annual variation of the heavy metal emissions by de-icing salts for one element and the metal content of the corresponding reference. Thus, the mean de-icing salt emissions presented in Table 7 can only be a very rough estimation. In contrast to the other traffic related emission sources, the variability of these emissions mainly depends on climatic factors.

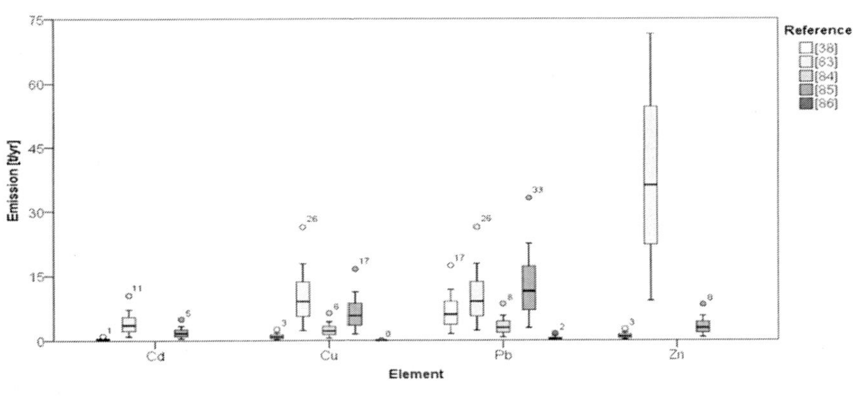

Figure 3. Variations of the traffic related heavy metal emissions by de-icing salts in Germany (1990–2013). Heavy metal contents of the de-icing salts differ largely within each reference.

For the year 2013 (year 2000 in brackets), the German Federal Environment Agency published national total annual emissions for road transportation as 0.78 (0.82) t Cd, 25.7 (25.1) t Cr, 2109 (1986) t Cu, 5.0 (4.7) t Ni, 88.7 (80.8) t Pb, and 1906 (1805) t Zn [88]. There is a slight increase of the emissions for five of the six heavy metals in the 21st century. In 2013, the percentages of the traffic related emissions based on the total emissions are 11% for Cd, 45% for Cr, 98% for Cu, 5% for Ni, 42% for Pb, and 95% for Zn. Considering these data, road transportation is an important source of total heavy metal emissions in Germany. With the exception of the data for Cu, these values are comparable with the ones presented in Table 7. Denier van der Gon et al. [89] presented annual road transportation emissions for Germany for the year 2000 of 0.88 t Cd, 4.0 t Cr, 102.9 t Cu, 5.0 t Ni, 0.40 t Pb, and 209.8 t Zn. Most of these emission values are lower than the ones proposed by Hillenbrand et al. [81], calculated in this study, and published by Umweltbundesamt [88]. These differences are quite high and this is not only related to the variability of the data but also to the fact that each study considers different traffic related sources, which can also vary because of substitutions.

Table 8. Variations of the traffic related heavy metal emissions (t/yr) by de-icing salts with different contents (cf. [38,83–86]) in Germany (1990–2013).

Parameter	Min					Median					Mean					Max				
	[38]	[83]	[84]	[85]	[86]	[38]	[83]	[84]	[85]	[86]	[38]	[83]	[84]	[85]	[86]	[38]	[83]	[84]	[85]	[86]
Cd	0.09	0.92	–	0.43	–	0.36	3.61	–	1.70	–	0.41	4.07	–	1.91	–	1.05	10.5	–	4.95	–
Cu	0.23	2.30	0.55	1.45	0.00	0.90	9.02	2.17	5.68	0.00	1.02	10.2	2.44	6.41	0.00	2.63	26.3	6.31	16.6	0.00
Pb	1.52	2.30	0.74	2.89	0.04	5.95	9.02	2.89	11.4	0.16	6.71	10.2	3.25	12.8	0.18	17.4	26.3	8.42	33.1	0.47
Zn	0.23	9.20	–	0.72	–	0.90	36.1	–	2.83	–	1.02	40.7	–	3.19	–	2.63	105	–	8.26	–

3.4.3. Transportation of Traffic Related Heavy Metals

A sustainable stormwater management combines the stormwater source control and the treatment of traffic area runoff loads to reduce the impact of heavy metals on receiving waters. Therefore, the influence of the traffic related heavy metal emissions on the runoff loads because of leaded gasoline, tire wear, brake lining wear, roadway abrasion, weights for tire balance, guardrails, lampposts/signs, and de-icing salts must also be considered to identify cost efficient mitigation strategies. This process was successfully implemented by the phase-out of Pb in recent decades (cf. Section 3.2.2). Nevertheless, not all traffic related emissions are found in the runoff because of wind turbulences, vehicles (traffic volume and vehicle speed), antecedent dry periods, and other site-specific factors (e.g., particle size distribution, land use, road surface roughness, or elevated configuration of the traffic area; [69]). For example, Hallberg et al. [90] found out that splashing corresponds with the average vehicle speed during storms. In Great Britain, Harrison et al. [91] measured that 90% of Pb emissions from fast-moving vehicles are deposited away from the immediate vicinity of a non-urban highway (up to several hundred meters). Consequently, most of the emitted Pb was not measured in the runoff. In contrast to Pb, Harrison and Johnston [92] showed that the Cd and Cu deposition fluxes were elevated close to the investigated highway section and decreased to background levels within 20–40 m. Langbein et al. [14] and Steiner et al. [93] measured the different fluxes of Cd, Cr, Cu, Pb, and Zn emissions at a non-urban road with approximately 17,000 vehicles per day in Switzerland (average vehicle speed between 60 km/h and 80 km/h). They evaluated that 17%–25% of the heavy metals were found in the runoff, 21%–38% in the splash water, 19%–45% were blown away and deposited in a distance of up to 25 m away from the road, and 17%–37% were deposited far away (>25 m). Thus, only some heavy metals were found in the runoff of a non-urban road and the fluxes of the emitted heavy metals varied between each element. Legret and Pagotto [38] calculated emission fluxes and estimated the percentages of Cd, Cu, Pb, and Zn that were removed by runoff waters. The following were considered as emission sources: tire and brake wear, gasoline, safety fences, de-icing salts, and atmospheric deposition (wet and dry). At the investigated rural highway bridge, 313% of the emitted Cd, 2% of Cu, 5% of Pb, and 37% of Zn were found in the runoff water and the rest was dispersed into the atmosphere as dust. For Cd, the authors assumed that some sources might be unrealized or underestimated. Hewitt and Rashed [94] also measured that the largest part of Pb (approximately 86%) was dispersed in the atmosphere. Most of these mass fluxes are proposed for non-urban sites and bridges. In contrast, a higher percentage of the traffic related emissions (e.g., 50%) might be found in urban runoff from traffic areas [14]. Thus, the heavy metal emissions in the urban areas influence the runoff concentrations and loads more directly [11]. Moreover, all traffic related emissions that were not found in the runoff (i.e., approximately 50%–98% according to the above data) would be deposited on non-traffic areas and consequently have an effect on the environment.

The influence of traffic related emissions on the runoff loads can be determined for de-icing salt applications on highways. Granato [95] calculated mean annual runoff loads as 4100 g/ha Cl, 1.8 g/ha Cr, 6.6 g/ha Cu, 2.0 g/ha Ni, and 3.6 g/ha Pb for the application of sodium chloride on highway runoff. Compared with the median runoff loads for highways calculated in this review (76 g/ha Cr, 355 g/ha Cu, 90 g/ha Ni, and 72 g/ha Pb), the de-icing salt emissions contribute to about 2.4%, 1.9%, 2.2%, and 5.0% of the total annual runoff loads for Cr, Cu, Ni, and Pb, respectively. Higher ratios were calculated for H01 with 38% for Cr, 56% for Ni, 2.2% for Pb, and 1.5% for Zn [32]. Thus, the use of de-icing salts with lower metal impurities can directly reduce the runoff loads by some percentages although de-icing salts are only used for several months per year.

In summary, the identification and subsequent reduction or substitution of relevant traffic related heavy metal sources (cf. Section 3.4.2) can help to fulfill the requirements of a sustainable stormwater management. Since the runoff loads are also influenced by special site-specific factors (e.g., nearby industries like Zn smelters) and non-traffic related emissions, a runoff treatment of distinct traffic area categories (e.g., highways, roads, and frequently used parking lots) is also necessary. This can be done by permeable pavements, grass swales, and technical stormwater treatment systems [96,97].

4. CONCLUSIONS

For stormwater management, an improved knowledge about the runoff loads of different traffic area categories is necessary. These loads were summarized for highways, roads, and parking lots. The highest loads were determined for highways, roads, and frequently used parking lots (in particular by trucks). In contrast to highways and roads, parking lots must be considered individually because of site-specific factors. The traffic related emissions in Germany were estimated for seven different sources (tire wear, brake lining wear, roadway abrasion, weights for tire balance, guardrails, lampposts/signs, and de-icing salts). Zn is mostly emitted by galvanized elements and tires, Cu and Pb by brakes, and Cd by de-icing salts. The calculated loads are comparable with the ones presented in other studies for most metals. However, a statistical analysis of traffic related metal mass fluxes (e.g., the contribution of metal emissions on the runoff loads, in particular for different traffic area categories and land uses) is not currently possible because of a lack of monitoring data (e.g., metal contents of different sources, emission factors, deposition rates for different traffic area categories and land use characteristics, and size characterizations of distinct catchment areas for different traffic area categories).

Nevertheless, the estimation of the runoff loads and the emission loads for Germany stated that the vehicles, the road design, and the winter services emit heavy metals in large quantities (0.93 t/yr Cd, 935 t/yr Cu, 84.4 t/yr Pb, and 2094 t/yr Zn) and the runoff also contains high amounts of metal loads per year. Currently, the most relevant metals are Cu and Zn because the annual Pb loads have decreased significantly in the last few decades and traffic related Cd and Ni contribute only 5% and 11% of the total emissions in Germany, respectively. However, the loads of the other metals can also be detrimental for receiving waters and aquatic biota. In particular, for the highly toxic metals Cd and Sb. Thus, efforts should be intensified to reduce traffic related loads (both emissions and runoff) and subsequently to minimize their input into receiving waters and soils.

For the future, more long-term data, which include a large variety of substances and a detailed description of climatic factors, must be measured and published to improve the determination of metal loads, mass balances, and fluxes and to reduce the large uncertainties of several values used in this review for the calculations.

ACKNOWLEDGMENTS

Special thanks to Alicia Dery and Antje Welker for their contributions, support, and valuable comments. This work was supported by the German Research Foundation (DFG) and the Technical University of Munich (TUM) in the framework of the Open Access Publishing Program.

REFERENCES

1. Dauber, L.; Novak, B.; Zobrist, J.; Zürcher, F. Schmutzstoffe im Regenwasserkanal einer Autobahn. Stuttgarter Berichte zur Siedlungswasserwirtschaft. In *Beeinflussung der Gewässergüte durch Regenabflüsse von Straßen*; Oldenbourg Verlag: München, Germany, 1979; Volume 64, pp. 41–57. (In German)
2. Huber, M.; Welker, A.; Helmreich, B. Critical review of heavy metal pollution of traffic area runoff: Occurrence, influencing factors, and partitioning. *Sci. Total Environ.* **2016**, *541*, 895–919.
3. Opher, T.; Friedler, E. Factors affecting highway runoff quality. *Urban Water J.* **2010**, *7*, 155–172.
4. Schueler, T. Stormwater pollution source areas isolated in Marquette, Michigan. *Watershed Prot. Tech.* **2000**, *3*, 609–612.
5. Bannerman, R.T.; Owens, D.W.; Dodds, R.B.; Hornewer, N.J. Sources of pollutants in Wisconsin stormwater. *Water Sci. Technol.* **1993**, *28*, 241–259.
6. Ball, J.E. *Stormwater Quality at Centennial Park, Sydney, Australia*; University of New South Wales, School of Civil and Environmental Engineering, Water Research Laboratory: Sydney, New South Wales, Australia, 2002; pp. 1–62.
7. Driscoll, E.; Shelley, P.E.; Strecker, E.W. *Pollutant Loadings and impacts from Highway Stormwater Runoff, Volume IV: Design Procedure*; Report no. FHWA-RD-88-009; US Department of Transportation: Washington, DC, USA, 1990; pp. 1–143.
8. Burton, G.A.; Pitt, R.E. *Stormwater Effects Handbook: A Toolbox for Watershed Managers, Scientists, and Engineers*; CRC Press: Boca Raton, FL, USA, 2001.
9. SN640347. *Strassenentwässerung. Belastung von Strassenabwasser (gültig ab 1. Februar 2009)*; Schweizerischer Verband der Strassen- und Verkehrsfachleute, Schweizerische Normen-Vereinigung SNV: Zürich, Switzerland, 2009; pp. 1–20. (In German)
10. Hamilton, R.S.; Revitt, D.M.; Warren, R.S.; Duggan, M.J. Metal mass balance studies within a small highway dominated catchment. *Sci. Total Environ.* **1987**, *59*, 365–368.
11. Barrett, M.; Zuber, R.D.; Collins, E.R.; Malina, J.F.; Charbeneau, R.; Ward, G.H. *A Review and Evaluation of Literature Pertaining to the Quantity and Control of Pollution from Highway Runoff and Construction*; CRWR Online Report 95-5; University of Texas at Austin: Austin, TX, USA, 1995; pp. 1–180.
12. Racin, J.A.; Howell, R.B.; Winters, G.R.; Shirley, E.C. *Estimating Highway Runoff Quality*; Report No. FHWA-CA/TL-82/11; California Department of Transportation: Sacramento, CA, USA, 1982; pp. 1–97.

13. Muschack, W. Pollution of street run-off by traffic and local conditions. *Sci. Total Environ.* **1990**, *93*, 419–431.
14. Langbein, S.; Steiner, M.; Boller, M. *Schadstoffe im Straßenabwasser einer stark befahrenen Straße und deren Retention mit neuartigen Filterpaketen aus Geotextil und Adsorbermaterial, Schlussbericht des Forschungsprojekts*; Eawag: Dübendorf, Switzerland, 2006; pp. 1–77. (In German)
15. Gan, H.; Zhuo, M.; Li, D.; Zhou, Y. Quality characterization and impact assessment of highway runoff in urban and rural area of Guangzhou, China. *Environ. Monit. Assess.* **2008**, *140*, 147–159.
16. Wiseman, C.L.S.; Zereini, F. Airborne particulate matter, platinum group elements and human health: A review of recent evidence. *Sci. Total Environ.* **2009**, *407*, 2493–2500.
17. Dai, H.L.; Zhang, K.L.; Xu, X.L.; Yu, H.Y. Evaluation on the effects of deicing chemicals on soil and water environment. *Procedia Environ. Sci.* **2012**, *13*, 2122–2130.
18. Eriksson, E.; Baun, A.; Scholes, L.; Ledin, A.; Ahlman, S.; Revitt, M.; Noutsopoulos, C.; Mikkelsen, P.S. Selected stormwater priority pollutants—A European perspective. *Sci. Total Environ.* **2007**, *383*, 41–51.
19. Folkeson, L.; Bækken, T.; Brenčič, M.; Dawson, A.; Frančois, D.; Kuřímská, P.; Leitão, T.; Ličbinský, R.; Vojtěšek, M. Sources and fate of water contaminants in roads. In *Water in Road Structures—Movement, Drainage and Effects*, 1st ed.; Dawson, A., Ed.; Springer Science+Business Media B.V.: Dordrecht, The Netherlands, 2009; pp. 107–146.
20. Roeva, N.N.; Rovinskii, F.Y.; Kononov, E.Y. Special features of the behavior of heavy metals in various natural environments. *J. Anal. Chem.* **1996**, *51*, 352–364.
21. Kayhanian, M.; Stransky, C.; Bay, S.; Lau, S.L.; Stenstrom, M.K. Toxicity of urban highway runoff with respect to storm duration. *Sci. Total Environ.* **2008**, *389*, 386–406.
22. Barbosa, A.E.; Hvitved-Jacobsen, T. Highway runoff and potential for removal of heavy metals in an infiltration pond in Portugal. *Sci. Total Environ.* **1999**, *235*, 151–159.
23. Spromberg, J.A.; Baldwin, D.H.; Damm, S.E.; McIntyre, J.K.; Huff, M.; Sloan, C.A.; Anulacion, B.F.; Davis, J.W.; Scholz, N.L. Coho salmon spawner mortality in western us urban watersheds: Bioinfiltration prevents lethal storm water impacts. *J. Appl. Ecol.* **2016**, *53*, 398–407.
24. McIntyre, J.K.; Davis, J.W.; Hinman, C.; Macneale, K.H.; Anulacion, B.F.; Scholz, N.L.; Stark, J.D. Soil bioretention protects juvenile salmon and their prey from the toxic impacts of urban stormwater runoff. *Chemosphere* **2015**, *132*, 213–219.
25. Scholz, M. Sustainable drainage systems. *Water* **2015**, *7*, 2272–2274.
26. Klein, H. Die Beschaffenheit der Niederschlagsabflüsse von Autobahnen. *Wasserwirtschaft* **1982**, *72*, 37–43. (In German)
27. Stotz, G. Investigations of the properties of the surface water run-off from federal highways in the FRG. *Sci. Total Environ.* **1987**, *59*, 329–337.
28. Wu, J.; Allan, C.; Saunders, W.; Evett, J. Characterization and pollutant loading estimation for highway runoff. *J. Environ. Eng.* **1998**, *124*, 584–592.
29. Rushton, B.T. Low-impact parking lot design reduces runoff and pollutant loads. *J. Water Resour. Plan. Manag.* **2001**, *127*, 172–179.
30. Barrett, M.E.; Malina, J.F.; Charbeneau, R.J.; Ward, G.H. Characterization of highway runoff in Austin, Texas, area. *J. Environ. Eng.* **1998**, *124*, 131–137.
31. Mosley, L.M.; Peake, B.M. Partitioning of metals (Fe, Pb, Cu, Zn) in urban run-off from the Kaikorai Valley, Dunedin, New Zealand. *N. Z. J. Mar. Freshw. Res.* **2001**, *35*, 615–624.
32. Hedley, G.; Lockley, J.C. Quality of water discharged from an urban motorway. *Water Pollut. Control* **1975**, *74*, 659–674.
33. Cathelain, M.; Friant, G.; Olie, J.L. Les eaux de ruissellement de chaussees autoroutieres: Evaluation des charges de pollution. *Paris Bull. Liaison Lab. Ponts Chaussees* **1981**, *116*, 9–24. (In French)
34. Chui, T.W.D.; Mar, B.W.; Horner, R.R. *Highway Runoff in Washington State: Model Validation and Statistical Analysis*; Report No. WA-RD-39.12; Washington State Department of Transportation: Olympia, WA, USA, 1982.
35. Balades, J.D.; Cathelain, M.; Marchandise, P.; Peybernard, J.; Pilloy, J.C. Chronic pollution of intercity motorway runoff waters. *Water Sci. Technol.* **1984**, *17*, 1165–1174.

36. Montrejaud-Vignoles, M.; Roger, S.; Herremans, L. Runoff water pollution of motorway pavement in Mediterranean area. In Proceedings of the 7th International Conference on Urban Storm Drainage, Hannover, Germany, 9–13 September 1996; pp. 247–252.
37. CH2MHILL. *Highway Stormwater Runoff Study*; The Michigan Department of Transportation: Lansing, MI, USA, 1998; pp. 1–48.
38. Legret, M.; Pagotto, C. Evaluation of pollutant loadings in the runoff waters from a major rural highway. *Sci. Total Environ.* **1999**, *235*, 143–150.
39. Lange, G.; Grotehusmann, D.; Kasting, U.; Schütte, M.; Dietrich, M.; Sondermann, W. *Wirksamkeit von Entwässerungsbecken im Bereich von Bundesfernstraßen. Forschungsberichte aus dem Forschungsprogramm des Bundesministers für Verkehr, Bau- und Wohnungswesen und der Forschungsgesellschaft für Straßen- und Verkehrswesen e.V*; Bundesministerium für Verkehr, Bau- und Wohnungswesen, Abteilung Straßenbau, Straßenverkehr: Bonn, Germany, 2003; Volume 861. (In German)
40. Suter, J. Schwermetallbilanz eines Autobahnabschnittes und die Behandlung des Strassenabwassers. Diploma Thesis, ETHZ/Eawag, Zürich/Dübendorf, Switzerland, 2000. (In German).
41. Scheiwiller, E. Schadstoffabschwemmungen: Am Beispiel von Hochleistungsstraßen. *GWA* **2008**, *88*, 539–546. (In German)
42. Grotehusmann, D.; Lambert, B.; Fuchs, S.; Graf, J. *Schlussbericht zum BASt Forschungsvorhaben FE-Nr. 05.152/2008/GRB "Konzentrationen und Frachten organischer Schadstoffe im Straßenabfluss"*; Ingenieurgesellschaft für Stadthydrologie mbH, Hannover: Bundesanstalt für Straßenwesen, Bergisch Gladbach, Germany, 2014; pp. 1–89. (In German)
43. Scheiwiller, E.; (Amt für Wasser und Abfall des Kantons Bern, Bern, Switzerland). Personal communication, 2014.
44. Gäth, S.; Sternheim, M.; Frede, H.-G. Einfluss des Kraftfahrzeugverkehrs auf den Schwermetallgehalt von Straßenabflußwasser. *Forum Städte—Hyg.* **1990**, *41*, 235–238. (In German)
45. Nadler, A.; Meißner, E. *Versickerung des Niederschlagswassers von befestigten Verkehrsflächen, Abschlussbericht, Entwicklungsvorhaben, Oktober 1996—Oktober 2005*; Bayerisches Landesamt für Umwelt: Augsburg, Germany, 2007; pp. 1–72. (In German)
46. Prestes, E.C.; dos Anjos, V.E.; Sodré, F.F.; Grassi, M.T. Copper, lead and cadmium loads and behavior in urban stormwater runoff in Curitiba, Brazil. *J. Braz. Chem. Soc.* **2006**, *17*, 53–60.
47. Hilliges, R.; Schriewer, A.; Helmreich, B. A three-stage treatment system for highly polluted urban road runoff. *J. Environ. Manag.* **2013**, *128*, 306–312.
48. Wichern, M.; Vesting, A.; Pecher, K.H.; Giga, A.; Wolke, M.; Helmreich, B.; Huber, M.; Li, Y. *Untersuchungs- und Entwicklungsvorhaben im Bereich Abwasser zum Themenschwerpunkt Niederschlagswasserbeseitigung: Vergabe-Nr. 08/058.2. Reduktion von Kohlenwasserstoffen und anderen organischen Spurenstoffen durch ein dezentrales Behandlungssystem für Verkehrsflächenabflüsse, gefördert durch das Ministerium für Klimaschutz, Umwelt, Landwirtschaft, Natur- und Verbraucherschutz des Landes Nordrhein-Westfalen, wissenschaftlich begleitet durch das Landesamt für Natur, Umwelt und Verbraucherschutz Nordrhein-Westfalen (IV-7-042 600 002E). Abschlussbericht Projektphase II*; Ruhr-Universität Bochum: Bochum, Germany, 2015; pp. 1–260. (In German)
49. Grotehusmann, D.; Kasting, U. *Schlussbericht: Untersuchung über die Beschaffenheit der Oberflächenabflüsse von Parkflächen bei Tank- und Rastanlagen*; Ingenieurgesellschaft für Stadthydrologie mbH, Hannover: Bundesanstalt für Straßenwesen: Bergisch Gladbach, Germany, 2002; pp. 1–62. (In German)
50. Ball, J.E.; Jenks, R.; Aubourg, D. An assessment of the availability of pollutant constituents on road surfaces. *Sci. Total Environ.* **1998**, *209*, 243–254.
51. Wilson, D. *Highway 520 Bridge Storm Water Runoff Study*; King County Water and Land Resources Division, Department of Natural Resources and Parks: Seattle, WA, USA, 2006; pp. 1–66.

52. Lau, S.-L.; Han, Y.; Kang, J.-H.; Kayhanian, M.; Stenstrom, M.K. Characteristics of highway stormwater runoff in Los Angeles: Metals and polycyclic aromatic hydrocarbons. *Water Environ. Res.* **2009**, *81*, 308–318.
53. Kayhanian, M.; Singh, A.; Suverkropp, C.; Borroum, S. Impact of annual average daily traffic on highway runoff pollutant concentrations. *J. Environ. Eng.* **2003**, *129*, 975–990.
54. Horstmeyer, N.; Huber, M.; Drewes, J.E.; Helmreich, B. Evaluation of site-specific factors influencing heavy metal contents in the topsoil of vegetated infiltration swales. *Sci. Total Environ.* **2016**, *560–561*, 19–28.
55. FHWA. *Is Highway Runoff a Serious Problem?* Office of Infrastructure R&D, Turner-Faribank Highway Research Center: McLean, VA, USA, 1999.
56. Berbee, R.; Rijs, G.; de Brouwer, R.; Van Velzen, l. Characterization and treatment of runoff from highways in the Netherlands paved with impervious and pervious asphalt. *Water Environ. Res.* **1999**, *71*, 183–190.
57. Hagner, C. European regulations to reduce lead emissions from automobiles—Did they have an economic impact on the German gasoline and automobile markets? *Reg. Environ. Chang.* **2000**, *1*, 135–151.
58. Mangani, G.; Berloni, A.; Bellucci, F.; Tatàno, F.; Maione, M. Evaluation of the pollutant content in road runoff first flush waters. *Water Air Soil Pollut.* **2005**, *160*, 213–228.
59. Sansalone, J.J.; Buchberger, S.G.; Al-Abed, S.R. Fractionation of heavy metals in pavement runoff. *Sci. Total Environ.***1996**, *189–190*, 371–378.
60. Davis, A.P.; Shokouhian, M.; Ni, S. Loading estimates of lead, copper, cadmium, and zinc in urban runoff from specific sources. *Chemosphere* **2001**, *44*, 997–1009.
61. Root, R.A. Lead loading of urban streets by motor vehicle wheel weights. *Environ. Health Perspect.* **2000**, *108*, 937–940.
62. Bleiwas, D.I. *Stocks and Flows of Lead-Based Wheel Weights in the United States, U.S. Geological Survey Open File Report 1111*; U.S. Department of the Interior, U.S. Geological Survey: Reston, VA, USA, 2006; pp. 1–6.
63. Shinya, M.; Funasaka, K.; Katahira, K.; Ishikawa, M.; Matsui, S. Lead isotope ratios in urban road runoff. *Water Sci. Technol.* **2006**, *53*, 185–191.
64. McKenzie, E.R.; Money, J.E.; Green, P.G.; Young, T.M. Metals associated with stormwater-relevant brake and tire samples. *Sci. Total Environ.* **2009**, *407*, 5855–5860.
65. Thorpe, A.; Harrison, R.M. Sources and properties of non-exhaust particulate matter from road traffic: A review. *Sci. Total Environ.* **2008**, *400*, 270–282.
66. Helmreich, B.; Hilliges, R.; Schriewer, A.; Horn, H. Runoff pollutants of a highly trafficked urban road—Correlation analysis and seasonal influences. *Chemosphere* **2010**, *80*, 991–997.
67. Stenstrom, M.K.; Kayhanian, M. *First Flush Phenomenon Characterization*; University of California: Berkeley, CA, USA, 2005; pp. 1–69.
68. Terzakis, S.; Fountoulakis, M.S.; Georgaki, I.; Albantakis, D.; Sabathianakis, I.; Karathanasis, A.D.; Kalogerakis, N.; Manios, T. Constructed wetlands treating highway runoff in the central Mediterranean region. *Chemosphere* **2008**, *72*, 141–149.
69. Liu, A.; Gunawardana, C.; Gunawardena, J.; Egodawatta, P.; Ayoko, G.A.; Goonetilleke, A. Taxonomy of factors which influence heavy metal build-up on urban road surfaces. *J. Hazard. Mater.* **2016**, *310*, 20–29.
70. Driscoll, E.; Shelley, P.E.; Strecker, E.W. *Pollutant Loadings and Impacts from Highway Stormwater Runoff, Volume III: Analytical Investigation and Research Report*; Report No. FHWA-RD-88-008; U.S. Department of Transportation: Washington, DC, USA, 1990; pp. 1–162.
71. Li, M.; Barrett, M.; Rammohan, P.; Olivera, F.; Landphair, H. Documenting stormwater quality on Texas highways and adjacent vegetated roadsides. *J. Environ. Eng.* **2008**, *134*, 48–59.
72. Dannecker, W.; Au, M.; Stechmann, H. Substance load in rainwater runoff from different streets in Hamburg. *Sci. Total Environ.* **1990**, *93*, 385–392.
73. Zessner, M. *Bedeutung und Steuerung von Nährstoff- und Schwermetallflüssen des Abwassers*; Wiener Mitteilungen: Wien, Austria, 1999; Volume 157. (In German)

74. Ilyin, I.; Gusev, A.; Rozovskaya, O.; Strijkina, I. *Transboundary Pollution of Germany by Heavy Metals and Persistent Organic Pollutants in 2010*; EMEP/MSC-E technical Report 5/2012; EMEP/MSC-E: Moscow, Russia, 2012; pp. 1–23.
75. Kocher, B.; Brose, S.; Chlubek, A.; Görg, C.; Klein, N.; Siebertz, I. *Stoffeintrag in Straßenrandböden: Messzeitraum 2006/2007*; Berichte der Bundesanstalt für Straßenwesen: Bergisch Gladbach, Germany, 2010; Volume 199, pp. 1–38. (In German)
76. Hullmann, H.; Kraft, U. Kupfer und Zink—Die Bedeutung der Abschwemmungen von Metalldächern. *gwf Wasser/Abwasser* **2003**, *144*, 127–133. (In German)
77. Pick, V.; Fettig, J.; Miethe, M.; Strüber, K. Dezentrale Behandlung des Niederschlagsabflusses von Verkehrsflächen in einer mehrstufigen Schachtanlage. *KA Abwasser Abfall* **2002**, *49*, 312–320. (In German)
78. Lippold, C. *Der Elsner 2015: Handbuch für Straßen- und Verkehrswesen (Planung, Bau, Erhaltung, Verkehr, Betrieb)*, 69 ed.; Otto Elsner Verlagsgesellschaft: Berlin, Germany, 2014. (In German)
79. DESTATIS. *Land-und Forstwirtschaft, Fischerei. Bodenfläche nach Art der tatsächlichen Nutzung. Fachserie 3 Reihe 5.1*; Statistisches Bundesamt: Wiesbaden, Germany, 2015. (In German)
80. Horwatich, J.A.; Bannerman, R.T. *Pollutant Loading to Stormwater Runoff from Highways Impact of a Highway Sweeping Program: Phase II, Madison, Wisconsin*; Wisconsin Department of Transportation: Madison, WI, USA, 2009; pp. 1–121.
81. Hillenbrand, T.; Toussaint, D.; Böhm, E.; Fuchs, S.; Scherer, U.; Rudolphi, A.; Hoffmann, M.; Kereißig, J.; Kotz, C. *Einträge von Kupfer, Zink und Blei in Gewässer und Böden—Analyse der Emissionspfade und möglicher Emissionsminderungsmaßnahmen*; Forschungsbericht 202 242 20/02, UBA-Texte 19/05; Umweltbundesamt: Dessau, Germany, 2005; pp. 1–329. (In German)
82. Kocher, B.; Brose, S.; Feix, J.; Görg, C.; Peters, A.; Schenker, K. *Stoffeinträge in den Straßenseitenraum—Reifenabrieb*; Berichte der Bundesanstalt für Straßenwesen: Bergisch Gladbach, Germany, 2010; Volume 188, pp. 1–33. (In German)
83. FGSV379/1. *Hinweise und Empfehlungen für die Beschaffung von Streustoffen—Kommentar zur TL-Streu*; FGSV Forschungsgesellschaft für Straßen- und Verkehrswesen e. V.: Köln, Germany, 2004. (In German)
84. Harned, D.A. *Effects of Highway Runoff on Streamflow and Water Quality in the Sevenmile Creek Basin, a Rural Area in the Piedmont Province of North Carolina, July 1981 to July 1982*; U.S. Geological Survey Water-Supply Paper 2329; U.S. Geological Survey: Denver, CO, USA, 1988; pp. 1–44.
85. Kobriger, N.P.; Geinopolos, A. *Sources and migration of highway runoff pollutants. Volume III: Research Report*; Report No. FHWA-RD-84/059; US Department of Transportation: Washington, DC, USA, 1984; pp. 1–368.
86. Gupta, M.K.; Agnew, R.W.; Kobriger, N.P. *Constituents of Highway Runoff. Volume I: State-of-the-Art Report*; Report No. FHWA-RD-81/042; US Department of Transportation: Washington, DC, USA, 1981; pp. 1–124.
87. Heshmati, A.; Vahidinia, A.; Salehi, I. Determination of heavy metal levels in edible salt. *Avicenna J. Med. Biochem.* **2014**, *2*, e19836.
88. Umweltbundesamt. *National Trend Tables for the German Atmospheric Emission Reporting (Heavy Metals) 1990–2013*; Final Version: 3 February 2015; Umweltbundesamt: Berlin, Germany, 2015.
89. Denier van der Gon, H.A.C.; van het Bolscher, M.; Visschedijk, A.J.H.; Zandveld, P.Y.J. *Study to the Effectiveness of the UNECE Heavy Metals Protocol and Costs of Possible Additional Measures. Phase I: Estimation of Emission Reduction Resulting from the Implementation of the HM Protocol*; TNO-Report B&O-A R 2005/193; TNO: Apeldoorn, The Netherlands, 2005; pp. 1–134.
90. Hallberg, M.; Renman, G.; Lundbom, T. Seasonal variations of ten metals in highway runoff and their partition between dissolved and particulate matter. *Water Air Soil Pollut.* **2007**, *181*, 183–191.
91. Harrison, R.M.; Johnston, W.R.; Ralph, J.C.; Wilson, S.J. The budget of lead, copper and cadmium for a major highway. *Sci. Total Environ.* **1985**, *46*, 137–145.

92. Harrison, R.M.; Johnston, W.R. Deposition fluxes of lead, cadmium, copper and polynuclear aromatic hydrocarbons (PAH) on the verges of a major highway. *Sci. Total Environ.* **1985**, *46*, 121–135.
93. Steiner, M.; Boller, M.; Schulz, T.; Pronk, W. Modelling heavy metal fluxes from traffic into the environment. *J. Environ. Monit.* **2007**, *9*, 847–854.
94. Hewitt, C.N.; Rashed, M.B. An integrated budget for selected pollutants for a major rural highway. *Sci. Total Environ.* **1990**, *93*, 375–384.
95. Granato, G.E. Deicing chemicals as source of constituents of highway runoff. *Transp. Res. Rec. J. Transp. Res. Board* **1996**, *1533*, 50–58.
96. Dierkes, C.; Lucke, T.; Helmreich, B. General Technical Approvals for Decentralised Sustainable Urban Drainage Systems (SUDS)—The Current Situation in Germany. *Sustainability* **2015**, *7*, 3031–3051.
97. Huber, M.; Welker, A.; Dierschke, M.; Drewes, J.E.; Helmreich, B. A novel test method to determine the filter material service life of decentralized systems treating runoff from traffic areas. *J. Environ. Manag.* **2016**, *179*, 66–75.

CHAPTER 2

Urban Stormwater Runoff: A New Class of Environmental Flow Problem

Christopher J. Walsh[1*], Tim D. Fletcher[1,2], Matthew J. Burns[2]

[1]Department of Resource Management and Geography, The University of Melbourne, Parkville, Victoria, Australia

2 Department of Civil Engineering and Monash Water for Liveability, Monash University, Victoria, Australia

ABSTRACT

Environmental flow assessment frameworks have begun to consider changes to flow regimes resulting from land-use change. Urban stormwater runoff, which degrades streams through altered volume, pattern and quality of flow, presents a problem that challenges dominant approaches to stormwater and water resource management, and to environmental flow assessment. We used evidence of ecological response to different stormwater drainage systems to develop methods for input to environmental flow assessment. We identified the nature of hydrologic change resulting from conventional urban stormwater runoff, and the mechanisms by which such hydrologic change is prevented in streams where ecological condition has been protected. We also quantified the increase in total volume resulting from urban stormwater runoff, by comparing annual streamflow volumes from undeveloped catchments with the volumes that would run off impervious surfaces under the same rainfall regimes. In catchments with as little as 5–10% total imperviousness, conventional stormwater drainage, associated with poor in-stream ecological condition, reduces contributions to baseflows and increases the frequency and magnitude of storm flows, but in similarly impervious catchments in which streams retain good ecological condition, informal drainage to forested hillslopes, without a direct piped discharge to the stream, results in little such hydrologic change. In urbanized catchments, dispersed urban stormwater retention measures can potentially protect urban stream ecosystems by mimicking the hydrologic effects of informal drainage, if sufficient water is harvested and kept out of the stream, and if discharged water is treated to a suitable quality. Urban stormwater is a new class of environmental flow problem: one that requires reduction of a large excess volume of water to maintain riverine ecological integrity. It is the best type of problem, because solving it provides an opportunity to solve other problems such as the provision of water for human use.

1. INTRODUCTION

Humanity faces a major challenge to provide the world's growing population with reliable and affordable water supplies, while protecting the ecological integrity of freshwater ecosystems [1]. The concept of environmental flows has developed to meet this challenge, with the aim of identifying the critical elements of flow regimes that should be retained or restored [2]. Environmental flow assessment and provision has to date focussed mainly on mitigation of the effects of water extraction, but increasingly, changes to the flow regime resulting from land-use changes are being considered [3], [4].

Of all land-uses, urbanization arguably causes the largest changes to volumes and patterns of flow running from catchments to streams and rivers. The urban water system has three elements; water supply, wastewater and stormwater. Managers of urban water resources focus primarily on water supply imported from distant sources (usually rural rivers), or from groundwater, distributed through a network of water supply pipes, and on wastewater exported through a second network to treatment plants. While the imported water supply and exported wastewater of urban areas substantially alters the water balance of cities [5], in most modern cities with separate stormwater drainage and sanitary sewerage systems, these elements of the urban water system bypass, and therefore have relatively little influence on, the flow regime of streams and rivers of the city themselves. It is the third element of the urban water system–stormwater (runoff from impervious surfaces during and immediately after wet weather)–, generally of a similar volume to imported water [6], but rarely considered by water resource managers, that has the largest effect on flow regimes of urban streams and rivers.

It has long been recognized that covering land with impervious surfaces, such as roofs and roads, reduces both the volume of water infiltrating into soils and the volume of water lost to the air through evapotranspiration, thus increasing the volume of runoff following rain events [7]. These changes to the water balance are exacerbated, and their effects transferred to streams and rivers, by stormwater drainage systems: a third network of pipes under cities (that have separate sewers), which is designed to minimize flood risk, by efficiently draining all runoff from impervious surfaces directly to the nearest receiving water [7]. Here we term such drainage systems, that directly connect impervious surfaces to a receiving water (via gutters, pipes and perhaps sealed channels), conventional stormwater drainage [8].

To date, urban stormwater runoff has primarily been considered a water quality problem [8]. The combined hydrologic and water quality perturbations caused by urban stormwater runoff make the management of urban stormwater an ideal case to develop and apply the new broad conception of environmental flow management [2].

Poff et al. [3] developed a framework for determining and implementing this new conception of environmental flow requirements. The framework requires initial hydrologic analyses to identify the nature and degree of flow alteration for each river type (based on hydrologic and geomorphic classifications) followed by an assessment of responses of river biota and ecological processes to flow alteration, a social process for determining ecological objectives, and adaptive implementation and monitoring [3].

A primary step in assessing and implementing environmental flows is to estimate how ecologically relevant components of the flow regime are altered by a human activity [3]. Typically, ecologically relevant indicators of hydrologic alteration have been selected by identifying those hydrologic indicators that are well correlated with changes in ecological indicators, and a few studies have applied such an approach in an urban setting. The general conclusion from such studies (e.g. [4], [12]) is that the primary urban-induced hydrologic changes driving ecological degradation centre on the increased frequency and magnitude, and reduced duration of storm flow events, and reduced low flows. While these studies did not explicitly consider water quality, both high and low flows in urban areas are associated with increased pollutant concentrations [13].

Conventional stormwater drainage has been identified as a primary driver of the commonly observed, severe degradation of stream ecosystems in urban catchments [9], [10]. Such systems send polluted stormwater to receiving waters every time there is sufficient rain to generate runoff from impervious surfaces, greatly increasing the frequency of hydraulic and water quality disturbances to streams [11]. The large changes to the volume and pattern of flow caused by urban stormwater drainage systems point to urban stream degradation being, in large part, a hydrologic problem. Its solution should therefore be informed by approaches to environmental flow management.

Recent considerations of environmental flow requirements for freshwater and estuarine systems have broadened the focus of environmental flow management to include all aspects of land and water use, and seek to integrate water quality with flow management [2].

When provision of environmental flow requirements is addressed solely through management of water storages and diversions, the mechanisms for hydrologic alteration are usually obvious. However, when the primary environmental flow problem is a result of altered land use, such as urbanization, then an additional step in an environmental flow assessment framework is required: determination of the mechanisms for the hydrologic alteration, and strategies for their mitigation. The ultimate mechanisms by which urbanization degrades streams are manifold, but can usually be attributed to a small number of landscape-scale land-use practices for which alternative management approaches are available [9]. Hydrologists have long recognised that conventional stormwater drainage is the dominant driver of urban-induced hydrologic changes [7], [8], [14], leading ecologists to posit the importance of such systems in driving the ecological degradation of urban stream ecosystems [9].

Only a few studies have attempted to distinguish the effects of urban stormwater drainage on stream ecosystems from the general effects of urban density. Most studies of ecological and hydrologic changes resulting from urbanization have used total impervious coverage or other correlated urban land-use measures as explanatory variables (e.g. [15], [16], [17]), and have therefore been unable to discern if variation in ecological response could have been caused by differences in stormwater drainage systems. Differences that could alter the response of stream ecosystems and their flow regimes include the proportion of impervious areas that are connected to conventional drainage systems, and the hydraulic efficiency of alternative drainage systems (i.e. the degree to which water and attached pollutants are retained or attenuated as they are transferred downstream). Studies of streams across urban areas with differing extent of conventional urban stormwater drainage have shown that streams with substantial catchment urbanization can retain ecological structure and function more typical of streams with undeveloped catchments, if their urban areas lack conventional stormwater drainage systems [18]. In contrast, streams with similar levels of catchment urbanization (as measured by total imperviousness), but in which most urban surfaces are drained by conventional stormwater drainage, exhibit poor ecological condition.

In this paper, we use evidence of ecological response to different stormwater drainage systems to develop methods for input to environmental flow assessment. We do not aim to conduct a full environmental flow assessment, but to demonstrate a process for identifying the nature of hydrologic change resulting from conventional urban stormwater runoff, and the mechanisms by which such hydrologic change is prevented in streams where ecological condition has been protected. We propose these as important steps in environmental flow assessment for stormwater management.

Unlike the usual environmental flow problem of needing to allocate a reduced volume of water to the environment, urban stormwater runoff presents a problem of increased runoff volume, which should be prevented from reaching receiving streams and thus could be used by humans. We quantify the magnitude of this problem by comparing streamflow volumes from a range of undeveloped catchments in the Melbourne region with the volumes that would run off impervious surfaces in those catchments, and use a published global analysis to demonstrate the generality of these findings. Our analyses reveal urban stormwater as a new class of environmental flow problem: one that requires reduction of a large excess volume of water to maintain riverine ecological integrity. Urban stormwater runoff is thus revealed as the best type of problem, because solving it provides an opportunity to solve other problems such as the provision of water for human use in cities.

2. MATERIALS AND METHODS

Our analysis comprises two parts. We first compare hydrographs of four streams of contrasting catchment urban density and stormwater drainage infrastructure. We use these data to contrast the severe alterations to flow patterns resulting from conventional stormwater drainage, to the conservation of more natural flow patterns afforded by informal drainage. We thus link these hydrologic patterns with previously reported strong correlation between stormwater drainage connection and ecological indicators to infer that the observed changes to flow regime are likely important drivers of observed ecological degradation. Because such a short-term hydrographic analysis does not allow a valid assessment of the urban water balance, we use a second approach to assess the influence of conventional urban stormwater drainage on streamflow coefficients. We compare streamflow coefficients for streams of the world [19], [20] as a function of catchment rainfall and vegetation, validated with coefficients calculated for streams of the Melbourne region with a long flow record. We compare these to impervious runoff coefficients estimated from long-term rainfall records.

2.1 Determining the effect of conventional stormwater drainage on flow regimes

We first consider four of the small streams in eastern Melbourne, Victoria, Australia, studied by Walsh et al. [18], each with some level of catchment urbanization, but two of which almost completely lack conventional stormwater drainage and two for which most urban areas are drained conventionally (Figure 1).

The two streams with near-zero connected imperviousness, Olinda and Sassafras creeks, were in good ecological condition: they shared comparably low pollutant concentrations and algal biomass, and supported algal and invertebrate assemblages comparable to undisturbed forested streams of the region (Figure 1). Little Stringybark Creek, which has only marginally higher total imperviousness than Sassafras Creek, but substantially higher connected imperviousness, was in poor ecological condition, like the even more urban Brushy Creek (Figure 1).

Most of the impervious surfaces of Little Stringybark and Brushy creeks drain to conventional drainage systems, as evidenced by the small differences between total and connected imperviousness (Figure 1A, B). In contrast, the very low levels of connected imperviousness in Sassafras and Olinda result from the buildings and roads of these catchments being almost all drained informally: roofs generally drain to surrounding gardens or to rainwater tanks, and almost all sealed roads lack curbs, and drain either into the surrounding forest or to earthen or vegetated swales.

The main road of Sassafras township, at the head of the Sassafras catchment, is conventionally drained by curb and channel, leading to a pipe which drains to the forested hillslope below, ~150 m upslope from the creek. In 2004, the outlet of the pipe distributed flow across the forest floor, permitting most flows from the pipe to infiltrate into forest soils.

The catchments of all four streams are dominated by urban land or forest, with little agriculture, except for a small area of horticulture in the Olinda Creek catchment, and low-intensity grazing in rural-residential areas of Little Stringybark Creek. Our conclusions are not influenced by abstraction patterns: there is no abstraction from Brushy or Little Stringybark creeks, and only very minor abstraction in Sassafras and Olinda Creek (mean licenced volume of 0.01 and 0.03 mm/d respectively: source Melbourne Water).

See [21] for further details on the catchment characteristics of the four streams.

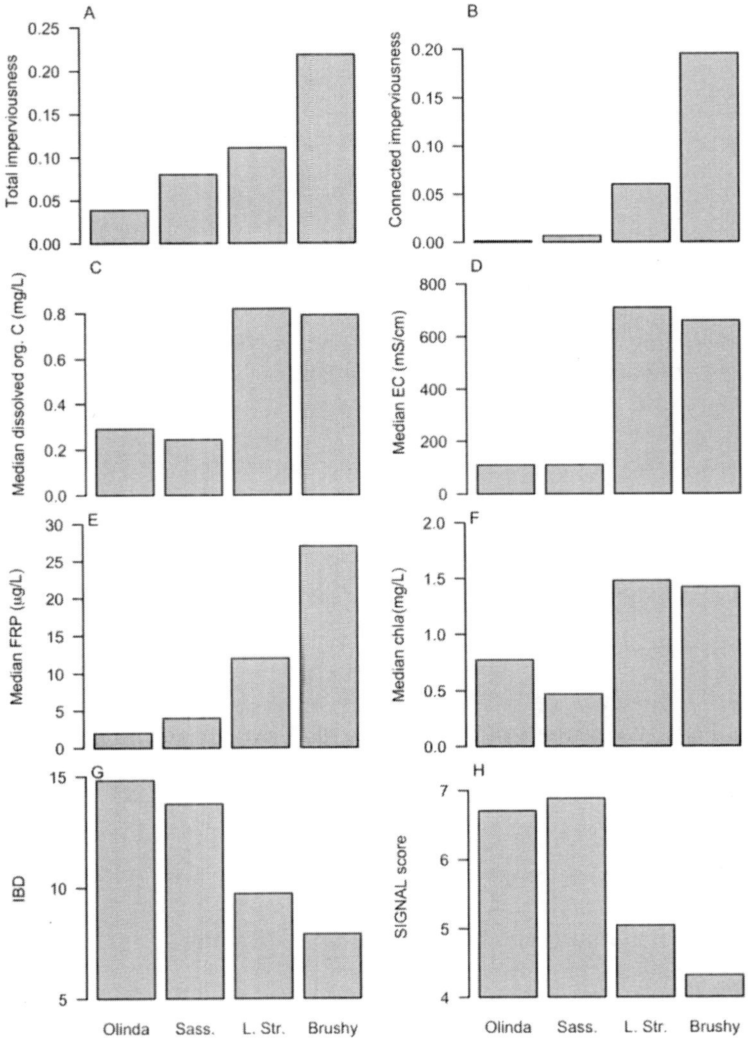

Figure 1. Catchment urbanization measures and in-stream ecological indicators for four contrasting streams in eastern Melbourne. (A) Total and (B) connected imperviousness (estimated for 2004). Median 2001–2002 (C) dissolved organic carbon, (D) electrical conductivity, (E) filterable reactive phosphorus. (F) Mean 2001–2002 median benthic chlorophyll *a*, as an estimate of algal biomass. (G) Indice biologique diatomée (IBD) [53]. (H) SIGNAL score [54] for stream-edge samples. IBD and SIGNAL are indices based on diatom species and macroinvertebrate families, respectively, weighting each taxon by their sensitivity to pollution. Adapted from Walsh et al. [18], with minor revision to Little Stringybark Creek values from 2004 aerial photos and ground-truthing). Sass. = Sassafras Creek, L. Str. = Little Stringybark Creek.

To assess the flow regime in the four sites we used 6-min time-step flow data from a permanent flow gauge installed ~100 m upstream of the sampling site on Brushy Creek (Melbourne Water station 229249: www.melbournewater.com.au), and a limited record of water depth collected at the other three sites (using Odyssey™ capacitance water level probes–Dataflow Systems, Christchurch, New Zealand–in 44-mm diameter polyvinyl-chloride stilling wells, logging at 5-min intervals). Concurrent data without gaps for all four streams were only available for the period 24 Nov 2004–6 Jan 2005 (43 days).

Discharge of Sassafras and Little Stringybark creeks was estimated manually using either a CMC 20 current meter counter or a Marsh-McBirney flow-mate™velocity meter, 6–10 times at each site over Nov 2004–Mar 2005, and relationships between discharge and simultaneous depth logger readings were derived. Discharge in Olinda Creek was estimated using a quadratic relationship between discharge recordings from a permanent flow station downstream (Melbourne Water Station 229260, catchment area 23 km^2 compared to 9 km^2 for our site, with similar catchment land use) and corresponding depth logger readings. All discharge estimates were normalized by catchment area, to be expressed as mm/d.

Discharge patterns in the four streams were compared to the daily rainfall recorded at the two Melbourne Water flow gauges and at the Bureau of Meteorology (www.bom.gov.au) pluviograph on Mt Dandenong (Station 086243). The three rainfall stations were within the area bounded by the four study catchments.

While the period of record considered here is short, it encompassed several rain events of a wide range of sizes (in which rain was relatively evenly spread across the catchments), and two inter-event periods that permit an assessment of differences and similarities in hydrologic response of the four streams.

2.2 Quantifying the increased volume of urban stormwater runoff

To assess the increase in runoff volume generated by impervious surfaces in the Melbourne region, we compared streamflow coefficients (mean annual discharge depth divided by mean annual rainfall) of undeveloped, unregulated streams in the region with impervious runoff coefficients (mean annual runoff depth from an impervious area divided by mean annual rainfall). The difference between the streamflow coefficient and impervious runoff coefficient of a site represents the loss of evapotranspiration and equates to the excess runoff generated by a given area of impervious surface above the volume that would have contributed to streamflow in the pre-urban state (the lost infiltration flows). The analyses were undertaken using streamflow and rainfall data from across the Melbourne region (Table 1), spanning the range of mean annual rainfall in the region (400–1800 mm/y, although the wettest urban areas receive 1200 mm/y).

The range of streamflow coefficients for streams and rivers of the Melbourne region in the absence of urbanization was estimated using the relationships derived by Zhang et al. [19], [20] that predict the proportion of mean annual rainfall lost to evapotranspiration (and conversely, the proportion that became stream flow, assuming no changes in soil water storage over the long-term and no loss to deep seepage [20]) as a function of catchment vegetation cover (grassland or forest) and rainfall (Figure 2).

Table 1. Flow and rainfall gauging stations used in the water balance analyses, with period of data used.

Gauge	Stream	Years	Vegetation[3]	Rainfall[4]
406213[1]	Campaspe32	1985–1994	Pasture, crops	
407214[1]	Creswick Creek	1985–1994	Mixed	
407253[1]	Piccaninny Creek	Long term	Mixed	
406224[1]	Mt. Pleasant Creek	1985–1994	Pasture	
406235[1]	Wild Duck Creek	1985–1994	Pasture, crops	
406262[1]	Axe Creek 33	1990–1994	Pasture, crops	
406200[1]	Coliban River	1985–1993	Pasture, crops	
228207A[2]	Bunyip River at Headworks	1994–1996	Forest	x
229627[2]	Merri Ck	1994–1996	Pasture	x
229215B[2]	Woori Yallock Ck	1994–1996	Mixed	x
229690[2]	Olinda Ck at Mt Evelyn	1994–1996	Forest	x
586097	Mt St Leonard	1954–1999		x

086027	Croydon	1965–1970, 1980–1997	x
086251	Upwey	1968–1976, 1978–2001	x
086071	Melbourne CBD	1970–2001	x
086282	Melbourne airport	1970–2001	x
587030	Werribee	1968–1990	x

[1]Streams outside the Melbourne region, for which streamflow coefficients were derived by Jolly et al. [22] and catchment vegetation classes were assigned by Zhang et al. [19].
[2]Streams in the Melbourne region, for which streamflow coefficients were derived using Melbourne Water flow gauge data and Bureau of Meteorology rainfall data for 1994–1996. Catchment vegetation classes were estimated from 2001 aerial imagery.
[3]Catchment vegetation class is indicated for flow-gauging stations used to calculate streamflow coefficients. [4]Stations from which rainfall data were used to estimate impervious runoff are indicated in the final column.

To confirm that measured streamflow coefficients for Melbourne streams fall within the predicted bounds for grassland and forested catchments, we plotted streamflow coefficients calculated by Jolly et al. [22], for seven undeveloped, unregulated streams within 200 km of Melbourne (Table 1). We also determined streamflow coefficients for four Melbourne Water flow gauges on streams with undeveloped, unregulated catchments within the Melbourne region. For each of these gauges, weighted mean catchment rainfall was estimated using dynamically assigned Thiessen polygons around all available daily rain gauge data (Bureau of Meteorology) for 1994–1996 (Table 1).

Runoff from impervious surfaces was estimated from daily rainfall data from 11 gauges across the Melbourne region (Table 1), assuming an initial loss of 1 mm/d (e.g. [23]). Impervious runoff coefficients were calculated by dividing the sum of estimated impervious runoff by the sum of rainfall over the period. To compare impervious runoff coefficients and streamflow coefficients in undeveloped catchments, the relationship between impervious runoff coefficient and mean annual rainfall (log-transformed to reduce leverage of high values) was determined by linear regression. The increase in impervious runoff was estimated using the range of differences between this line of best fit for impervious runoff coefficients and the two curves derived by Zhang et al. [20] for streamflow coefficients in grassland and forested catchments (Figure 2).

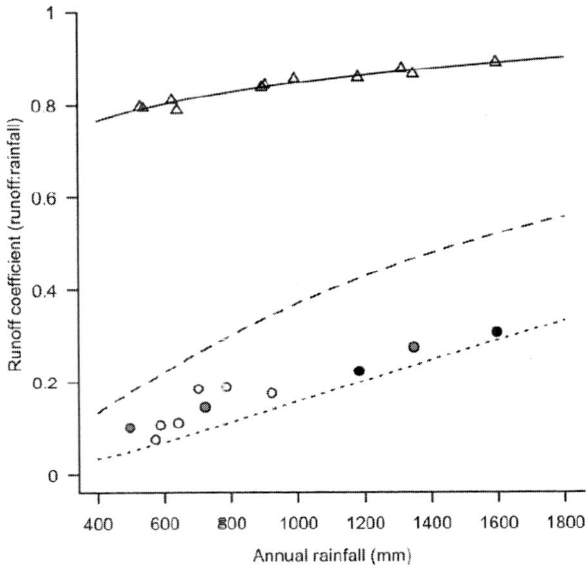

Figure 2. Impervious runoff and streamflow coefficients for the Melbourne region. Estimated annual runoff coefficients (C) from impervious surfaces (open triangles) from sites across the Melbourne region as a function of mean annual rainfall (MAR). Regression line: $C=0.230+0.206\times\log_{10}(MAR)$. $R^2=0.94$. Annual streamflow coefficients from 11 streams with forested (closed circles), grassland (open circles) or mixed forested and grassland catchments (grey circles) across the Melbourne region as a function of mean annual rainfall. The lines surrounding these stream points are the relationships between streamflow derived by Zhang et al. [20] for grassland (dashed curve) and forested catchments (dotted curve) of the world.

3. RESULTS

3.1 Determining the effect of conventional stormwater drainage on flow regime

Three major differences in the four hydrographs (Figure 3) are likely to signify ecologically important changes to the flow regime:

1) Olinda and Sassafras creeks, which had similarly low connected imperviousness (Figure 1), had similarly high baseflow, while Little Stringybark Creek, which had substantially higher connected imperviousness, but similar total imperviousness to Sassafras Creek

(Figure 1), had lower baseflow. Brushy Creek, which had higher connected imperviousness again (Figure 1), had even lower baseflow;.

2) Discharge increased only slightly in response to rain events of 10–20 mm in Olinda and Sassafras creeks (e.g. events A and C, Figure 3), with a slightly larger response in the latter stream, usually not exceeding an increase in discharge of 2–3 times. In contrast, such events caused an increase in discharge of an order of magnitude in Little Stringybark Creek, and of 2–3 orders of magnitude in Brushy Creek.

3) Discharge did not change in Olinda and Sassafras creeks following rain events of <5 mm/d (e.g. event B, Figure 3), compared to large increases in discharge in Little Stringybark and Brushy creeks.

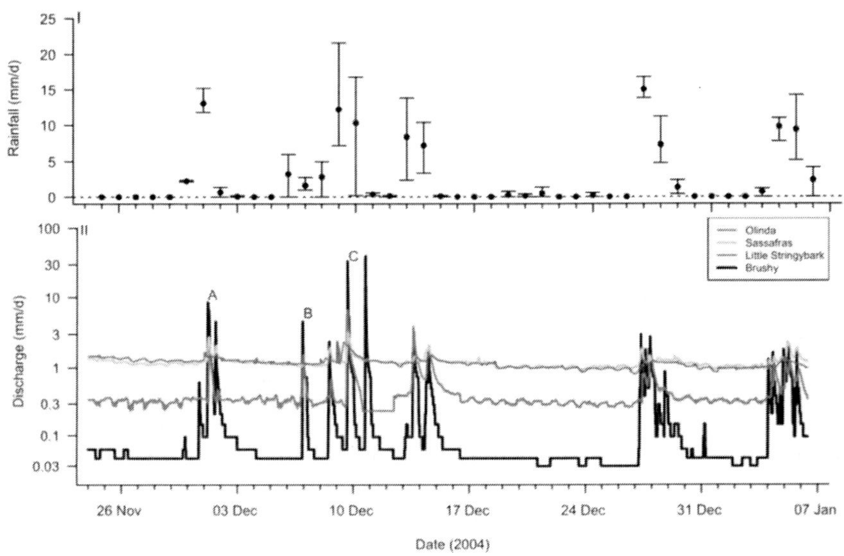

Figure 3. Rainfall and discharge of the four study streams over 43 days in 2004. I. Mean ± range daily rainfall recorded in the three rain gauges that fall within the area bounded by catchments of the streams. II. 5- or 6-minute hydrographs for each stream. Three flow events (A, B and C) are discussed in the text.

3.2 Quantifying the increased volume of urban stormwater runoff

Streamflow coefficients in and around the Melbourne region consistently fell within the bounds of the relationships derived by Zhang et al. [20] for forested (lower bound) and grassland (upper bound) catchments (Figure

2). While the predicted effect of vegetation on streamflow coefficient is not evident from the small number of points for the Melbourne region, the two curves provide useful bounds for estimating streamflow from undeveloped, unregulated catchments.

Impervious runoff coefficients were well predicted by mean annual rainfall (R^2=0.94), and were consistently much higher than streamflow coefficients in undeveloped streams (Figure 2). More importantly, the difference between streamflow and impervious runoff coefficients equates to an excess volume of runoff from 1 ha of impervious surface of 2.6–3.0 ML/y in catchments with mean annual rainfall of 400 mm rising to 5.1–7.8 ML/y in catchments with 1200 mm/y (Figure 4).

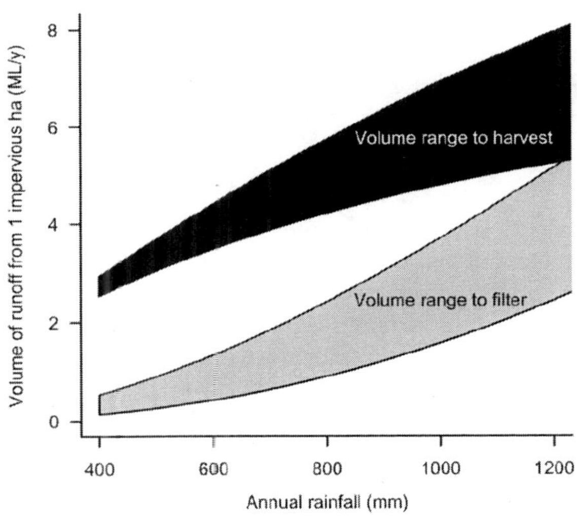

Figure 4. Impervious runoff volume partitioned into lost subsurface flows and lost evapotranspiration. Annual volume of runoff from 1 ha of impervious surface (from the relationship between impervious runoff coefficient and annual rainfall shown in Figure 2), partitioned into two parts: the volume that needs to be passed through infiltration systems (or catchment soils) to restore lost subsurface flows (grey polygon), and the volume that needs to be retained in the catchment and not delivered to the stream (through evapotranspirational loss or through use and export from the catchment through the wastewater stream). For each part, a range is indicated between situations in which the target streamflow is predicted by the grassland curve (more stream flow, less retention in catchment) or by the forest curve (less streamflow, more retention in the catchment) of Zhang et al. [20] (Figure 2).

4. DISCUSSION

4.1 Mechanisms by which informal drainage protects streams

The similarity of flow regimes in Olinda and Sassafras creeks (Figure 3) and their good water quality and ecological condition (Figure 1) demonstrates the potential for streams with substantial catchment urbanization to retain important elements of the flow regime and water quality that are likely to be required for the protection of in-stream ecological values.

The lack of conventional stormwater drainage in these two catchments points to several mechanisms that can be replicated by appropriate stormwater management technologies to retain or restore components of the flow regime. The clear driver of increased frequency and magnitude of high flows observed in Little Stringybark and Brushy creeks, but not the other two streams, is impervious runoff directed to the stream through pipes.

The informal drainage that dominates the Sassafras and Olinda catchments permits the retention (including some harvesting) and infiltration (and subsequent uptake and loss by the large area of downslope plants) of water from frequent, small-to-moderate storm flows [18], explaining the lack of streamflow response to events of <5 mm/d in Sassafras and Olinda creeks. In contrast, the conventional stormwater drainage systems of Little Stringybark and Brushy creeks pass runoff and its associated pollutants directly to the streams in all events large enough to elicit runoff from impervious surfaces (typically >1 mm/d, e.g. [23], although runoff from pitched roofs is typically initiated after as little as 0.5 mm [24]), explaining both the increased frequency and magnitude of high flows observed in these two streams.

Reduced infiltration resulting from coverage of impervious surfaces and lined drainage systems in Little Stringybark and Brushy creeks is most likely exacerbated by soil changes that occur through urbanization. The residential gardens of Little Stringybark and Brushy catchments are typically less treed than those of Sassafras and Olinda, and compaction of topsoils is likely to be more widespread. The resultant loss of soil storage capacity is likely to be another important driver of reduced baseflows in these catchments [25].

However, pervious areas with reduced soil storage capacity are unlikely to be large contributors to increased frequency or magnitude of high flows. Like informally drained impervious surfaces, runoff from such surfaces is unlikely to find its way into stormwater drains in most rain events. Instead, water is likely to flow to a downslope area with greater infiltration capacity, augmenting subsurface flow. Studies of hillslope storm hydrology of forested catchments have generally found subsurface flow to be the dominant pathway (e.g. [26]). Hillslopes form complex mosaics with differing soil porosity, meaning that any overland flow generated from intense rain events on a hillslope is likely to be re-absorbed into the soils downslope before reaching a stream [27].

In contrast, all urban stormwater delivered through conventional drainage systems is delivered to the receiving water unfiltered, through pipes, resulting in more frequent, larger flood peaks of shorter duration and of poor quality. The challenge for urban stormwater managers is therefore to mimic the natural hillslope hydrology of forested catchments, so that impervious runoff is delivered to streams with the appropriate temporal pattern, volume and quality.

4.2 Environmental flow standards for stormwater management

This assessment of the hydrologic differences between streams with informal catchment drainage (that retain good ecological condition) and streams with conventional stormwater drainage (that are degraded ecologically), allows the identification of environmental flow standards for the protection of streams from urban stormwater runoff. The first requirement of such standards is the provision of flow through filtration systems (or catchment soils) that mimics the temporal pattern, quantity and quality of subsurface flows in undeveloped catchments.

Secondly, the systems should be sized so that they overflow and spill to conventional drainage systems or streams only infrequently: ideally at a frequency similar to that of storm events large enough to initiate widespread overland flow on hill slopes of the undeveloped catchment.

In sparsely-to-moderately urbanized catchments such as Olinda and Sassafras creeks, informal drainage provides protection to the streams because the catchments retain sufficient areas of intact forest soils downslope that can absorb the additional runoff from urban surfaces. In more densely urbanized areas, the magnitude of flow increases and a lack of downslope soil storage (and coupled evapotranspiration through vegetation) will likely require constructed retention and treatment technologies to adequately replicate forested hillslope processes. Such techniques are likely to involve some form of infiltration.

Infiltration systems can be designed to retain and treat storm flows to substantially reduce the frequency of unfiltered runoff from impervious surfaces, while helping to restore baseflows [28], [29]. However, because impervious surfaces increase the volume of stormwater runoff and reduce vegetation cover (reducing capacity to lose water to evapotranspiration), very large infiltration systems will be needed to adequately retain storm flows, unless a large proportion of stormwater runoff is harvested. Without harvesting runoff for uses that either result in the water being lost to the wastewater stream through the (separate) sanitary sewer, or to the air through evapotranspiration by urban vegetation, the amount of infiltration required to match pre-development runoff frequency will result in infiltrated (filtered) flows greater than the volume that fed streamflow in the pre-urban state, resulting in increased streamflow volumes. Ideally, across urban Melbourne where mean annual rainfall varies from 400 to 1200 mm/y, only a minority of stormwater runoff should be delivered to the stream (as filtered flows), with the rest being harvested or lost to evapotranspiration (Figure 4).

This therefore points to a third requirement of an environmental flow standard for stormwater management: harvesting of an appropriate volume of stormwater runoff to prevent it from reaching the receiving water. This marks urban stormwater as a unique water resource: one for which human use provides environmental benefit.

4.3 Stormwater management technologies to meet environmental flow standards

The volume of excess urban stormwater runoff in Melbourne corresponds to a large proportion of the city's total water demand, which is currently met almost completely by water from rivers in forested catchments outside the city. The total water demand for Melbourne in 2004–2005 was equivalent to 0.14 ML/y per person [30], while the mean impervious area per person in 2004 was 266 m^2 (total impervious area 964 km^2, Melbourne Water, unpubl. data, population 3.63 million [31]). Excess runoff from an impervious area of this size in the driest part of Melbourne equals 57% of the total demand (the upper bound of the black polygon for annual rainfall of 400 mm in Figure 4=3 ML/y excess runoff from 1 ha, or 0.08 ML/y from 266 m^2), while in the wettest part of Melbourne excess runoff from an equivalent impervious area produces 147% of total demand (upper bound of black polygon for annual rainfall of 1200 mm in Figure 4). Such large volumes of excess runoff relative to demand, in the presence of existing water supplies, make adequate retention to protect stream ecosystems challenging.

In many cities of the world, with higher population densities than Melbourne and inadequate water supply systems, water demand is likely to be sufficient to allow harvesting of all excess urban stormwater runoff. In low-density cities such as Melbourne, use of all excess runoff is more challenging, but new and existing technologies for stormwater retention make such retention feasible, while the potential social and environmental benefits of using stormwater to improve the urban landscape and amenity provide incentives to create new uses for stormwater.

Household-scale harvesting (typically of roofwater) is already widely practiced in many places around the world. Collection, storage, distribution and treatment of urban stormwater at larger scales will be necessary to allow adequate harvesting of excess runoff from areas with low demands and distribution to areas with high demands. Such technologies are well developed, requiring storage volumes that are feasible in most urban settings [32]. Substantial challenges remain, however, in the development of cost-effective techniques for the treatment of stormwater [33], [34].

Technologies for stormwater infiltration, such as infiltration basins and trenches, soakaway pits, bioretention systems (rain gardens) and porous pavements are well developed [28], [35]. Rain gardens use vegetated soil media to improve stormwater quality [36], attenuate flow rates [37] and, depending on the design, promote infiltration and evapotranspiration [38]. Where infiltration is not possible, they can be built with an underdrain which discharges treated water directly to the stormwater system (and thus the receiving water), at a rate controlled to match the pre-development low-flow regime [37]. However, to mimic the type of highly attenuated flow regimes observed for Sassafras and Olinda Creek (Figure 3), large detention storage is required.

Typically, infiltration-based systems need to have an area of 2–5% of the upstream impervious area, depending on climate and soil type, to retain enough stormwater to permit replication of infiltration flows that would have occurred in the pre-urban context [38]. Infiltration systems can provide other benefits by forming part of green open space, while porous pavements can form other useable urban surfaces, but with less evapotranspiration loss.

Importantly, infiltration techniques alone are unlikely to achieve the stormwater retention necessary to return streamflow volumes to near natural levels (evapotranspiration losses in infiltration systems will not be enough to mimic pre-development catchment-wide evapotranspiration losses). Integration of harvesting and infiltration-based techniques to restore both high- and low-flow hydrology towards their natural levels is necessary, and can be achieved by simple methods such as using a proportion of harvested storage volumes for passive landscape irrigation. Such systems can reduce overflow frequency and increase evapotranspiration losses and infiltration, with little effect on harvested yields [39].

Existing stormwater management technologies are therefore capable of providing environmental flows to protect stream ecosystems in urban areas. The primary impediments to implementation of such technologies for the protection of streams are social and political [40]. A particular barrier is that the unique nature of the stormwater problem (as an environmental flow) and opportunity (as a water resource) is largely unrecognised in the environmental flow literature, and among urban water managers.

4.4 Urban stormwater as an environmental flow problem and an unrealized water resource

Most environmental flow problems arise from water being extracted for human use: the challenge for environmental flow researchers and practitioners in such situations is how to distribute the remainder for maximum environmental benefit [2]. This focus on extraction of water from aquatic ecosystems, compounded with the tendency for water resource managers to prefer centralized systems [41], [42], leads to a tendency of urban water managers to first consider extraction from urban rivers and drains when identifying urban stormwater harvesting projects (e.g. [43], [44]). Such a conception of stormwater harvesting has led to a misconception that urban stormwater runoff has some environmental flow benefit [45]. Studies showing consistent degradation in the face of increasing urban stormwater drainage [9], [10] indicate that the reverse is true. Urban stormwater runoff, delivered through conventional drainage systems, is a complex environmental flow problem that can, in large part, be solved by harvesting stormwater *before* it reaches aquatic ecosystems.

A recent review of large-scale urban stormwater harvesting projects funded in Australia [46] showed that almost all projects extracted water from waterways or large drains (rather than directly harvesting from impervious areas). Such schemes can at best harvest a small proportion of the damaging stormwater flows (their ability to divert and store flows is small compared to the large flow rates that accumulate with catchment area), do nothing to restore lost dry-weather flows, and fail to protect upstream waters. Of greater concern, about a quarter of the projects identified harvested dry-weather flows from drains or waterways. These projects only serve to exacerbate the reduction of dry-weather flows caused by reduced infiltration, as observed in this study.

The protection and restoration of urban streams has thus been hampered by a lack of understanding of the unique nature of urban stormwater runoff as an environmental flow problem that could be solved by using stormwater as a water resource. Stormwater harvesting defies the dominant conception of water resource management that extraction of water from ecosystems must result in a monotonic decline in the ecological condition of that ecosystem (Figure 5.I) [47].

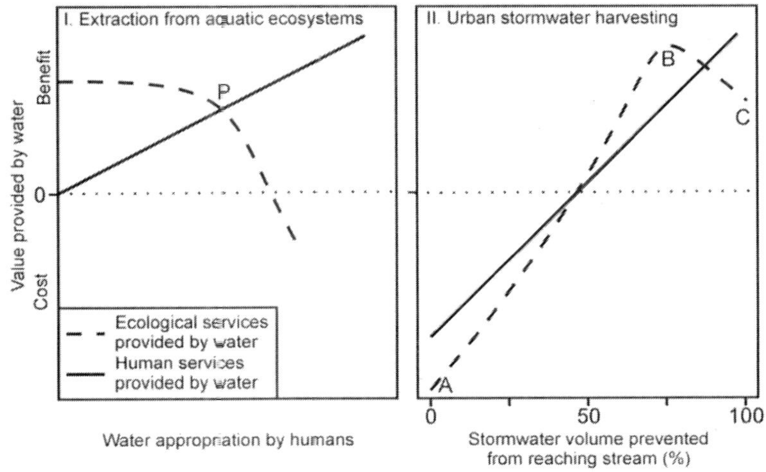

Figure 5. Conceptual graphs of ecological and human value of water. I. The model proposed by Gleick and Palaniappan [47] assumes that any extraction from aquatic ecosystems has a negative ecological impact, predicting a monotonic decline of increasing gradient with greater extraction. The benefits accrued by the human population rise linearly with the volume extracted. Beyond peak ecological water (P) [47], any increase in human benefit is outweighed by reduced ecological benefit. II. illustrates different trends in ecological and human cost and benefit with increasing retention and use of stormwater before it reaches aquatic ecosystems. No stormwater use (A) results in ecological degradation of receiving waters. It also presents greater costs in urban microclimate control and flood mitigation than if stormwater was harvested. Using a volume of stormwater equivalent to the volume lost to evapotranspiration in the pre-urban state (B), if coupled with infiltration systems to restore lost sub-surface flows, provides maximum environmental benefit. Using all available stormwater runoff (C) has an environmental cost by reducing subsurface flow delivery to stream.

By not harvesting stormwater to keep an appropriate proportion of it out of receiving waters, we not only forego the benefits to society of this large water resource, but also contribute to the degradation of waterways (Figure 5.II), resulting in a loss of biodiversity and ecological function provided by healthy streams. In addition, the use of stormwater for landscape irrigation can mitigate against the urban heat island effect [48],

[49]. The wider use of urban stormwater would reduce demand on potable water supplies, potentially freeing water to provide environmental flows below water supply storages. In growing cities, urban stormwater harvesting could allow water managers to avoid or delay the augmentation of other potable water supply options that have negative environmental impacts. Retention and use of stormwater also contributes to mitigation of urban flooding, thus potentially reducing costs for flood protection [50].

Using a volume of stormwater equivalent to the volume lost to evapotranspiration in pre-urban state (Figure 5.II B), if coupled with infiltration to restore lost sub-surface flows, provides maximum environmental benefit, by optimizing the retention of water in the catchment and restoration of downstream flow regimes. The integration of harvesting with baseflow restoration techniques such as infiltration is thus important. In contrast, using all available stormwater runoff (Figure 5.II C) has an environmental cost by reducing subsurface flow delivery to stream. However, in many urban settings this loss can be compensated by increased infiltration in non-treed open spaces, or by leakage of water supply systems (e.g. [51]).

Degradation of stream biotic assemblages occurs at very low levels of (connected) imperviousness [17], [52]. Therefore, protection of the ecological integrity of stream ecosystems is likely to require interception and treatment of runoff from almost all catchment impervious surfaces, including the prevention of excess runoff from reaching streams. The provision of environmental flow standards for the protection of streams from urban stormwater runoff will therefore require a universal change from conventional stormwater management in any catchment where the decision is made to protect the stream from stormwater impacts (or to remove existing impacts). Such a fundamental change in practice will require the addressing of technical challenges (concerning design standards, available space), economic barriers (costs of construction and maintenance) and social and institutional impediments. Realizing such change will require that managers of rivers, of stormwater drainage systems, and of urban water resources, recognize the rare advantage of urban stormwater as a water resource that can provide both human and environmental benefit.

ACKNOWLEDGMENTS

We thank the following for their critical discussions which have helped us to develop our arguments: Tony Ladson, Belinda Hatt, Phil Johnstone, Geoff Vietz, Mike Stewardson and Perrine Hamel. The paper was greatly improved by the comments of Liz Martin.

REFERENCES

1. Millennium Ecosystem Assessment (2005) Fresh water. Ecosystems and human well-being: current state and trends Findings of the condition and trends working group. Washington, DC, USA: Island Press.
2. Arthington AH, Naiman RJ, McClain ME, Nilsson C (2010) Preserving the biodiversity and ecological services of rivers: new challenges and research opportunities. Freshwater Biol 55: 1–16.
3. Poff NL, Richter BD, Arthington AH, Bunn SE, Naiman RJ, et al. (2010) The ecological limits of hydrologic alteration (ELOHA): a new framework for developing regional environmental flow standards. Freshwater Biol 55: 147–170.
4. Kennen JG, Kauffman LJ, Ayers MA, Wolock DM, Colarullo SJ (2008) Use of an integrated flow model to estimate ecologically relevant hydrologic characteristics at stream biomonitoring sites. Ecol Model 211: 57–76.
5. Kennedy C, Cuddihy J, Engel-Yan J (2007) The changing metabolism of cities. J Ind Ecol 11: 43–59.
6. Mitchell VG, McMahon TA, Mein RG (2003) Components of the total water balance of an urban catchment. Environ Manage 32: 735–745.
7. Leopold LB (1968) Hydrology for urban land planning – a guidebook on the hydrologic effects of urban land use. Washington DC: US Geological Survey. 554. 18 p.
8. Burns MJ, Fletcher TD, Walsh CJ, Ladson AR, Hatt BE (2012) Hydrologic shortcomings of conventional urban stormwater management and opportunities for reform. Landscape Urban Plan 105: 230–240.
9. Walsh CJ, Roy AH, Feminella JW, Cottingham PD, Groffman PM, et al. (2005) The urban stream syndrome: current knowledge and the search for a cure. J N Am Benthol Soc 24: 706–723.
10. Wenger SJ, Roy AH, Jackson CR, Bernhardt ES, Carter TL, et al. (2009) Twenty-six key research questions in urban stream ecology: an assessment of the state of the science. J N Am Benthol Soc 28: 1080–1098.
11. Walsh CJ, Fletcher TD, Ladson AR (2009) Retention capacity: a metric to link stream ecology and stormwater management. J Hydrol Eng 14: 399–406.
12. .Roy AH, Freeman MC, Freeman BJ, Wenger SJ, Ensign WE, et al. (2005) Investigating hydrologic alteration as a mechanism of fish assemblage shifts in urbanizing streams. J N Am Benthol Soc 24: 656–678.
13. Hatt BE, Fletcher TD, Walsh CJ, Taylor SL (2004) The influence of urban density and drainage infrastructure on the concentrations and loads of pollutants in small streams. Environ Manage 34: 112–124.
14. Shuster W, Bonta J, Thurston H, Warnemuende E, Smith D (2005) Impacts of impervious surface on watershed hydrology: a review. Urban Water J 2: 263–275.
15. Brown LR, Cuffney TF, Coles JF, Fitzpatrick F, McMahon G, et al. (2009) Urban streams across the USA: lessons learned from studies in 9 metropolitan areas. J N Am Benthol Soc 28: 1051–1069.
16. Wang LZ, Lyons J, Kanehl P (2003) Impacts of urban land cover on trout streams in Wisconsin and Minnesota. T Am Fish Soc 132: 825–839.
17. King RS, Baker ME, Kazyak PF, Weller DE (2010) How novel is too novel? Stream community thresholds at exceptionally low levels of catchment urbanization. Ecol Appl 21: 1659–1678.
18. Walsh CJ, Fletcher TD, Ladson AR (2005) Stream restoration in urban catchments through re-designing stormwater systems: looking to the catchment to save the stream. J N Am Benthol Soc 24: 690–705.
19. Zhang L, Dawes WR, Walker GR (1999) Predicting the effect of vegetation changes on catchment average water balance. Canberra: Cooperative Research Centre for Catchment Hydrology. Technical Report 99/12.

20. Zhang L, Dawes WR, Walker GR (2001) Response of mean annual evapotranspiration to vegetation changes at catchment scale. Water Resour Res 37: 701–708.
21. Taylor SL, Roberts SC, Walsh CJ, Hatt BE (2004) Catchment urbanisation and increased benthic algal biomass in streams: linking mechanisms to management. Freshwater Biol 49: 835–851.
22. Jolly ID, Dowling TI, Zhang L, Williamson DR, Walker GR (1997) Water and salt balances of the catchments of the Murray-Darling Basin. Canberra: CSIRO Land and Water.
23. Boyd MJ, Bufill MC, Knee RM (1993) Pervious and impervious runoff in urban catchments. Hydrolog Sci J 38: 463–478.
24. Ragab R, Bromley J, Rosier P, Cooper JD, Gash JHC (2003) Experimental study of water fluxes in a residential area: 1. Rainfall, roof runoff and evaporation: the effect of slope and aspect. Hydrol Process 17: 2409–2422.
25. Konrad CP, Booth DB (2005) Hydrologic changes in urban streams and their ecological significance. Am Fish S S 47: 157–177.
26. Tanaka T, Yasuhara M, Sakai H, Marui A (1988) The Hachioji experimental basin study–storm runoff processes and the mechanism of its generation. J Hydrol 102: 139–164.
27. Kirchner JW (2003) A double paradox in catchment hydrology and geochemistry. Hydrol Process 17: 871–874.
28. Mikkelsen PS, Jacobsen P, Fujita S (1996) Infiltration practice for control of urban stormwater. J Hydraulic Res 34: 827–840.
29. Holman-Dodds JK, Bradley AA, Potter KW (2003) Evaluation of hydrologic benefits of infiltration based urban storm water management. J Am Water Resour As 39: 205–215.
30. Melbourne Water (2006) Water supply-demand strategy for Melbourne 2006–2055. Melbourne: Melbourne Water.
31. Australian Bureau of Statistics (2010) Population growth: past, present and future. 41020 Australian Social Trends, Jun 2010. Canberra: Australian Bureau of Statistics.
32. Mitchell VG, McCarthy D, Deletic A, Fletcher TD (2008) Sensitivity of urban stormwater harvesting storage capacity-reliability-yield relationships to behaviour analysis method selection. Environ Modell Softw 23: 782–793.
33. Hatt BE, Deletic AB, Fletcher TD (2006) A review of integrated stormwater treatment and re-use in Australia. J Environ Manage 76: 102–113.
34. Fletcher TD, Deletic A, Mitchell V, Hatt BE (2008) Reuse of urban runoff–a review of recent Australian advances and remaining challenges. J Environ Qual 37: 116–127.
35. Pagotto C, Legret M, Le Cloirec P (2000) Comparison of the hydraulic behaviour and quality of highway runoff water according to the type of pavement. Water Res 34: 4446–4454.
36. Bratières K, Fletcher TD, Deletic A, Alcazar L, Le Coustumer S, et al.. (2008) Removal of nutrients, heavy metals and pathogens by stormwater biofilters. 11th International Conference on Urban Drainage (ICUD); Edinburgh. International Water Association.
37. Li HW, Sharkey LJ, Hunt WF, Davis AP (2009) Mitigation of impervious surface hydrology using bioretention in North Carolina and Maryland. J Hydrol Eng 14: 407–415.
38. Hamel P, Fletcher TD, Walsh CJ, Beringer J, Plessis E. (2011) Water balance of infiltration systems in relation to their operating environment. 12th International Conference on Urban Drainage (ICUD); Porto Allegre, Brazil.
39. Burns MJ, Fletcher TD, Duncan HP, Hatt BE, Ladson AR, et al.. (2012) The stormwater retention performance of rainwater tanks at the land-parcel scale. In: Wong THF, McCarthy DT, editors. 7th International Conference on Water Sensitive Urban Design. Melbourne.
40. Roy AH, Wenger SJ, Fletcher TD, Walsh CJ, Ladson AR, et al. (2008) Impediments and solutions to sustainable, watershed-scale urban stormwater management: lessons from Australia and the United States. Environ Manage 42: 344–359.
41. Harremoes P (1997) Integrated water and waste management. Water Sci Technol 35(9): 11–10.
42. Mitchell VG (2006) Applying integrated urban water management concepts: a review of Australian experience. Environ Manage 37: 589–605.

43. Anon (2007) Stormwater recycling feasibility study: high level assessment of option report. Melbourne: Sinclair Knight Merz.
44. Newton D, Ewert J (2009) A simple method to estimate stormwater harvesting entitlements from urban streams. 32nd Hydrology and Water Resources Symposium; Newcastle, Australia. 102–113.
45. Victorian Government: Department of Sustainability and Environment (2006) Sustainable water strategy, Central Region: action to 2055. Melbourne: Department of Sustainability and Environment, Victoria.
46. Knights D, McAuley A (2009) What makes a sustainable stormwater harvesting project? Stormwater Industry Association of NSW and Victoria Joint Annual Conference; Albury, NSW, Australia http://equatica.com.au/pdf/Knights and McAuley SIA 2009_final_.pdf. Accessed 2012 March 16.
47. Gleick PH, Palaniappan M (2010) Peak water limits to freshwater withdrawal and use. Proc Natl Acad Sci U S A 107: 11155–11162.
48. Coutts AM, Beringer J, Tapper NJ (2007) Impact of increasing urban density on local climate: spatial and temporal variations in the surface energy balance in Melbourne, Australia. J Appl Meteorol 47: 477–493.
49. Carlson TN, Arthur ST (2000) The impact of land use – land cover changes due to urbanization on surface microclimate and hydrology: a satellite perspective. Global Planet Change 25: 49–65.
50. Tourbier J, White I (2007) Sustainable measures for flood attenuation: sustainable drainage and conveyance systems SUDACS. In: Ashley R, Garvin S, Pasche E, Vassilopoulis A, Zevenbergen C, editors. Advances in Urban Flood Management. London: Taylor & Francis. 13–28.
51. Price K (2011) Effects of watershed topography, soils, land use, and climate on baseflow hydrology in humid regions: a review. Prog Phys Geog 35: 465–492.
52. Walsh CJ, Kunapo J (2009) The importance of upland flow paths in determining urban effects on stream ecosystems J N Am Benthol Soc. 28: 977–990.
53. Lenoir A, Coste M (1996) Development of a practical diatom index of overall water quality applicable to the French National Water Board Network. In: Whitton BA, Rott E, editors. Use of algae in monitoring rivers. Innsbruck, Switzerland: Institut fur Botanik, Universität Innsbruck. 29–43.
54. Chessman BC (1995) Rapid assessment of rivers using macroinvertebrates: a procedure based on habitat-specific sampling, family level identification and a biotic index. Aust J Ecol 20: 122–129.

CHAPTER 3

Eco-Revelatory Design

Nurgül Konaklı Arısoy

Selçuk University, Agricultural Faculty, Department of Landscape Architecture, Turkey

1. INTRODUCTION

Built environments ignore people's need and their potential for learning. The negative effects related to the ignorance of natural systems in human development are evident. Making natural cycles and processes visible bring the designed environment back to life. Effective design helps inform us of our place within nature. Landscape architects have developed theories and methodologies which represent a new, ecologically oriented approach to design.

Eco-revelatory design (ERD) is an ecological design concept in the field of landscape architecture means "a design strategy that attempts to enhance site ecosystems as well as engage users by revealing ecological and cultural phenomena, processes and relationships affecting a site " [1]. Landscape architects reveal nature through their form, materials and formation, and they also reveal the nature of the person who designed them.ERD is a new approach to landscape architecture, one where ecological processes and the environment is a fundamental determinant of the design.

ERD is a different exposition and interpretation –updated version- of design. It is an integrative and ecologically responsible design. It is a partnership between people and nature. ERD attempts to enhance site ecosystems and engage users by revealing ecological and cultural phenomena, processes, and relationships affecting a site. It aspires to reveal endemic ecological process and affords a more direct connection between fundamental ecological process and the phenomenological experience of landscape. It is important to involve using knowledge about how interact with environment to form objects and spaces with skill and artistry [4]. This approach to design should be applied not only to urban sides, but also to non-urban sides such as wetlands, arboretums. The theory has received heavy criticism about its ability to absorb an audience in ecological understanding or improve site conditions [2,3].

Another criticism about ERD is some designs make tangible improvements in local ecological health while others are symbolic gestures [2]. It is adequate for people to develop ecological perception by means of the visibility of ecological process in design. The theory has received heavy criticism about its ability to absorb an audience in ecological understanding or improve site conditions.

It aspires to reveal endemic ecological process and affords a more direct connection between fundamental ecological process and the phenomenological experience of landscape. It is important to involve using knowledge about how interact with environment to form objects and spaces with skill and artistry (Ndubisi, 2002). This approach to design should be applied not only to urban sides, but also to non-urban sides such as wetlands, arboretums.

They are typically designed landscapes that elucidate natural phenomenon such as the cleansing action of wetlands. Ecorevelatory landscapes have also been referred to as "educational and enlightening". They are reference sites for what we understand about our environment and its workings. Designs can convey knowledge through direct experience as well as by interpretation. Interpretation seeks to create connections between the resources that are being interpreted and relevant everyday knowledge that everyone has by engaging emotions, creating experiences, and entertaining ideas through engagement [5]. By highlighting the particular ecological relationships at any given site, such design can punctuate and enliven our environment and sensitize us to what is known about its interlocking complexities" [1]. These sites often still use traditional interpretive media such as wayside exhibits and publications to fully convey their meaning and function, but eco-revelatory design can bring to light processes that usually remain unseen and forgotten.

2. HISTORY

This study explores descriptive theory, principals, techniques and practice of eco-revelatory design, can help to plan a sustainable development, which uses and reveals natural systems to reconciling human systems and its effects on the surrounding environment. The goal of this study, therefore, is to generate ideas and to begin a discussion about how the design within an eco-revelatory framework will be.

Most of the "eco" prefixes such as eco-city, eco-technique, eco-efficient was began to used nineties; also ERD emerged in 1998 as a new theory within the field of landscape architecture.

The term and practice of ERD was coined from an exhibit Nature Constructed/Nature Revealed sponsored by the University of Illinois at Urbana-Champaign. Two predominant schools of thought, one insistently cultural and the other assertively ecological, reigned over the conceptual and theoretical dialog in landscape design and planning. In 1998 a group of practitioners and landscape scholars published a special issue of Landscape Journal as a catalogue and record of the exhibition Eco-Revelatory Design: Nature Constructed/Nature Revealed. Brenda Brown, Terry Harkness, and Douglas Johnston chaired the exhibition and served as guest editors of the journal (Figure 1). The exhibit opened at the University of Illinois in 1998 and closed at the Washington DC's National Building Museum in 2000.

Brown identifies three areas of investigation on landscapes as/of sound: listening gardens, sound, or listening trails, and sound designs [3]. She is particularly interested in what she calls "the reciprocal revelations of landscapes and sounds, how sounds can reveal landscapes and how landscapes can reveal sounds." She is therefore also concerned with how people perceive, understand and engage with landscapes.

ERD roots are based on ecologically minded landscape architects such as Ian McHarg and Frederick Law Olmsted but they didn't use the term as ERD. These landscape architects created works to foster function in natural systems and processes aesthetically. Ian McHarg was accepted as the first person who to apply his ecological knowledge into design. He began advocating the use of ecology as a basis for design in the early 1960s. He accomplished his goal of merging design with ecology. He gave importance of ecological principles in design.

Thirty years ago, in *Design with Nature*, McHarg proposed a system of ecological inventories to help explain the way natural processes may influence regional and urban planning and design [6]. Also, it makes the case for an ecological approach to design. McHarg gave a new dimension to the historical goal of imitating nature' (mimesis). He was concerned both with the practicalities of 'design with nature' and with the aesthetic results of a naturalist approach to landscape and garden design. It was demanded that local landscapes should follow ecological principles that implied that landscape could not represent some particular claim of social identity.

In the second half of the 19th century, Frederick Law Olmsted completed a series of parks which continue to have a huge influence on the practices of Landscape Architecture today. Among these were Central Park in New York City (Figure 2), Prospect Park in Brooklyn (Figure3), New York and Boston's Emerald Necklace park system (Figure 4).

Figure 1. Eco-revelatory design exhibit

Figure 2. Central Park, NYC

Figure 3.Prospect Park in Brooklyn

Figure 4. Boston's Emerald Necklace park system

The Emerald Necklace consists of a 1,100-acre (4.5 km^2) chain of parks linked by parkways and waterways in Boston and Brookline, Massachusetts.

Several components of the Emerald Necklace pre-date the plan to unite them. Some links of the Emerald Necklace not only offer an opportunity for recreation in a wooded environment, but are also ecologically important urban wilds that provide nesting places for migratory birds, fishes and other animals and improve the air quality of the city. The Emerald Necklace

Project successfully ties together conservation, land restoration, sewage treatment, solid waste disposal, recreation, transportation, and water and visual quality [7]. The park system provides opportunities for people to learn about natural systems by enjoying, observing, and appreciating these systems.

Today, landscape architects recognize the effects of ignorance on natural systems in human development. So they coined a new design approaches.

Arcata Marsh is one of the ERD examples of today design approach. The Arcata Marsh and Wildlife Sanctuary was constructed in 1981 (Figure 5). The City of Arcata incorporated wastewater treatment to the system in 1986. The City of Arcata's unique wastewater treatment facility, marsh, and wildlife sanctuary attracts approximately 150,000 visitors per year [8]. Arcata's wastewater treatment plant is an example of a community involvement in environmental politics, innovative uses of land, and applications of appropriate technology in a small urban community. The Arcata Wastewater Treatment Plant combined with the Arcata Marsh and Wildlife Sanctuary has multiple uses, including wastewater treatment, recreation, wildlife habitat, education, and research. The residents of Arcata who stroll by the wetland can, for instance; see that wastewater treatment wetlands can be important habitat for fish and birds, as well as an energy-efficient, biologically based method of controlling water pollution. The experience of the Arcata wetland shows that ecological processes can be brought into a constructive partnership with human settlements.

Figure 5. The Arcata Marsh and Wildlife Sanctuary (source: http://www.humboldt.edu/engineering/graduate/facilities)

Students are centrally involved in the original design and development of the Marsh's constructed wetland water treatment system, and they continue to play a key role through projects and research geared towards the continued optimization of the system.

The second example for ERD is the Gateway Business Center - post-industrial area- is a 25-acre site consisting of 10 industrial tilt-up buildings located at the San Gabriel Mountain foothills on a great alluvial fan that crosses the San Gabriel Valley on its way to the Pacific Ocean. The site concept transformed an industrial building environment into a sustainable and nature-inspiring experience.

The layout of cairns act as way finding and directional markers and add to the overall landscape display and experience of bioswales, infiltration zones, rain harvesting, solar canopies and an array of recycled materials.

The site landscape is designed to mimic the natural feel of the local native landscape character and capture it into the renovated industrial complex. The goal is to create "biophilia" (an instinctive bond between man and nature).

The bioswale has taken the place of the previous concrete swale, while the concrete catch basin remains to collect any additional runoff in a deluge, and as a reminder of the previous form of site storm water drainage. More than 95 percent of the runoff water in this parking lot is captured.

The storm garden is irrigated by the adjacent rain harvesting tanks. A natural wash picks up excess rainwater and roof water not captured by the parking lot bioswale. [9] (Figure 6).

Figure 6. The Storm Garden of Gateway Business Center

3. DESIGN AND ECOLOGY RELATIONSHIP

Design is the intention that redefines how we relate to each other and our environment. The process of designing places and artefacts are opportunities to reimagine a new relationship with our environment, especially when it seems remote or difficult to create. Design that reveals hidden systems, whether ecological or economic, is a powerful way to meet the challenges of ecologically benign communities. Landscape design is a discipline which transfer the knowledge developed in landscape ecology to application [10].

The term ecology, like the term landscape, has multiple meanings. Ecology has historically focused on pristine, natural environments, however, by the 1970's many ecologists began to turn their interest towards the ecological interactions taking place in, and caused by urban environments. Urban ecology is recognized as a diverse and complex concept which differs in application between North America and Europe. The European concept of urban ecology examines the biota of urban areas while to the North American concept which has traditionally examined the social sciences of the urban landscape as well as the ecosystem fluxes and processes.

Environmental knowledge have been part of the intent of design in landscape architecture and various conceptions and methodologies have been improved for measuring the environmental consequences of design.

Nassauer and Opdam define design as intentional change of landscape pattern, for the purpose of sustainably providing ecosystem services while recognizably meeting societal needs and respecting societal values [11]. Design is both a product, landscape pattern changed by intention, and the activity of deciding what that pattern could be.

Ecology plays a big role in ERD. Disciplines in applied ecology such as urban ecology, landscape ecology etc. helps planning and design practice in landscape architecture. These disciplines help put theories into practice. The most important thing is to find how to use ecological thinking in design.

4. LANDSCAPE ARCHITECTURE AND LANDSCAPE ECOLOGY

Landscape architecture is informed by scientific knowledge and aspires to provide aesthetic expressions in landscapes across a range of spatial scales. Landscape ecology has been defined as the study of the effect of landscape pattern on process, in heterogeneous landscapes, across a range of spatial and temporal scales [12]. The logical reasons for integrating these two fields are clear and compelling, with a great potential to support sustainable landscapes through ecologically based planning and design.

Proponents of ERD recognized that landscape architecture alters and directs both cultural and ecological systems. Furthermore, they acknowledged landscape architects' capacity to direct human experience and reveal, through design, aspects of ecology and culture. This integrated approach provides opportunity for people to place themselves in and as part of an interconnected socio-ecologic world, reinforcing the relationships between humans and the bio-geosphere.

Landscape ecology is defined as a problem-oriented science [13]. It has developed from the growing awareness of environmental problems since the nineteen seventies. Spatial planning and landscape design are disciplines which transfer the knowledge developed in landscape ecology to application. To optimize this process of knowledge transfer, landscape ecology must co-evolve with spatial planning [14]. The development of ecologically sustainable landscapes requires that patterns of future landscapes sustain the necessary ecological processes in the landscape. Therefore, we must know how landscape patterns relate to these processes.

Humans are the driving force behind urban ecology and influence the environment in a variety of ways, such as modifying and altering land surfaces and waterways, introducing foreign species. In this context, changes between human-environment affect the styles of landscape design directly.

However, the present concept of ecological design and its interpretation in this sense does not refer to information charts or written explanations, which people encounter at places like zoological gardens or arboretums.

Generally, "ecological designs blend with their contexts and results in a diffuse visual pattern" [15]. Consequently, this perceptual subtlety can make ecological landscapes difficult for inhabitants to recognize and care about. ERD make ecological considerations perceivably a "visible part of landscape experience". It achieve this is by benefiting from the contrast between cultural and ecological domains. One way to achieve this is by exploiting the power of contrast, particularly the contrast between cultural and ecological domains. Ecological design has tended to diffuse edges to provide transition.

We can use basic principles of landscape ecology in landscape design. Such as linear parks, neighbourhood parks, playgrounds as for patch, greenways for corridor.

Every human directly or indirectly contributes towards enriching as well as degrading the quality and experience of cities. In order to create a successful ecological design it is important to recognize and interpret the historic and cultural significance of the landscape. In other words, "eco-revelatory design expands by hitching human habitat and their inevitable cultural determinants in to an environmentally inclusive vision" [16].

Farina observed that landscape design is an important component in practical landscape ecology as it expresses the relationship of spatial patterns and processes in a practical manner [17]. It provides in depth understanding on wildlife habitats and movements and biological interdependency within a region. The two disciplines should be complementary [17].

People use land for its scenic and recreational uses. Thus, the aesthetic use is important for them. However, natural systems are the important part of the designing decisions.

It is important to understand the social and cultural backgrounds and expectations of the society who use the land are determinant. It depends on two factors [18], in how people experience and use their landscapes, and their understanding of ecological processes.

It is important to organize a linkage between cultural expectations and ecological process. Man made land uses affect the ecological system activities such as wildlife crossing and subsurface water movement. Lyle has explained the following six basic ecological processes that are vital for operational integrity of natural systems [19].

In order to participate creatively in natural processes and to do so with reasonable hope of success, we need to include as subject of design the inner workings of the landscape, the systems that motivate and maintain it, and reveal them through creative, imaginative, and visible form of the landscape [19].

5. THEORY

A design theory is a procedure for how to set about a design project. The classical twentieth century approach to landscape design has been Survey-Analysis- Design (SAD). It has been elevated to the status of a design methodology and cruelly overworked. The resultant places lacked clarity of intention. The intention of ERD is to "connect people with the natural environment". A dynamic balance between natural environment and society, intended to reveal and interpret (resolve and educate the relationship between user and the designed area) and finally to provide awareness on ecological understanding.

Most of the landscape designers' style ignores natural systems and have only one purpose, primarily focused on aesthetics. ERD has to combine aesthetics and science. It is one of the endeavours that contribute to rediscovering aesthetics. It is hard to measure aesthetics outcome statistically or collectively. But can measure the environmental effects in numbers. So you can measure the sustainability by its effects on environment not by aesthetics.

6. METHOD

It is important to choose a method that is most compatible design strategy. It depends on the design intent, the place and the designer. How methods fit into design process should be determined clearly.

Generally two methods are used at ERD as follows;

a) Mimic the nature

b) Use the nature

a. We need to design a natural area to show the ecological processes and relationships. Artificial waterfalls can perfectly mimic the nature a way that highlight the system of the real one (Figure 7). Native trees and shrubs mimic the adjacent natural areas. The site landscape is designed to mimic the natural feel of the local native landscape character.

b. ERD utilizes the natural capabilities of the environment. In an ecorevelatory approach the pre-existing natural, structural and functional characteristics of the place are integrated into design and made them observable and understandable.

There are a lot of methods for ERD that landscape architecture can use such as environmental method, man centred method, evaluation method, interpretive analogy method. The design method is up to designer's idea and knowledge.

Through our senses, we form concrete relationships, we reconciliate, with the world. With this understanding, phenomenology can be used as a method for eco-revelatory design. Suitability analysis is also one of the approaches for ERD.

Figure 7. Artificial Waterfalls

7. DESIGN PROCESS

When we examine the literature about ERD we can't see any knowledge about the design process. Landscape ecology principles integrate to design process. How can we adopt the ecology approach and principles into design process? We can use basic principles of landscape ecology in landscape design such as linear parks, neighbourhood parks, playgrounds as for patch, greenways for corridor. Landscape ecology principles influence every stage of landscape design process such as site planning.

One of the design processes is traditional design process that landscape architects undertake. It typically starts with the selection of a site based on a set of criteria. Once a site is selected, the typical design process will move through a series of phase including site inventory, site analysis, conceptual design, design development, construction documentation and finally implementation [20].

Landscape designers need to understand how natural and human systems work and design as an integral part of a nature and establish relationship with nature. They can achieve this by combining those with new technology to meet changing cultural and ecological needs.

8. PRINCIPLES (DESIGN STRATEGY)

The hosts of the exhibit, Barbara Brown, Terry Harkness and Doug Johnston, take great care in describing what deserves to be called eco-revelatory design, and stressed stringent and ambitious goals for the competition. Those chosen for the exhibit represent rigorous application of eco-revelatory design, and utilize some or all of the following strategies [1]:

- Abstraction and simulation of natural processes
- New uses of landscapes producing deeper caring for life and ecological processes
- Signifying features that speak for natural/cultural processes that might otherwise remain invisible
- Expose infrastructure and process
- Reclaim landscapes so that the past is remembered
- Change perspectives by structuring how we interact with the Landscape.

Interpretation of ecological processes refers to the ability of the design to reveal ecological processes at work. This process of revealing can only be successful if the environments created are visible, observable, legible, and have the ability to raise curiosity in visitors to explore and understand the complexity of the landscape. Principles for ERD are described generally as follows:

9. VISIBILITY

Visibility is a cardinal point for ERD. Many designers and planners have become concerned in recent years with "revealing" ecological processes in their designs so that the users of the environment may experience, learn about, and appreciate those processes. In practice, "revelation" of ecological process has meant everything from capturing stormwater on the surface of the land before it drains away to the storm sewers. In addition, the ecological processes that are revealed may themselves be truly "natural," in the sense that they could continue to exist without the management of humans, or they may be highly artificial, engineered systems that need constant supervision if they are to persist in an urbanized context.

"Most of the time, natural systems themselves are not visible and readily engaging. What are visible are the surface manifestations and the material conclusions of these natural systems, for example layers of rocks are not ecological process, but the result of it" [21]. Thus the most important challenge for designers is to recognize which ecological processes can actually be made visible and how they can interpret these dynamic processes or their material conclusions to form and inform landscapes.

The Arcata Marsh and Wildlife Sanctuary, a real-world example of eco-revelatory design, highlights some of the positive potential of using visibility as a design strategy (Figure 8).

Some ERDs, for example, have sought to bring ecological processes (such as water flows) into the open, but then blend them in with the surrounding landscape as much as possible. Many proposals to capture rainfall in grassy areas and infiltrate it into the soil before it runs off into the storm sewers would use either parks or front lawns for this purpose. Although this strategy reveals an ecological process occurring during and just after a rainfall, the rest of the time these spaces would simply look like what they have always looked like--large grassy expanses--and would forfeit an opportunity to communicate a clear, consistent, and meaningful landscape message. Making natural cycles and processes visible bring the designed environment back to life.

In reference [22] the quantitative interpretation challenges aesthetics by rendering it negative, segregated and unstable. Under this "hegemony" various endeavours in which eco-revelatory design played a part involved in. It is important for people to "read" or "experienced" the ecological knowledge tried to given. This attempt is achieved by visibility.

Figure 8. An Oxidation Pond at the Arcata Wastewater Treatment Plant

10. SUSTAINABILITY

The second concern for ERD is to provide sustainability. The term sustainability was first used in 1980 in IUCN's World Conservation Strategy. The sustainability concept is arguably relevant to systems from the global to the local scale. Sustainable ecologically-based approaches to design are desirable but their application is not widely seen.

Sustainability reshapes environmental ethics, available technologies, planning techniques, and assessment criteria, which in turn influence environmental design disciplines. This approach suggests a need to contemplate spiritual aspects of sustainable design.

A set of criteria is developed for sustainability to evaluate the environmental performance. In this context, sustainability becomes the key consideration in design.

If we propose to use ecological design of urban places to promote cultural change in the human relationship to the environment, then, we should be thinking about how to create physical settings with cues for sustainable behaviour.

The design of sustainable systems is consistent with ecological principles, which integrate human society with its natural environment for the benefit of both.

There are many compelling reasons why environmental and resource problems should be placed in a dynamic perspective. It is important to provide sustainability by using moderate and efficient resource use. ERD strives for moderation and efficiency in resource use.

11. THE USE OF NATIVE PLANTS (NATURALNESS)

Naturalness provide to "sustainability" and "knowledge" about native plants. It also engages people and nature as it used to be. We can see the usage of native plants example on stormwater garden (Figure 9).

It gives opportunity to people to have information about their native plants. Native trees and shrubs mimic the adjacent natural riparian areas, native plant communities and wildlife habitat. Natural water sources and planting zones will be considered when choosing plant material to minimize the need for irrigation.

Figure 9. Stormwater management- the use of native plants

You can help stop the exotic plant invasion by using and nurturing native plants around your home and on your property. Native plants generally grow well and require less care than exotic species when grown on the proper soils under the right environmental conditions.

- Protective cover for most animals.
- Seeds, nuts, and fruits for squirrels
- and other mammals.
- Seeds, fruits, and insects for birds.
- add beauty to the landscape and preserve our natural heritage
- provide food and habitat for native wildlife
- serve as an important genetic resource for future food crops or other plant-derived products

If you enjoy observing nature, are concerned about the environment, or wish to make a long-term contribution to your community's ecosystem, then using native plants is a responsible, money-saving, long-term, positive investment to both your property and your community.

Landscaping with native plants improves the environment. Native plants are hardy because they have adapted to the local conditions. The native plants increase our connection to nature, help educate our neighbours, and provide a beautiful, peaceful place to relax.

The interest in the preservation and restoration of native plant communities increases as the public becomes more concerned about the environment. Native plants are valued for their economic, ecological, genetic, and aesthetic benefits in addition to the growing societal belief in their intrinsic value as living species.

12. OBSERVABILITY

It is relevant with visibility. People observe the design by hearing, listening and seeing. It makes awareness about the nature.

When ecological design incorporates "visibility" [23] and "observability" [24] it reveals ecological phenomena and processes and can be referred to as ERD.

13. MULTIFUNCTIONALITY

Multifunctionality is the most important characteristics of this design strategy. It serves people multifunctional activities such as recreation, protection and education together. For instance, The Arcata Wastewater Treatment Plant combined with the Arcata Marsh and Wildlife Sanctuary has multiple uses, including wastewater treatment, recreation, wildlife habitat, education, and research.

Some people come to learn about the innovative wastewater treatment that enhances the community. Other people come to see the more than 270 species of birds that make use of the habitat provided by the Marsh [8]. Even more people come to exercise on the trails while enjoying the natural experience.

Multifunctionality is fundamental for sustainability. Multi-functionality is generally desirable, as it encourages efficient use of land, delivers wider public benefit and builds partnerships of user groups, leading to better stewardship. [25].

A designed land has many layers for visitors such as place for leisure activities, resting areas, conservation areas, fish and wildlife areas, sports activities and educational areas together. People gain social, environmental, economical advantage from multifunctional places.

14. LEGIBILITY

The design gives its mission to the visitor clearly. ERD purpose is "educate" and illumine". In this context, the messages have to be understandable. Effective design helps inform us about our place within nature.

Legibility was interpreted as the understanding of how the landscape worked from its manifestation based on coherence between climate, soil, water and human occupation; or "understanding relationships between process and material, form and space" [26].

The spatial organization of design was utilized to clarify the ecological and cultural phenomena for people. People accompany the space by their senses. It gives people to create a spiritual connection with their environment and to have an educational consciousness.

15. ABILITY TO RAISE CURIOSITY

If the design does not take attention it can not to carry out its mission. To make people curious about the design shows its success. The perception is the most important element of design that its comprehensibleness and arouse curiosity.

It is important for people to "read" or "experienced" the ecological knowledge tried to given. This attempt is achieved by rising curiosity. If people wonder, they can learn the ecological awareness. Designers should enhance visitor's experience by encouraging interaction to interpret the ecological processes.

Effective design solutions or distinctive features are chosen as the general publics' ability to gain ecological information from their environment is limited. By using and revealing natural systems to spread consciousness and earn attention and care for our environment.

ERD overlap cultural needs and ecological processes to raise curiosity in visitors to explore and understand the complexity of the landscape.

16. CONCLUSION

ERD is a new approach in subdiscipline of design for landscape architecture, one where ecological process, the environment and the cultural awareness of people is a fundamental determinant of the design. Landscape architects should enhance visitor's experience by encouraging interaction to interpret the ecological processes. In this context, the ecological knowledge of them should be enough for interpretation of ecological processes on their designs. Also, it is important to be understood that the ecological phenomena helps people understand their environment and its cycles by their senses.

The present concept of ERD enriches the landscapes by incorporating visibility and observability. In this context, it reveals ecological phenomena and process. This process of revealing can only be successful if the environments created are visible, observable, legible, and have the ability to raise curiosity in visitors to explore and understand the complexity of the landscape.

In order to create a successful ecological design it is important to recognize and interpret the historic and cultural significance of the landscape. In other words, "ERD expands by hitching human habitat and their inevitable cultural determinants in to an environmentally inclusive vision" [16].

The *Eco-Revelatory Design Exhibit*, summarized in the 1998 special issue of *Landscape Journal* [1] clearly articulates an open challenge for landscape architects to work as environmental educators and to help heal the relationship between society and natural systems.

Explaining this new philosophy as distinctly different and complementary to ecological design, a more technical approach for designing with natural processes, ERD reveals the significant ecological aspects of a site and helps visitors build meaning and connection between the landscape and their own lives. In this sense, eco-revelatory designs are educative landscapes, pushing their visitors to think, gain perspective, and internalize new information [1].

ERD serves as a lens for reading the landscape's story. This design idea was resulted from environmental and ecological degradation, and the erosion of spiritual connections with the land.

Most of the projects have only one purpose: they are planned for recreation that ignores natural system usually or planned for protection. However, ERD is a multifunctional design that reveals natural systems and meets the cultural and ecological needs of human systems. By revealing natural systems earn people consciousness attention, and care for their environment. The basis for ERD is to create a landscape that is ecologically as well as culturally sustainable. It provides people educational and recreational opportunities.

The technology and the conditions of life break people's connections with nature. Landscape architecture can play a major role in the mission to bring together again nature and people and reveal the ecological processes by design.

REFERENCES

1. Brown B Harkness T Johnston D. Eco-Revelatory Design: Nature Constructed/Nature Revealed: Guest Editors' Introduction. Landscape Journal. 1998 Special Issue: x–xi.
2. Eisenstein W. Ecological design, urban places and the culture of sustainability: Can citybuilding foster a culture of sustainability? Retrieved 8/23/2009, from SPUR. 2001. http://www.spur.org/publications/library/article/ecologicaldesign09012001
3. Brown B. Holding moving landscapes. Landscape Journal 1998 (Special issue):53-68.
4. Ndubisi F.. Ecological planning: a historical and comparative synthesis. Johns Hopkins University Press, Baltimore; 2002
5. Cox Von Ins R E. Designing for Interpretation: Nanny's Mountain Park, Master Of Landscape Architecture. Maureen Grasso Dean of the Graduate School The University of Georgia , Athens; 2006.
6. McHarg I L. Design with Nature. Doubleday/Natural History Press, Garden City, NJ; 1969
7. Forman R T. The Missing Catalyst: Design and Planning with Ecology Roots. Washington Covelo London:Island Press; 2002
8. The Ashford Borough Council. 2008
9. http://www.landscapeonline.com/research/article/15739
10. Opdam P Verboom J and Reijnen R.. Landscape cohesion assessment: determining the conservation potential of landscapes for biodiversity (submitted); 2002
11. Nassauer J I Opdam P.. Design in science: extending the landscape ecology paradigm, Landscape Ecol, Springer 2008: 23:633–644, DOI 10.1007/s10980-008-9226-7
12. Turner M G. Landscape ecology: The effect of pattern on process. Annu. Rev. Ecol. Syst 1989: 20: 171-197.
13. IALE Executive Committee, 1998. IALE Mission Statement. IALE Bulletin 16:1
14. Ahern J. Integration of landscape ecology and landscape design: An evolutionary process. Edited by J. A. Wiens and M. R. Moss. International Association for Landscape Ecology, Guelph, Ontario, Canada.Landscape Ecology 1999: 119–123.
15. Lyle J T. Regenerative Design for Sustainable Development. New York: John Wiley & Sons;1994.
16. Mozingo L A. The Aesthetics of Ecological Design: Seeing Science as Culture Landscape Journa,l Spring, 1997 (16) 1 46-59 DOI: 10.3368/lj.16.1.46
17. Farina A. Principles and methods in landscape ecology. London: Chapman & Hall; 1998
18. Deshpande A M. Design Process to Integrate Natural and Human Systems, Landscape Architecture, State University, Ms Thesis, Blacksburg, VA.; 2003
19. Lyle J T. Design for human ecosystems. Landscape, land use, and natural resources. New York: Van Nostrand Reinhold Company Inc ; 1985
20. LaGro J A. Site analysis. Linking program and concept in land planning and design. New York: John Wiley and Sons: 2001
21. Nassauer J I. The appearance of ecological systems as a matter of policy. Landscape Ecology 1992: (6) 239–250
22. Zhang B. Technicalization of Environmental Aesthetics and a Resolution of Spirituality, Architecture, Culture, and Spirituality Symposium. JUne29-July 1 2011, Serenbe, Georgia, USA

23. Hough M. Cities and Natural Process, London: Routledge; 1995
24. Thayer R L. The experience of sustainable landscapes. Landscape Journal 1989: 8 (2):101-110.
25. Selman P. Community Essay. Department of Landscape, University of Sheffield, Crookesmoor Building, Conduit Road, Sheffield S10 1FL United Kingdom, Planning for landscape multifunctionality, Sustinnability: Science, Practice &Policy 2009: 5 (2)
26. Spirn A W. The language of landscape. New Haven and London, Yale University Press; 1998.
27. Lyle J T. Regenerative Design for Sustainable Development. New York: John Wiley & Sons; 1994

CHAPTER 4

Impact of Short Duration Intense Rainfall Events on Sanitary Sewer Network Performance

Tasnim Nasrin [1], Ashok K. Sharma [1,2] and Nitin Muttil [1,2,*]

[1]College of Engineering and Science, Victoria University, P.O. Box 14428, Melbourne, VIC 8001, Australia

[2]Institute for Sustainability and Innovation, Victoria University, P.O. Box 14428, Melbourne, VIC 8001, Australia

ABSTRACT

Short duration intense rainfall causes an increase in rainfall derived infiltration and inflow (RDII) into aging sewer networks, which leads to Sanitary Sewer Overflows (SSOs). This study presents a generalised framework for assessing and mitigating the impacts of intense rainfall on sanitary sewer networks. The first part of the proposed framework involves a detailed hydraulic modelling to evaluate the performance of the sewer network. The second part deals with the development of SSO mitigation strategies based on Water Sensitive Urban Design (WSUD) approaches. This paper also demonstrates the application of the first part of the proposed framework for a case study catchment in Melbourne, Australia. The hydraulic performance of the case study sewer network during a wet and a dry year is presented. The analysis found that for the wet year, 11 manholes had sewer overflows, whereas 53 of 57 manholes in the network of 3.2 km had surcharges. Such a study will benefit the water authorities to develop mitigation strategies for controlling SSOs in their sewer systems.

Keywords: intense rainfall events; rainfall derived infiltration and inflow; sanitary sewer overflows; mitigation strategies; water sensitive urban design

1. INTRODUCTION

The negative impacts of climate change on urban water infrastructure have been an essential part of an intensive scientific discussion over the last couple of decades [1,2]. The Intergovernmental Panel on Climate Change (IPCC) reported an increase in the frequency of intense rainfall as a consequence of global climate change, which will continue to alter hydrologic regimes across the world [3,4,5]. This increasing intensity of extreme rainfall combined with increasing urbanization (resulting in more impervious areas) are making conventional drainage systems more vulnerable due to increased peak flow volumes and shorter times to peak flow. Recent studies have pointed out that increased intense rainfall events and increased urbanization have increased the risk of widespread urban flooding and sewage overflow hazards [6,7,8]. It was reported that short duration intense rainfall events (with durations of 12, 18, 30 min and 1 h) have become more frequent in recent years [9]. This paper focuses on the effects of this increasing intensity of extreme rainfall events on the performance of the sewer network and presents a generalised framework for assessing and mitigating the impact of such rainfall events on the sanitary sewer network.

Intense rainfall increases flow into the sewer system and this increased portion of flow that occurs during and after a rainfall event is called Rainfall Derived Infiltration and Inflow (RDII). Sanitary sewers are designed to accommodate a certain amount of inflow and infiltration. During intense rainfall events, this amount of inflow and infiltration is exceeded and hence may lead to Sanitary Sewer Overflows (SSOs) [10,11,12]. The SSOs occur when the sewage overflows from the manholes to the surface level due to sewers running under pressure, while manhole surcharge is a situation when sewage rises in the manhole shaft but does not overflow as in the case of SSOs. These SSOs carry inherent risks to human health as well as to the environment. It is necessary to have a better understanding about the sources of RDII in planning a sewer system and propose mitigation strategies to reduce SSOs. As the name indicates, RDII is made up of stormwater entering the sanitary sewer system in terms of inflow as well as rainfall derived infiltration. Inflow is stormwater which enters the sewer pipes through direct connections: roof downpipes which are illegally connected to the sanitary sewers, broken manhole covers and cross-connections between stormwater and sewer pipes. On the other hand, infiltration is the runoff that filters through the soil and then enters the sewer network through cracked pipe sections, defective joints and damaged manhole walls. It can also occur due to rise in the water table.

This study presents a generalised framework for assessing and mitigating the impacts of intense rainfall on the sanitary sewer network. The first part of this framework aims to evaluate the possible impacts of short duration intense rainfall on the performance of the sanitary sewer system. This will include determining RDII flows and conducting hydraulic modelling of a sewer system for performance assessment. The performance of the sewer network is assessed based on various indicators that were used to quantify the SSOs and manhole surcharges. These indicators will help to investigate the hydraulic performance of the existing sewer system under short duration intense rainfall events. The second part of this framework proposes sustainable mitigation strategies for controlling the SSO and surcharge problems. Earlier studies have recommended the implementation of Water Sensitive Urban Design (WSUD) strategies as sustainable and cost-effective approaches for managing stormwater runoff [13,14,15]. There are different types of sustainable WSUD strategies available in the literature: rainwater tanks, rain gardens, bio-retention cells, porous pavements, green roofs, infiltration trenches and vegetative swales [16,17]. These techniques are also known with different terminologies: low impact development (LID), sustainable urban drainage system (SUDS) and, most recently, green infrastructure [18,19]. These WSUD strategies can also reduce SSOs and surcharges by controlling excess stormwater runoff entering the sewer network in terms of RDII. However, there are few studies about the benefits of WSUD for reducing sewer overflows as these strategies are commonly used for stormwater management. Common approaches of SSO mitigation focus largely on structural actions and are well documented in literature [20,21,22,23]. Hence, the framework presented in this study includes a detailed modelling of WSUD approaches for reducing rainfall induced sewer overflows. Such a framework will benefit the water industry as it will improve the sustainability of the sewer network.

Furthermore, the generalised framework proposed in this study has been applied for a case study residential catchment in Melbourne, Australia. However, this paper covers only the first part of the framework. This includes the detailed hydraulic modelling to analyse the performance of the case study sewer network during a wet and a dry year. The chosen wet year was 2010, which was identified by the Australian Bureau of Meteorology (BoM) as the third wettest year on record for Australia [24]. For the case study area, the total annual rainfall in 2010 was 681 mm, which was well above the annual mean rainfall of 588 mm. Such an analysis will help to provide cost-effective remedial solutions for controlling existing SSO and surcharge problems. This will help the relevant water authorities to take adequate measures to minimize the environmental and human health impacts at locations identified to be at risk. The number of sewer overflows and surcharges at those locations are expected to increase because short duration intense rainfall events are becoming more frequent as a consequence of climate change [9]. The second part of the proposed framework will be applied at a later stage.

2. PROPOSED FRAMEWORK

Figure 1 presents the generalised framework used for assessing and mitigating the impacts of short duration intense rainfall on the sewer network.

The proposed framework is divided into two parts. The first part evaluates the hydraulic performance of the existing sewer network in terms of SSOs and surcharges. The second part proposes sustainable mitigation strategies for reducing the sewer overflows.

2.1. Part 1: Evaluation of the Hydraulic Performance of the Sewer Network

The modelling steps involved in evaluating the performance of the sewer system are described in the following sections:

- Selection of suitable modelling tools—Two modelling tools are required (1) Model for hydraulic evaluation of the sewer network and (2) Model to evaluate RDII parameters.
- PCSWMM which uses Stormwater Management Model (SWMM) as its basic engine was used in this study for the sewer hydraulic modelling. PCSWMM (CHI software) was selected because EPA SWMM is a widely used platform to investigate sewer overflows [25]. PCSWMM is used to simulate sewer flows and compute surcharge and overflow through manholes. It also provides hydrographs at each surcharged manhole including duration and surcharge depth, which can be used to assess the hydraulic performance of the existing sewer system [26,27].

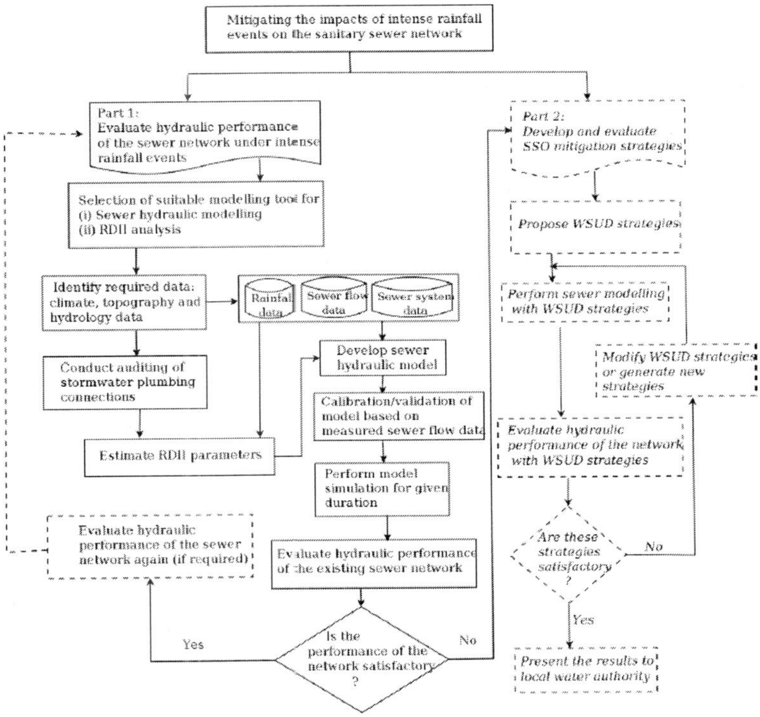

Figure 1. Generalised framework for assessing and mitigating the impacts of intense rainfall on the sanitary sewer network.

- The Sanitary Sewer Overflow Analysis and Planning (SSOAP) Toolbox [28] has been selected to estimate RDII parameters as input into PCSWMM. The SSOAP software applies the synthetic unit hydrograph method (SUH) to estimate the RDII parameters. This procedure is recommended in the literature as accurate and the industry standard methodology to quantify RDII parameters [11,29].
- Data collection—Data required for the hydraulic performance analysis are rainfall, measured sewer flow at point of interest and physical sewer system network geometry and layout. The SSOAP toolbox requires rainfall and measured sewer flow to estimate RDII flows. The sewer system data are needed for the hydraulic modelling for the selected case study area.
- Auditing of households stormwater plumbing connections—The connections of properties in the study area should be checked (if possible) for any cross-connections between the stormwater plumbing and existing sewer pipes. This is because the inflow component of the

RDII depends on the stormwater that enters the sewer network through direct connections and it plays a significant role in generating peak RDII flows.

- RDII parameters estimation—The SSOAP toolbox identifies the unit hydrograph parameters or the RDII parameters (R, T, K) through a systematic analysis of measured sewer flow data and rainfall data. This method contains three unit hydrographs and each unit hydrograph is characterized by a set of R, T and K parameters. Here, R parameter represents the fraction of rainfall volume that enters the sanitary sewer system as RDII during a rainfall event. T parameter is the time from the onset of rainfall to the peak of the RDII hydrograph in hours, and K parameter represents the ratio of time to recession of the RDII hydrograph to the time to peak; this analysis is undertaken in two steps. In the first step, the measured wastewater flow is analysed to generate the RDII hydrograph which is then used (in the second step) to identify the RDII parameters (R, T, K).
- Developing a hydraulic model of the sewer network—The rainfall, flow, sewer system data and RDII parameters are used as the input to the sewer hydraulic model to estimate the RDII flows. After assigning the input data to the model, the next step is to divide the total catchment area into the sub-catchment areas to prepare a hydrologic model for a sanitary sewer system and divert wastewater flows to various loading point of sewer system. Geographic Information System (GIS) has been used to delineate the sewer-shed area into small sub-catchments leading to the flow monitoring locations (manholes). Here, each sub-catchment area is associated with a flow loading point (i.e., manhole) in the developed sewer network model.
- Model calibration and validation—The successful application of a model depends on how well it is calibrated. During calibration, model parameters are adjusted within their physically meaningful ranges until the simulated flow matches the observed flow. Validation is then conducted to confirm that the calibrated parameters provide a consistent prediction. Model calibration and validation can be conducted using peak rainfall events from the available data. Sewer system performance evaluation—After calibration and validation, the next step is to perform a continuous simulation of the sewer network for a given duration. This simulation will help to assess the impacts of intense rainfall events on the performance of the sewer network in terms of SSOs and surcharges. In addition, a set of performance indicators needs to be developed for evaluating the performance of the sewer system. The performance indicators that could be selected can include number of overflowing manholes, number of overflow days, total overflow volume, peak overflow rates, manhole with maximum

volume of overflow, manhole with maximum hours flooded, number of surcharging manholes and manhole with maximum hours surcharged. Based on the local conditions, other performance indicators can also be selected. Earlier studies such as [30,31] have recommended performance indicators as assessment tools because simulation results do not adequately reflect the sewer network performance. Moreover, these indicators can also be used for describing and comparing the possible impacts of intense rainfall on the sewer system. Mitigation strategies are proposed based on the results of these performance indicators. If the system experiences SSO and surcharges, the next step is to develop mitigation strategies for controlling these problems.

2.2. Part 2: Define Optimal Strategies for Sewer Overflow Mitigation

In the second part of this framework, mitigation strategies are developed and assessed. These strategies are selected based on their sustainability and cost efficiency. WSUD strategies have shown to be sustainable, innovative and cost-effective approaches for controlling stormwater runoff. These strategies can also reduce sewer overflow volume and peak overflow rate by capturing excess stormwater runoff entering the sewer network during intense rainfall events [17,32,33,34,35,36,37]. The steps involved in the modelling of WSUD strategies for reducing sewer overflows are presented in brief below.

- Selection of suitable WSUD approaches based on literature review and discussion with local water professionals.
- Development of WSUD strategies for sewer modelling.
- Perform sewer hydraulic modelling with the developed strategies.
- Evaluation of hydraulic performance of the existing sewer network with WSUD strategies.
- Check if the proposed strategies meet the desired objectives. If the proposed strategies fail to provide satisfactory results, then modify or generate new strategies.

This paper presents the detailed modelling results for the first part of the framework and only a brief discussion is provided for the second part. Another publication will present the detailed outcomes of modelling for the second part of the framework. The first part of the framework has been applied to a residential catchment in Melbourne, which is described in the next section.

3. CASE STUDY AREA

The selected case study area was a residential catchment in Glenroy (a suburb in northern Melbourne), which comes under the jurisdiction of Yarra Valley Water, a water retailer servicing 1.6 million people. The study area is located within the larger Pascoe Vale catchment and consists mainly of residential households. The study area is serviced by a separate sewer and stormwater drainage system. The total contributing sewershed area of the catchment is 6.88 km^2 and has 3750 sewer connections. The length of the main sewer pipe is approximately 3.2 km and the pipe material is concrete. A flow meter was temporarily installed in the sewer network at a downstream manhole (named GLN8) (Yarra Valley Water sewer network identification number) for flow data collection from the whole catchment. The location of the study area and the layout of the Glenroy sewer network indicating the main sewer pipe and the location of the flow meter are shown in Figure 2.

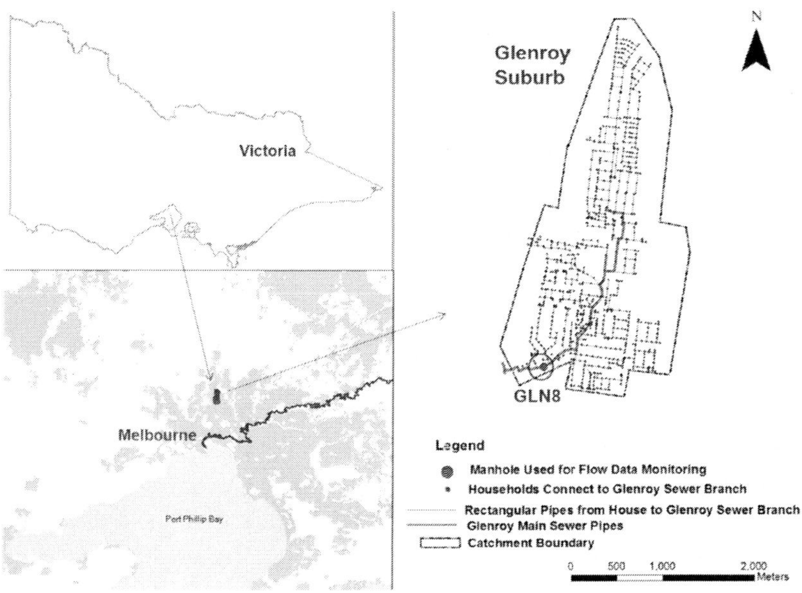

Figure 2. Location of the Glenroy sewershed in Melbourne.

At the GLN8 manhole, flow data was measured at a six-minute time-step during the period November–December of 2010. Flow was measured continuously at the manhole using a Sigma 940 flow meter. This flow meter is widely used for long-term flow monitoring and sanitary sewer evaluation studies. It has high accuracy levels of data measurement in low-flow, full-pipe or reversed-flow conditions [38]. Six-minute time-step rainfall data was obtained from the Bureau of Meteorology, Australia, for a nearby rain gauge station (Essendon Airport Melbourne; station no 086038). There were several intense rainfall events in 2010 and a number of them occurred during the months of November and December. Few events occurred on 8 December 2010 (23.4 mm of rainfall occurred on that day) with 9.8 mm falling in an interval of 18 min (from 9.48 a.m. to 10.06 a.m.) with an Average Recurrence Interval (ARI) of about 1.5 years.

The ARIs were calculated using the 2016 Intensity–Frequency–Duration (IFD) design rainfalls that are provided by the Bureau of Meteorology for use in conjunction with the 2016 edition of Australian Rainfall and Runoff (ARR) [39]. This intense rainfall over a short duration had caused the downstream manhole GLN8 to overflow (which was observed during the flow measurements).

4. SEWER SYSTEM HYDRAULIC PERFORMANCE ASSESSMENT

The proposed framework (presented in Figure 1) for assessing the hydraulic performance of the sewer network has been systematically implemented for the case study sewershed area. The detailed modelling of the hydraulic performance analysis is described in the following sections.

4.1. Data Collection

Climate, topography, and hydrology data are needed for this hydraulic performance assessment. The data used for this study were collected from various Australian government authorities. The climate data was collected from the Bureau of Meteorology. The GIS data for the catchment and the sewer network data were collected from Yarra Valley Water. All datasets used for the sewer performance analysis are presented in Table 1.

The investigation of stormwater plumbing connections at the properties were acquired from discussions with the water utility (YVW) and were not performed for this study due to the budgetary and time constraints. However, the utility is planning to conduct auditing of household stormwater connections in some other catchment.

4.2. RDII Analysis

To quantify RDII flow, the SSOAP toolbox determines the RDII hydrograph parameters (R, T, K) through unit hydrograph curve fitting analysis. As stated earlier, this analysis is undertaken in two steps. These two steps of the analysis are presented in the following sub-sections and a detailed description can be found in [26].

Table 1. Input data for model development.

Data Type	Specifications	Data Source
Meteorological Data	Rainfall time series data of 6-min resolution (from January 1 to December 31) for the wet year (2010) and for the dry year (2008)	Bureau of Meteorology (Essendon Airport Melbourne; station no 086038)
Observed Flow Data	Sewer flow data of 6 min interval at the downstream manhole (GLN8) during wet weather period (24 November–16 December, 2010)	Yarra Valley Water
Sewer Network Data	Sewer pipes, junctions, outlets with manhole and pipe properties	Yarra Valley Water
Subcatchment	GIS layers of the catchment area and subcatchment properties	Yarra Valley Water
Impervious Area Map	Vector polygons	Yarra Valley Water

4.2.1. Hydrograph Decomposition

The SSOAP toolbox applies the Hydrograph Decomposition method to decompose the measured wastewater flows into Dry Weather Flow (DWF) and RDII components. A description of the steps involved in the hydrograph decomposition is given below.

- Step 1—Dry Weather Flow analysis: The DWF analysis determines the typical characteristics of DWF diurnal patterns in each flow meter location.

- Step 2—Wet Weather Flow analysis: After computing the average DWF hydrograph, the model performs wet weather flow (WWF) analysis to calculate the RDII hydrograph for each rainfall event. WWF is the combination of DWF (identified in Step 1 above) and RDII. The latter component for each rainfall event is calculated by deducting the average DWF hydrograph from the total WWF hydrograph (which is the measured flow). This difference is the rainfall induced RDII volume entering the sewer network during the rainfall event.

Figure 3 shows the RDII hydrograph determined by subtracting the DWF from the measured wastewater flows. The figure also indicates the two significant rainfall events selected for the RDII analysis at the GLN8 flow meter.

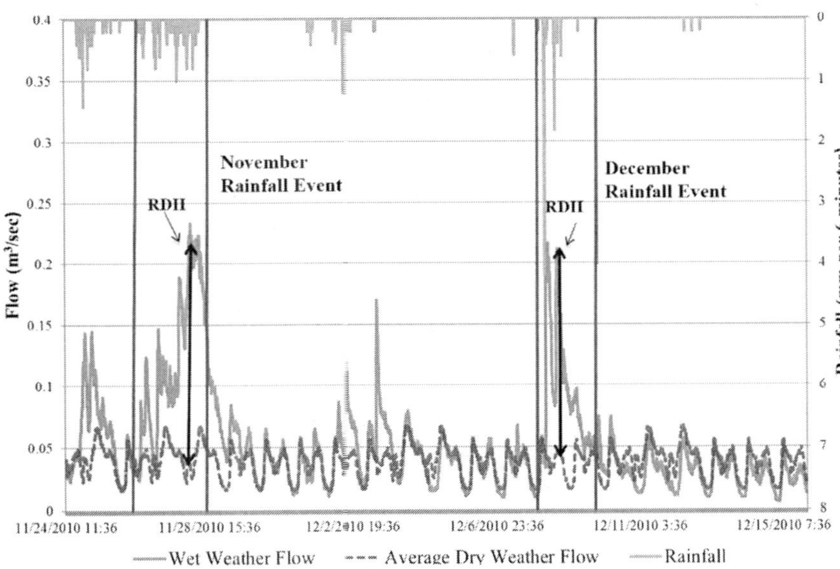

Figure 3. The RDII hydrograph determined using the Hydrograph Decomposition method.

4.2.2. Triangular Unit Hydrograph Curve Fitting

The SSOAP software defines the triangular unit hydrograph (UH) curve fitting method to simulate the RDII. This method is based on fitting three triangular unit hydrographs to an actual RDII hydrograph derived (in the previous step) from the flow data. As stated earlier, each hydrograph has three parameters of R, T, K. Total R value is the sum of three unit hydrograph parameters (R1, R2, and R3) and it denotes the percentage of rainfall volume which enters the sewer network in terms of RDII. A high value of R1 suggests that the inflow is a major component of RDII. If the proportion of R2 and R3 dominants in the total R, it denotes that infiltration is a major component of RDII. These three unit hydrographs also describe fast, medium and slow response of RDII. Figure 4 shows estimation of the R,

T, K parameters based on the triangular unit hydrograph method (using the SSOAP software).

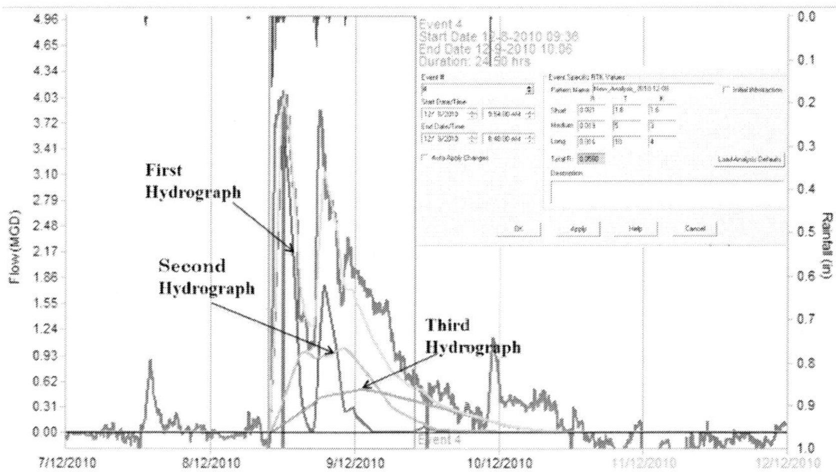

Figure 4. Estimation of the R, T, K parameters based on the triangular unit hydrograph method in the SSOAP toolbox.

The first hydrograph (Figure 4) indicates the fast response of the rainfall derived inflow component. The second hydrograph represents a medium response component of the RDII hydrograph, which contains both rainfall derived infiltration and inflow. The third hydrograph denotes rainfall derived infiltration and represents the slow response of the RDII.

A previous study was undertaken for calibrating the R, T and K parameters using the SSOAP toolbox [40]. This software simplifies the calibration procedure by using a built-in graphical tool (shown in Figure 4) that provides a visual-based automatic calibration approach. The model uses a FPS (foot–pound–second) unit system and the same parameters and their units are depicted in Figure 4. Table 2 presents the three sets of calibrated R, T and K parameters (nine parameters in total) for the two intense rainfall events during the period November–December 2010 at the GLN8 flow meter location.

RDII responses within the sanitary sewer network vary from event to event because different rainfall events could lead to different RDII responses throughout the year. Therefore, the ideal condition for determining the characteristics of the relationship between rainfall and RDII responses needs accurate long-term (a year or more) rainfall and measured sewer flow data. Total R, which is the fraction of rainfall volume entering the network as RDII, depends on many factors including total event rainfall, event rainfall intensity, ground water infiltration and antecedent moisture conditions. Hence, total R is the main parameter causing the variability of the RDII responses. The other parameters, T and K, are not changing significantly between rainfall events since they depend on the geometry and sewer system layout. These parameter values are only used for the iterative process in the triangular unit hydrograph curve fitting method. For continuous simulation of the hydraulic routing, monthly varying R, T, K parameters were required for evaluating the different RDII responses throughout the year. Therefore, we sought to establish a multi-variable linear regression equation to predict the remaining month's (January–October) RDII responses based on the limited data (measured during November–December, 2010). It is recommended in the literature to apply multi linear regression to predict better responses of RDII when long-term data is unavailable [26,41]. In the multi-variable linear regression analysis, the dependent variable is total R and the selected independent variables are total event rainfall, peak rainfall intensity and 7-day rainfall total before the event. For developing the multi-variable linear regression model, six significant rainfall events from the wet months of November and December were considered for calibration (including one large rainfall event in November, 2010).

Table 2. R, T and K parameters at GLN8 manhole for the November and December intense rainfall events.

Downstream Manhole, GLN8		R	T	K
November rainfall event	Short (R1, T1, K1)	0.021	1.5	1.6
	Medium (R2,T2, K2)	0.028	3	3
	Long (R3,T3, K3)	0.035	4	8
	Total R = (R1 + R2 + R3)	0.084	-	-
December rainfall event	Short (R1, T1, K1)	0.021	1.8	1.6
	Medium (R2, T2, K2)	0.019	5	3
	Long (R3,T3, K3)	0.016	10	4
	Total R = (R1 + R2 + R3)	0.0560	-	-

The coefficient of determination (R2) was used to identify the best regression equation and a satisfactory value of 0.99 was obtained for the calibration. The regression model was validated for two significant rainfall events in December 2010 (including one large rainfall event). The coefficient of determination (R2) for the validation was 0.96, which indicates a good predictive capability of the regression equation. The equation (with regression coefficients) developed from this multi-variable linear regression analysis is as follows.

Total R = 0.0244 + [total event rainfall × (0.095)] + [peak rainfall intensity × (−0.1465)] + [7 days total rainfall before the event × (−0.00622)]

After calculating the monthly varied total R value, the fast, medium and slow RDII responses (R1, R2 and R3) were determined using the RDII analysis tool of the SSOAP software over the analysis period total months. These calibrated R, T and K parameters will now be used as inputs to the sewer hydraulic model for continuous simulation.

4.3. Sewer Hydraulic Modelling

This task involves the simulation of the sewer model using PCSWMM for assessing the performance of the existing sewer network.

4.3.1. Model Development

In this part of the study, GIS has been applied to delineate the study area into 38 sub-catchments with areas varying from 2 ha to 101 ha based on the overall layout of the sewer network. The sewer modelling has been undertaken using the dynamic wave routing approach. A set of model parameters has been used in the study. The following parameters were used in the hydraulic model: Percentage of imperviousness (*Imp*); Catchment slope (*S*); Manning's *n* for overland flow (n_{Imp} and n_{Per}); Manning's *n* for closed conduit (n_{Conduit}); Depression storage for impervious and pervious areas (DS_{Imp} and DS_{Per}); and the Infiltration parameters. The parameters calibrated, the calibration procedure and the parameter values are discussed in the next sub-section.

4.3.2. Model Calibration and Validation

Model calibration and validation have been conducted using two intense rainfall events during the wet months of November and December, 2010. The sewer model has been calibrated by using the measured sewer flow at the GLN8 flow meter location for the November rainfall event, 27–28 November. Then the calibrated model has been validated for another intense rainfall event that occurred from 8 to 9 December. The time series plots of the calibration and validation rainfall events are shown in Figure 3.

In PCSWMM, calibration is done by using the Sensitivity Radio Tuning Calibration (SRTC) tool [42], which was adopted for this study. However, the automatic calibration is also popular in rainfall-runoff modelling to estimate the best combination of model parameters [43,44]. Especially, the population-evolution-based Genetic algorithm (GA) has been extensively used for evaluating the optimal values of model parameters [45,46]. As the SRTC tool provided acceptable accuracy for calibration and validation (discussed in the next sub-section), automatic calibration tools such as the GA were not employed in this study. The different model parameters and their calibrated values are shown in Table 3, which also presents typical values for the model parameters obtained from literature or calculated from the catchment data.

Table 3. Model parameters and their values.

Parameters	Parameter Values Calculated from Catchment Data	Calibrated Value
Percentage of imperviousness (*Imp*) (%)	51%	60%
Catchment Slope (S) (%)	0.21	Actual data available
Parameters	**Typical Parameter Values Obtained from Literature**	**Calibrated Values**
Manning's n for overland flow over impervious area (n_{Imp})	0.01–0.015 [47,48]	0.01
Manning's n for overland flow over pervious area (n_{Per})	0.02–0.8 for pervious surfaces [47]	0.1
Manning's n for closed conduits ($n_{Conduit}$)	0.011–0.015 to 0.017 [49,50]	0.0125
Depression storage for impervious areas (DS_{Imp}) (mm)	0.03–0.25 [48,51]	0.05
Depression storage for pervious areas (DS_{Per}) (mm)	0.25–0.5 for pervious surfaces [48,51]	0.25
Infiltration parameters (Green-Ampt)	SWMM Manual [50] Suction Head: 316 mm Conductivity (K): 0.6 mm/hr Initial Deficit: 0.21	Not calibrated for this study

Other than the parameters presented in Table 3, the unit hydrograph parameters (R, T and K) are also important parameters to be calibrated. These parameters have already been calibrated in the RDII analysis presented earlier using the SSOAP toolbox.

4.3.3. Calibration/Validation Results

The fitness evaluation of the calibration and validation hydrographs has been undertaken based on the Nash–Sutcliffe coefficient of efficiency (E_{NS}), which is a commonly used goodness-of-fit measure in hydrological models. The E_{NS} is suitable for reflecting the trends and overall fit of a flow hydrograph [52]. The E_{NS} is represented by Equation (1) (taken from [53]), where Q_{obs} and Q_{simu} refer to the measured sewer flows and model simulated flows, respectively and N defines the number of observations.

$$E_{NS} = 1 - \left(\frac{\sum_{i=1}^{N} \left| Q_{obs}^{i} - Q_{simu}^{i} \right|^{2}}{\sum_{i=1}^{N} \left| Q_{obs}^{i} - \overline{Q_{obs}^{i}} \right|^{2}} \right) \quad (1)$$

If the value of the E_{NS} is close to 1, it denotes that the prediction of the model simulated flow is as accurate as the measured flow. A comparison of the measured and simulated hydrograph for the calibration period is presented in Figure 5, whereas the validation hydrographs are presented in Figure 6. The E_{NS} values are also indicated in both these figures.

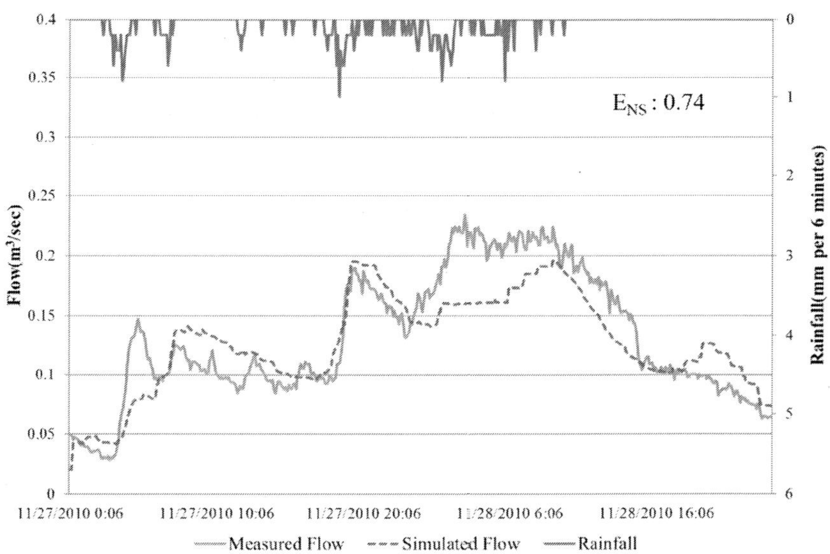

Figure 5. Hydrographs for the calibration period (November rainfall event) at GLN8 manhole.

From Figure 5, it can be seen that the calibration results are satisfactory with an E_{NS} coefficient of 0.74 and, although the model is not able to accurately capture some of the peak flows. An error of 13% was observed between the measured and simulated values over the calibration period, whereas the error over the validation period was 5%. Good performance is also achieved for the validation results, as can be seen in Figure 6. The validation results again indicate an accurate match between the measured and simulated hydrographs and a much better simulation of the peak flows (when compared to that in the calibration).

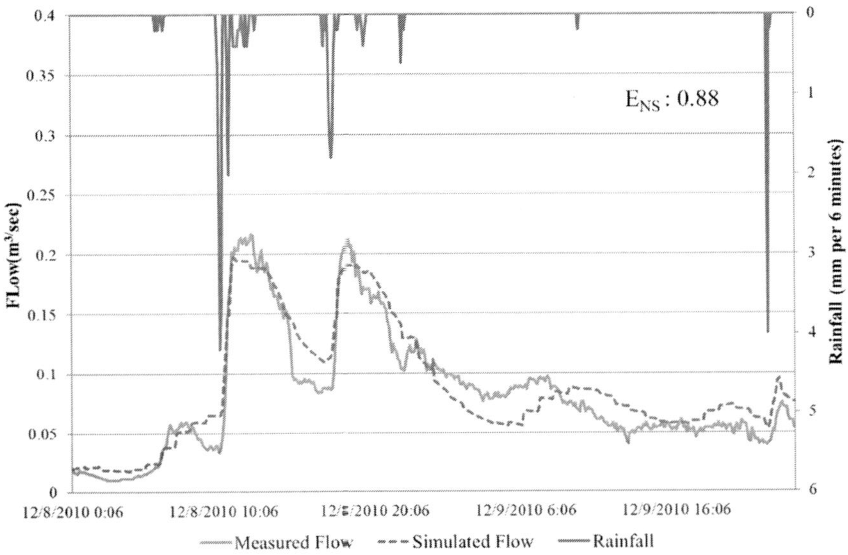

Figure 6. Hydrographs for the validation period (December rainfall event) at GLN8 manhole.

4.4. Sewer Network Performance Evaluation

As stated earlier, a set of performance indicators were developed that would help in assessing the hydraulic performance of the sewer network under intense rainfall events. The performance indicators were compared for two years, one of which (2010) was a wet year in Melbourne with many intense rainfall events, whereas the other (2008) was a relatively dry year. According to the Bureau of Meteorology, if the daily rainfall exceeds 10 mm, the day is known to be a heavy precipitation day [54]. The year 2010 had 19 heavy precipitation days whereas 2008 had nine such days. Moreover, the total annual rainfall for 2010 was 681.2 mm, which exceeds the long-term average of 587.9 mm and was accordingly denoted as a wet year [55]. However, the total annual rainfall for 2008 was 369.8 mm, which was well below the long-term average (587.9 mm) and defined as a dry year. A continuous simulation of the sewer model was undertaken individually for both years using a six-minute time-step.

Figure 7 presents a hydraulic profile plot that indicates the manholes which had SSOs and surcharges due to the most intense rainfall event in 2010, which occurred on 30 October. On that day, the total daily rainfall was 55.2 mm, which was the most for a day in 2010. A rainfall of 29.2 mm occurred from 1:00 p.m. to 3.30 p.m., which corresponds to an ARI of 5 years. As a consequence of the intense rainfall over a short duration, 11 manholes had sewer overflows and 53 manholes had sewer surcharges (out of a total of 57 manholes in the network). Figure 7 shows the location of the overflowing manholes, which in turn can be used by the water authorities to implement sewer overflow mitigation strategies.

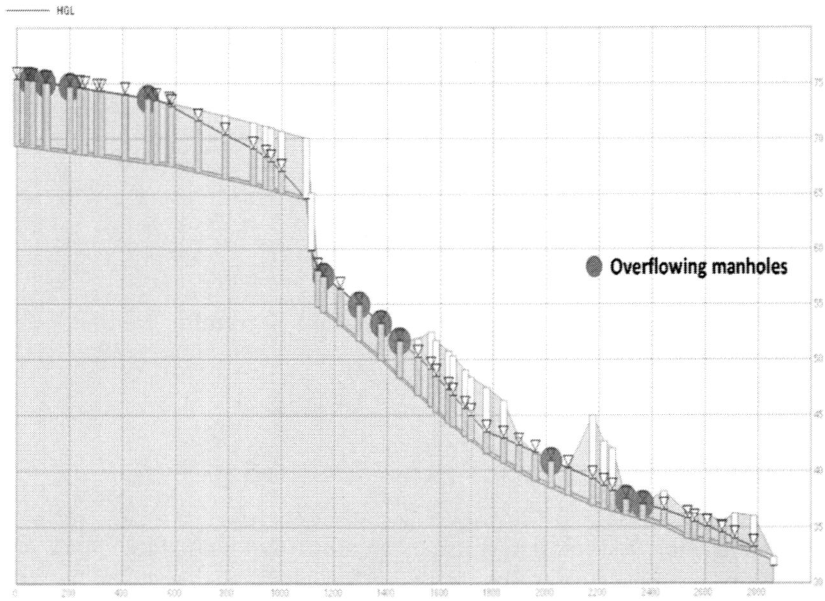

Figure 7. Hydraulic profile plot indicating the locations of SSOs and surcharges that occurred at 3:15 p.m. on 30 October 2010 (units for x- and y-axis are metres)

The performance indicators used in this study to assess the performance of the sewer network in terms of SSOs and sewer surcharges and their values for 2008 and 2010 are presented in Table 4.

Table 4. Sewer network performance indicators for a dry (2008) and a wet (2010) year.

Type	Performance Indicators	Year	
		2008	2010
Manhole Overflows	Number of overflowing manholes	Six manholes	11 manholes
	Number of overflow days	Manholes overflowed 48 h	Manholes overflowed 264 h
	Total overflow volume	3.42 ML	23 ML
	Peak overflow rates	0.14 m^3/sec	0.4 m^3/sec
	Manhole with maximum volume of overflow	GLN8A (1.8 ML)	GLN8A (8.8 ML)
	Manhole with maximum hours flooded	GLN8 (17.89 h)	GLN8 (79.70 h)
Manhole Surcharges	Number of surcharging manholes	52 manholes	53 manholes
	Manhole with maximum hours surcharged	GLN6 (80.94 h)	GLN6 (174.39 h)

The performance indicators presented in Table 4 for the wet year (2010) showed that 11 manholes had sewer overflows, whereas 53 had sewer surcharges. There were several intense rainfall events in 2010 and sewer overflows occurred on 11 days in that year. However, for the dry year (2008), there was an intense rainfall event on 13 December (14 mm rainfall in two hours) and it caused six manholes to overflow and 52 manholes to surcharge. It can also be seen that 23 ML of sewage overflowed in 2010 (wet year) but only 3.42 ML in 2008 (dry year), which could have serious aesthetic, environmental and health issues for the selected residential catchment.

5. DISCUSSION ON SEWER NETWORK HYDRAULIC PERFORMANCE ASSESSMENT

The aforementioned results describe and compare the simulated impact of short duration intense rainfall events on an existing sewer network. In this analysis, we sought to establish a set of sewer network performance indicators which will help to evaluate the current situation of the existing sewer system in terms of its hydraulic performance. Since the sewer network in this case study area was quite old, RDII would be a major problem, in addition to pipe collapse, debris build-up and blockages. These in turn will lead to SSOs and surcharges.

Comparing the sewer network's performance for the representative wet and dry years showed that the wet year with frequent intense rainfall events had five times the number of sewer overflow days than the dry year.

The sewer overflow volume on the other hand was 23 ML for the wet year as compared to 3.42 ML for the dry year. These overflows release many harmful contaminants and spread pollutants, nutrients, and hazardous substances into the suburban creeks and waterways. Thus, these sewage overflows affect the ecosystem and biota in the receiving waters.

Another point to note is that an increase in rainfall intensity or extended surcharge conditions will eventually lead to more of the surcharging manholes to overflow. The existing network fails to cope with the intense rainfall events and as a consequence, the wet year indicates a significant amount of overflow volume, more overflowing days and a large number of critical locations where overflows and surcharges occur frequently. The results (presented in Table 4) seem to indicate that the existing system in the case study catchment will fail to cope with the increased rainfall intensities. Therefore, the framework also proposes sustainable and cost-effective WSUD approaches for mitigating these negative impacts of sewer overflows. A future study will undertake detailed hydraulic modelling with various WSUD strategies for reducing rainfall induced sewer overflows for the same sewershed area. The proposed framework and these findings will serve as the basis for future studies. As stated in the framework, the performance indicators will also be used for comparing the performance of the existing system with that when sustainable mitigation strategies have been implemented.

6. SUMMARY AND CONCLUSIONS

Intense rainfall has an adverse impact on the performance of the sewer network by causing sanitary sewer overflows (SSOs) and surcharges. The aim of this paper is to investigate the negative impacts of the intense rainfall events on the performance of the existing sewer network. Therefore, this study has presented a framework for evaluating and mitigating the impacts of intense rainfall on the performance of the sewer network. The framework for the hydraulic performance assessment has been applied to a residential catchment in Melbourne and has assessed the performance of the sewer system under intense rainfall events. The application of the framework indicated that that overflows and surcharges were very frequent in the case study sewer network, especially for the wet year.

The framework undertaken for the hydraulic performance assessment would provide a good indication to the water managers in local councils about the performance of the sewer network, especially during intense rainfall events. The performance assessment modelling identifies the most critical locations (manholes) where overflows and surcharges frequently occur as well as providing the overflow volumes, and thus suitable measures could be taken to alleviate the problem.

ACKNOWLEDGMENTS

The authors wish to thank Yarra Valley Water and the Bureau of Meteorology for providing the necessary data used in this study.

REFERENCES

1. Willems, P. Revision of urban drainage design rules after assessment of climate change impacts on precipitation extremes at Uccle, Belgium. *J. Hydrol.* 2013, *496*, 166–177.
2. Gamerith, V.; Olsson, J.; Camhy, D.; Hochedlinger, M.; Kutschera, P.; Schlobinski, S.; Gruber, G. *Assessment of Combined Sewer Overflows under Climate Change-Urban Drainage Pilot Study Linz*; IWA World Congress on Water, Climate and Energy: Dublin, Irland, 2012; Availableonline: http://www.smhi.se/polopoly_fs/1.24829.1347460676!/Paper_ WCE_ Dublin_Gamerith_etal.pdf (accessed on 9 May 2014).
3. Berggren, K.; Olofsson, M.; Viklander, M.; Svensson, G.; Gustafsson, A.M. Hydraulic impacts on urban drainage systems due to changes in rainfall caused by climatic change. *J. Hydrol. Eng.* 2011, *17*, 92–98.
4. Nie, L.; Lindholm, O.; Lindholm, G.; Syversen, E. Impacts of climate change on urban drainage systems—A case study in Fredrikstad, Norway. *Urban Water J.* 2009, *6*, 323–332.
5. IPCC. Summary for policymakers. In *Climate Change 2007: The Physical Science Basis. Contribution of Working Group I to the Fourth Assessment Report of the Intergovernmental Panel on Climate Change*; Solomon, S., Qin, D., Manning, M., Chen, Z., Marquis, M., Averyt, K.B., Tignor, M., Miller, H.L., Eds. Cambridge University Press: Cambridge, UK, 2007; pp. 1–18.
6. Astaraie-Imani, M.; Kapelan, Z.; Fu, G.; Butler, D. Assessing the combined effects of urbanisation and climate change on the river water quality in an integrated urban wastewater system in the UK. *J. Environ. Manag.*2012, *112*, 1–9.
7. Willems, P.; Arnbjerg-Nielsen, K.; Olsson, J.; Nguyen, V.T.V. Climate change impact assessment on urban rainfall extremes and urban drainage: Methods and shortcomings. *Atmos. Res.* 2012, *103*, 106–118.
8. Semadeni-Davies, A.; Hernebring, C.; Svensson, G.; Gustafsson, L.G. The impacts of climate change and urbanisation on drainage in Helsingborg, Sweden: Combined sewer system. *J. Hydrol.* 2008, *350*, 100–113.
9. Yilmaz, A.G.; Perera, B.J.C. Extreme Rainfall Nonstationarity Investigation and Intensity-Frequency-Duration Relationship. *J. Hydrol. Eng.* 2014, *19*, 1160–1172.
10. Pawlowski, C.W.; Rhea, L.; Shuster, W.D.; Barden, G. Some factors affecting inflow and infiltration from residential sources in a core urban area: Case study in a Columbus, Ohio, neighborhood. *J. Hydraul. Eng.* 2013, *140*, 105–114.
11. Karuppasamy, E.; Inoue, T. Application of USEPA SSOAP Software to Sewer System Modeling. In Proceedings of the ASCE's World Environmental and Water Resources Congress 2012, Crossing Boundaries, Albuquerque, NM, USA, 20–24 May 2012; pp. 3494–3504.
12. Zhang, Z. Estimating rain derived inflow and infiltration for rainfalls of varying characteristics. *J. Hydraul. Eng.* 2007, *133*, 98–105.

13. Yazdanfar, Z.; Sharma, A.K. Urban Drainage System Planning and Design—Challenges with Climate Change and Urbanization: A Review. *Water Sci. Technol.* 2015.
14. Blecken, G.T.; Hunt, W.F., III; Al-Rubaei, A.M.; Viklander, M.; Lord, W.G. Stormwater control measure (SCM) maintenance considerations to ensure designed functionality. *Urban Water J.* 2015, *14*, 278–290.
15. Elliott, A.H.; Trowsdale, S.A. A review of models for low impact urban stormwater drainage. *Environ. Model. Softw.* 2007, *22*, 394–405.
16. Arnbjerg-Nielsen, K.; Fleischer, H.S. Feasible adaptation strategies for increased risk of flooding in cities due to climate change. *Water Sci. Technol.* 2009, *60*, 273–281.
17. Abi Aad, M.P.; Suidan, M.T.; Shuster, W.D. Modeling techniques of best management practices: Rain barrels and rain gardens using EPA SWMM-5. *J. Hydrol. Eng.* 2009, *15*, 434–443.
18. Fletcher, T.D.; Shuster, W.; Hunt, W.F.; Ashley, R.; Butler, D.; Arthur, S.; Trowsdale, S.; Barraud, S.; Semadeni-Davies, A.; Bertrand-Krajewski, J.L. SUDS, LID, BMPs, WSUD and more—The evolution and application of terminology surrounding urban drainage. *Urban Water J.* 2014, *12*, 525–542.
19. Fryd, O.; Dam, T.; Jensen, M.B. A planning framework for sustainable urban drainage systems. *Water Policy*2012, *14*, 865–886.
20. Beeneken, T.; Erbe, V.; Messmer, A.; Reder, C.; Rohlfing, R.; Scheer, M.; Schuetze, M.; Schumacher, B.; Weilandt, M.; Weyand, M. Real time control (rtc) of urban drainage systems—A discussion of the additional efforts compared to conventionally operated systems. *Urban Water J.* 2013, *10*, 293–299.
21. Dirckx, G.; Schütze, M.; Kroll, S.; Thoeye, C.; De Gueldre, G.; Van De Steene, B. Cost-efficiency of rtc for cso impact mitigation. *Urban Water J.* 2011, *8*, 367–377.
22. Fu, G.; Khu, S.T.; Butler, D. Optimal distribution and control of storage tank to mitigate the impact of new developments on receiving water quality. *J. Environ. Eng.* 2009, *136*, 335–342.
23. Butler, D.; Schütze, M. Integrating simulation models with a view to optimal control of urban wastewater systems. *Environ. Model. Softw.* 2005, *20*, 415–426.
24. BoM, Australia. 2015. Available online: http://www.bom.gov.au/climate/enso/lnlist (accessed on 5 May 2015).CHI. PCSWMM. 2016. Availableonline: http://www.chiwater.com/Software/PCSWMM/index.asp (accessed on 12 December 2016).
25. Vallabhaneni, S.; Camp, D. *Computer Tools for Sanitary Sewer System Capacity Analysis and Planning*; US Environmental Protection Agency, Office of Research and Development: Washington, DC, USA, 2007.
26. Vallabhaneni, S.; Lai, F.H.; Chan, C.; Burgess, E.H.; Field, R. SSOAP—A USEPA toolbox for SSO analysis and control planning. In Proceedings of the World Environmental and Water Resources Congress 2008, Honolulu, HI, USA, 12–16 May 2008; pp. 1–10.
27. Bennis, S.; Bengassem, J.; Lamarre, P. Hydraulic performance index of a sewer network. *J. Hydraul. Eng.* 2003, *129*, 504–510.
28. Hsu, M.H.; Chen, S.H.; Chang, T.J. Inundation simulation for urban drainage basin with storm sewer system. *J. Hydrol.* 2000, *234*, 21–37.
29. Engelhard, C.; De Toffol, S.; Rauch, W. Suitability of CSO Performance indicators for Compliance with Ambient Water Quality. *Urban Water J.* 2008, *5*, 43–49.
30. Berggren, K. Indicators for urban drainage system-assessment of climate change impacts. In Proceedings of the 11th International Conference on Urban Drainage, Edinburgh, UK, 31 August–5 September 2008; Available online: http://web.sbe.hw.ac.uk/ staffprofiles/ bdgsa/11th_International_Conference_on_Urban_Drainage_CD/ICUD08/pdfs/394.pdf (accessed on 11 June 2014).
31. Llopart-Mascaró, A.; Farreny, R.; Gabarrell, X.; Rieradevall, J.; Gil, A.; Martínez, M.; Puertas, J.; Suárez, J.; Río, H.D.; Paraira, M. Storm tank against combined sewer overflow: Operation strategies to minimise discharges impact to receiving waters. *Urban Water J.* 2015, *12*, 219–228.
32. Lucas, W.C.; Sample, D.J. Reducing combined sewer overflows by using outlet controls for Green Stormwater Infrastructure: Case study in Richmond, Virginia. *J. Hydrol.* 2015, *520*, 473–488.

33. Liao, Z.L.; Zhang, G.Q.; Wu, Z.H.; He, Y.; Chen, H. Combined sewer overflow control with LID based on SWMM: An example in Shanghai. *Water Sci. Technol.* 2015, *71*, 1136–1142.
34. Shamsi, U.M. Modeling to Quantify the LID Benefits for CSO Reduction. In Proceedings of the 5th EWRI National Low Impact Development Conference and Low Impact Development Symposium: Greening the Urban Environment, 2011, Philadelphia, PA, USA, 25–28 September 2011.
35. Montalto, F.; Behr, C.; Alfredo, K.; Wolf, M.; Arye, M.; Walsh, M. Rapid assessment of the cost-effectiveness of low impact development for CSO control. *Landsc. Urban Plan.* 2007, *82*, 117–131.
36. Sharma, A.K.; Pezzaniti, D.; Myers, B.; Cook, S.; Tjandraatmadja, G.; Chacko, P.; Chavoshi, S.; Kemp, D.; Leonard, R.; Koth, B.; et al. Water sensitive urban design: An investigation of current systems, implementation drivers, community perceptions and potential to supplement urban water services. *Water*2016, *8*, 272.
37. Hach Sigma 940 (2007): Hach Sigma 940 intrinsically safe area velocity flow meter. Hach Company: Loveland, CO, USA. Available online: www.e-d-s.com.au/documents/sigma_940.pdf (accessed on 20 November 2015).
38. ARR Australia. IFD Design Rainfalls for Use with the 2016 Edition of Australian Rainfall and Runoff. 2016. Available online: http://www.bom.gov.au/water/designRainfalls/ifd/index.shtml (accessed on 15 December 2016).
39. Nasrin, T.; Tran, H.D.; Muttil, N. Modelling Impact of Extreme Rainfall on Sanitary Sewer System by Predicting Rainfall Derived Infiltration/Inflow. In Proceedings of the MODSIM 2013: 20th International Congress on Modelling and Simulation, Adelaide, Australia, 1–6 December 2013; Available online: http://www.mssanz.org.au/modsim2013/ L12/nasrin.pdf (accessed on 16 May 2014).
40. Loehlein, M.D.; Meeneghan, T.J.; Prevost, T. A continuous simulation approach for separate sewered areas. *J. Water Manag. Model,* 2005, *13*, 435–450.
41. CHIWater Support. Calibrating a SWMM5 Model Using the SRTC Tool. 2014. Available online: http://support.chiwater.com/support/solutions/articles/29894-calibrating-a-swmm5-model-using-the-srtc-tool (accessed on 29 June 2014).
42. Barco, J.; Wong, K.M.; Stenstrom, M.K. Automatic calibration of the US EPA SWMM model for a large urban catchment. *J. Hydraul. Eng.* 2008, *134*, 466–474.
43. Muttil, N.; Jayawardena, A.W. Shuffled Complex Evolution model calibrating algorithm: Enhancing its robustness and efficiency. *Hydrol. Processes* 2008, *22*, 4628–4638.
44. Fang, T.; Ball, J.E. Evaluation of spatially variable control parameters in a complex catchment modelling system: A genetic algorithm application. *J. Hydroinform.* 2007, *9*, 163–173.
45. Liong, S.Y.; Chan, W.T.; ShreeRam, J. Peak-flow forecasting with genetic algorithm and SWMM. *J. Hydraul. Eng.* 1995, *121*, 613–617.
46. Huber, W.C.; Dickinson, R.E.; Barnwell, T.O., Jr.; Branch, A. *Storm Water Management Model*; version 4; US Environmental Protection Agency, Environmental Research Laboratory: Washington, DC, USA, 1988.
47. Sun, N.; Hall, M.; Hong, B.; Zhang, L. Impact of SWMM Catchment Discretization: Case Study in Syracuse, New York. *J. Hydrol. Eng.* 2012, *19*, 223–234.
48. Chow, V.T.; Maidment, D.R.; Mays, L.W. *Applied Hydrology*; McGraw-Hill Education: New York, NY, USA, 1988.
49. Rossman, L.A. *Storm Water Management Model User's Manual*; version 5.0; National Risk Management Research Laboratory, Office of Research and Development, U.S. Environmental Protection Agency (U.S. EPA): Cincinnati, OH, USA, 2010.
50. Tsihrintzis, V.A.; Hamid, R. Runoff quality prediction from small urban catchments using SWMM. *Hydrol. Processes* 1998, *12*, 311–329.
51. Coutu, S.; Del Giudice, D.; Rossi, L.; Barry, D.A. Parsimonious hydrological modeling of urban sewer and river catchments. *J. Hydrol.* 2012, *464*, 477–484.
52. Nash, J.; Sutcliffe, J.V. River flow forecasting through conceptual models part I—A discussion of principles. *J. Hydrol.* 1970, *10*, 282–290.

53. BoM, Australia. 2016. Available online: http://www.bom.gov.au/climate/ change/about/extremes.shtml (accessed on 5 December 2016).
54. Walsh, T.C.; Pomeroy, C.A.; Burian, S.J. Hydrologic modeling analysis of a passive, residential rainwater harvesting program in an urbanized, semi-arid watershed. *J. Hydrol.* 2014, *508*, 240–253.

CHAPTER 5

Framework for the optimization of operation and design of systems with different alternative water sources

Lisa J. Blinco[1,2*] , Martin F. Lambert[1,2], Angus R. Simpson[1,2] and Angela Marchi[1,2]

[1]*School of Civil, Environmental and Mining Engineering, The University of Adelaide, Engineering North N136, North Terrace Campus, Adelaide, South Australia 5005, Australia*

[2]*Cooperative Research Centre for Water Sensitive Cities, Monash University LPO, PO BOX 8000, Clayton, Victoria 3800, Australia*

ABSTRACT

Water security has become an increasing concern for many water system managers due to climate change and increased population. In order to improve the security of supply, alternative sources such as harvested stormwater, recycled wastewater and desalination are becoming more commonly used. This brings about the need for tools to analyze and optimize systems that use such sources, which are generally more complex than traditional water systems. Previous methodologies have been limited in their scope and cannot be applied to all types of water sources and systems. The framework presented in this paper has been developed for holistic analysis and optimization of water supply and distribution systems that use alternative water sources. It includes both design and operational decision variables, water and energy infrastructure, simulation of systems, analysis of constraints and objectives, as well as policies and regulations which may affect any of these factors. This framework will allow users to develop a comprehensive analysis and/or optimization of their water supply system, taking into account multiple types of water sources and consumers, the effect of their own design and operational decisions, and the impact of government policies and different energy supply options. Two case study systems illustrate the application of the framework; the first case study is a harvested stormwater system that is used to demonstrate the importance of simulation and analysis prior to optimization, the second utilizes four different water sources to increase security of supply and was optimized to reduce pump energy use.

Keywords: Water distribution systems, Integrated water resources management, Decision-making, Conceptual framework, Alternative water sources Optimization

1. INTRODUCTION

A changing climate and increasing population have put a strain on traditional water resources, which typically rely on natural catchment water. This has made water security an increasing concern for many water system managers, who have investigated options for reducing demand and supplementing supply. Alternative water sources, such as harvested stormwater, recycled wastewater and desalination, are increasingly being used to improve water security of cities and towns. Methods for simulation, analysis and optimization of traditional potable water distribution systems (WDSs) cannot necessarily be directly transferred to systems that use alternative water sources. Therefore there is a need to develop a methodology specifically for alternative water source systems, which includes both hydraulic and hydrologic considerations, as well as the many additional parameters and variables associated with alternative water sources. There are many modelling tools used in current practice for integrated water management, such as eWater Source, WEAP (Water Evaluation and Planning System) and Mike Basin. These modelling tools do not include hydraulic simulation, and therefore may not accurately represent performance of urban water networks. Moreover, this framework is not software, rather its purpose is to guide water system managers in how to best simulate and optimize their systems, particularly those that integrate multiple water sources, and natural and human-made systems.

The framework should be used to determine which system components need to be modelled, which type of modelling tools are most appropriate, what regulations and policies need to be taken into account and how to evaluate the performance of the system.

The framework introduced in this paper can be applied to the optimization of the design and operation of water supply and distribution systems from source to consumer, considering multiple traditional and alternative sources, multiple uses and multiple objectives. Electrical energy sources and their effect on electricity prices and greenhouse gas (GHG) emissions are included, as are several types of government policies that may affect the design, operation, data and evaluation of the system. The objectives of this paper are to (1) develop a generalized framework that could be applied to any water supply and/or distribution system optimization problem and (2) outline the application of this framework to two case study systems with a focus on optimizing their operation.

2. LITERATURE REVIEW

Since 2000, there has been significant consideration of the concept of water security (Cook and Bakker 2012) as water is increasingly seen as a fundamental and finite resource (Bogardi et al. 2012). Consequently, the use of alternative sources, such as harvested stormwater, desalination, recycled wastewater and rainwater, has gained traction (Fielding et al. 2015). Harvested stormwater schemes are often decentralized and used for non-potable supply such as household gardening and irrigation of public reserves (Naylor et al. 2012), however, in some cases are also used for potable supply (McArdle et al. 2011). While desalination is a climate independent (and therefore more reliable) source, is often not the most cost effective or environmentally sensitive option (Becker et al. 2010; Miller et al. 2015). Recycled wastewater is also climate independent, and generally used for large scale non-potable applications (Muga and Mihelcic 2008; Oron et al. 2014), however, it can also be used for indirect or direct potable supply (Rodriguez et al. 2009; Nagal 2015). Domestic rainwater tanks are increasing in popularity and have benefits of reducing water usage from utilities and reducing stormwater runoff from houses (Campisano and Modic 2012; Umapathi et al. 2013). Demand management strategies have also been used to reduce per capita consumption and therefore reduce the pressure on limited water supplies (Dawadi and Ahmad 2013; Friedman et al. 2014).

Some alternative sources, such as harvested stormwater, introduce additional complexity to the problem of modeling and optimization than has been previously considered for traditional water systems (Marchi et al.

2016). There is, for example, the need to consider the supply and distribution systems together, rather than separately, as it is less likely that there will be large storages isolating the supply side from the distribution side. When including the supply side, longer simulation times often need to be used, requiring rainfall and evaporation scenarios to be taken into account. The security of supply with regard to climate change needs to be considered (Paton et al. 2014; Cai et al. 2015), as some sources are climate dependent and some are climate independent. The social acceptability of using particular sources for particular applications and the willingness of consumers to pay more for alternative source systems to be constructed and maintained may need to be incorporated (Hwang et al. 2006; Londoño Cadavid and Ando 2013; Fielding et al. 2015). The perception of risks associated with alternative water source systems by water system managers may also present a barrier to the implementation and success of such systems (Dobbie and Brown 2012; West et al. 2016). Many alternative sources also have associated externalities that result in either cost or benefit to the user, such as reduced effluent flow to the ocean or receiving water body by reusing wastewater and reduced urban stream flows by harvesting stormwater (Marchi et al. 2016).

The increased use of alternative water sources then raises the question of how such systems should by analyzed and optimized to ensure they are implemented as effectively as possible. Stokes et al. (2014) developed a framework for optimizing the cost and GHG emissions of WDSs, taking into account both the design and operation of the system, energy sources and GHG emission factors. This study, however, was applicable only to traditional WDSs, with no consideration of the supply side and alternative water sources. Chung et al. (2008) developed a mathematical model for evaluating integrated water supply systems with decentralized treatments. Multiple sources, uses, transportation and treatment systems can be considered, however only surface water, groundwater and recycled wastewater sources are included. This model does not incorporate any optimization procedure, only analysis of different options developed by the user. Makropoulos et al. (2008), with further developments in Rozos and Makropoulos (2013), produced a decision-support tool for modeling the urban water system from source to tap. The software can be used to select combinations of water saving strategies and technologies, including how much water from each type of demand (for example domestic, commercial) is obtained from each source and how the system is operated. It uses a demand-oriented, water balance approach and does not include capability for other types of simulation models such as hydraulic and hydrologic modeling.

Uncertainty, particularly with regard to climate change, is an important consideration that has been taken into account in several methodologies. Paton et al. (2014) developed a framework for water supply system planning with alternative sources and climate change considerations, while Beh et al. (2014, 2015) developed two methods for optimal sequencing of urban water supply augmentation options under deep uncertainty regarding demands and climate. The research by both Paton et al. (2014) and Beh et al. (2014, 2015) considered only the planning of water supply projects, and did not optimize the specific design or operation of the systems. Sequencing is also considered in Cai et al. (2015), however, in this case it is applied to planning of drought mitigation strategies in agricultural systems. They consider multiple decision stages in which options such as infiltration ponds, parallel terraces, irrigation triggering threshold and irrigation water sources can be implemented. Marchi et al. (2016) developed a methodology for optimizing the design of harvested stormwater systems taking into account future climate scenarios; however, it does not apply to other types of alternative sources or optimization of system operation. It does include a detailed analysis of the associated externalities, such as reduced peak flows and improved economic value of properties near stormwater schemes. Ashbolt et al. (2014) introduced a framework for planning of short-term operations for water systems using surface water, groundwater, desalination, and recycled wastewater with multiple objectives and multiple inflow replicates to account for uncertainty. Long-term operating strategies and the design of the system were not included and the operating strategies considered were limited to bulk water transfers and not the operation of pumps and smaller storages.

3. FRAMEWORK FOR THE OPTIMIZATION OF ALTERNATIVE WATER SOURCE SYSTEMS

The framework presented in the current paper was developed to guide the modeling and optimization of water supply and distribution systems that use alternative water sources. It is comprised of several components and sub-components that fit within an optimization structure, for example, a multi-objective evolutionary algorithm (Fig. 1). The options component [OPT] describes the potential 'decision variables' that are available in an optimization problem, that is, the factors that can be changed in order to produce a different outcome. This includes both the initial design of the water supply and distribution infrastructure and the long- and short-term rules that govern the operation of the system once it has been commissioned. The infrastructure component [INF] describes the physical components of the system that need to be modeled and the data associated with each, including both water infrastructure and energy infrastructure, which may affect the evaluation of electrical energy cost and life-cycle GHG emissions. There is also a government policy component [G] that covers the policies from regulating bodies that may affect other aspects of the framework. The analysis component [ANL] describes the simulation of each potential system configuration and evaluation against objectives and constraints. The optimization algorithm [OA] investigates different possible combinations of decision variables from the options component, models the system according to the infrastructure component and evaluates it using the analysis component to find the optimal solution(s).

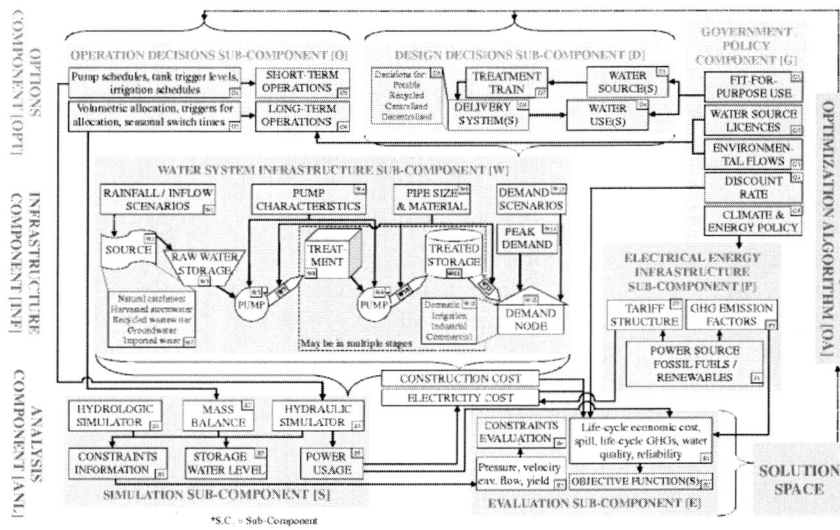

Figure. 1 Schematic of the framework for optimizing the design and operation of alternative water source systems

Details of the components and sub-components are shown in Fig. 1 and described in Sections 'Options component [OPT]', 'Infrastructure component [INF]', 'Government policy component [G]' and 'Analysis component [ANL]'. Table 1 summarizes the parameters that need to be considered in the optimization and simulation of alternative water source systems with respect to the different items that are presented in Fig. 1 and in the following sections. There are three (non-exclusive) categories that each parameter may be placed in – decision variables, parameters that are set, and uncertain parameters. Decision variables are parameters that the user may be able to examine using optimization. It is important to note that in most optimization problems, not all of these parameters will be available as decision variables at once, and it is likely that only a small number will be considered. For example, when optimizing pump operations for an irrigation system, only the first three 'decision variables' shown in Table 1 (pump schedules, tank trigger levels, and demand scheduling) may be considered. The remaining parameters that are designated as decision variables in Table 1, particularly those relating to the design of the system (for example, delivery system layout and pump sizing) would already be set and not able to be optimized if the existing infrastructure cannot be modified. The parameters that are set are those that very rarely, if ever, are able to be optimized by the user. These include parameters that would be controlled by external sources, for example consumers of domestic or commercial demands, pipe manufacturers and higher level government and regulatory bodies; and also parameters that need to be predefined to a known or assumed value before optimization or simulation can be performed, for example, fire demand/reserve, hydrologic/hydraulic variables and objective and constraint selection and definition. The final category, uncertainty, designates those externally set or predefined variables that are not well known or may be subject to change in the future and therefore may need to be considered in a sensitivity analysis. While the selected values of decision variables have an impact on the performance of a system, they are generally within the control of the decision maker, and therefore are not classed as 'uncertain'. It is important to note that the categorization in this table is indented as an indication of how each parameter is typically treated There are, of course, exceptions to this, as almost all of the parameters could be considered as decision variables if desired and have some associated uncertainty. For example, environmental flows have been designated as an externally set parameter, as it is likely that the operator of a system will have to meet requirements set by an external organization such as the Environmental Protection Agency. They may, however, want to investigate providing greater environmental flows, or show the benefits of reducing their environmental flow requirements and being able to supply more water elsewhere.

4. OPTIONS COMPONENT [OPT]

The options component covers the potential decision variables (and the range of possible choices for the decision variables) for an optimization problem. This component is split into two sub-components; the operational decisions sub-component [O] and the design decisions sub-component [D]. Design decisions include elements that can be changed before a system is constructed, such as the layout and capacities, materials and other properties of the various infrastructure components. Operational decisions include elements that can be changed after construction during the daily management of the system, such as the operating rules for pumps and valves and allocation of water from different sources.

Table 1 Summary of parameters for the design and operation of alternative water source systems

Parameter	Decision variable[a]	Parameter that is set	Uncertain parameter	Relevant items in Fig. 1
Operational inputs [O]				
Pump schedule	X			O1
Tank trigger levels	X			O1
Tank/storage maximum and minimum allowable levels		X		O1, W3, W11
Demand pattern (irrigation, agriculture)	X			O1, D4, W13
Demand pattern (domestic, commercial, industrial)		X	X	O1, D4, W13
Demand flow rate (peak, average, peak day)		X	X	O1, D4, W14
Valve settings or operating rules	X			O1
Pump pressure settings	X			O1
Volumetric allocation of water	X			O2
Reservoir trigger levels	X			O2
Switch time between operating regimes	X			O2
Priority ranking of operating rules		X		O3, O4
Design inputs [D] and water infrastructure [W]				
Water source selection	X			D1, W2
Water source infrastructure (layout, capacity)	X			D1, W2
Treatment type selection	X			D2, W8
Treatment infrastructure (layout, capacity, treatment rate/level)	X			D2, W8
Delivery system type selection	X			D3
Delivery system layout (lengths, elevations, junctions, tank locations)	X			D3, W7, W10, W12, W15
Pipe material and diameters	X			D3, W7, W10, W12
Pipe parameters (unit cost, pipe wall roughness (ε), wall thickness, embodied energy)		X	X (ε)	D3, W6, W7, W10, W12
Pump sizing	X			D3, W5, W9
Pump performance characteristics and cost		X		D3, W4
Tank sizing (capacity, height, diameter)	X			D3, W3, W11
Fire demand/reserve		X		D3, W11
Water user type selection	X			D4
Rainfall/streamflow series		X	X	W1
Reservoir capacity and volume curve	X			W3
Pond (e.g. wetland) capacity and volume curve	X			W3
Prioritization rules for demands types		X		W15
Other inputs [P], [G] and [S]				
Power source selection		X	X	P1, P3, G5
Electricity tariff structure and cost		X	X	P2
GHG emission factors		X	X	P3, G5
Fit-for-purpose requirements		X		G1
Water license amounts		X		G2
Environmental flow amounts		X		G3
Discount rate		X	X	G4
Hydrologic variables (e.g. permeability)		X		S1
Hydraulic variables (e.g. water temperature)		X		S3
Optimization problem formulation [E]				
Objective selection		X		E1
Objective function(s)		X		E2
Constraint selection		X		E3
Constraint limits (maximum and minimum)		X		E4
Penalty costs		X		E4

[a]Note: Parameters specified as decision variables are shown in bold throughout Sections 'Options component [OPT]' and 'Infrastructure component [INF]'

5. OPERATIONAL DECISIONS SUB-COMPONENT [O]

Both short- and long-term operations are considered in the operational decisions sub-component. The critical aspects of this sub-component (items in **bold** can be optimized), as shown in Fig. 1 and Table 1 are:

*[O1]*the specific short term operating strategies including **pump schedules** (when pumps are turned on or off based on time), **trigger levels** (water levels in tanks or other storages that determine when pumps or valves turn on or off), **irrigation or demand schedules** (for systems where they can be pre-determined), **valve settings and operating rules**, and **pressure settings for pumps** (to maintain the set pressure at a particular point).

*[O2]*the specific long term operating strategies including **volumetric allocation of water** from different alternative sources, **trigger levels** (for example in reservoirs) that determine allocations from different sources or water demand restriction levels, **switch times between different operating regimes** (for example between different trigger level sets for different seasons) and power source selection.

*[O3]*the overall short-term operating strategy, including operating rules that are optimized in [O1] and operating rules that are pre-set and are not to be optimized (acting as constraints). Where there are multiple operating rules, the priority of each rule and order they are enforced in is important to consider.

*[O4]*the overall long-term operating strategy, including operating rules that are optimized in [O2] and operating rules that are pre-set and are not to be optimized. Again, the priority and order of the rules is important to consider.

Most systems have multiple operating conditions to meet and therefore multiple operating rules will be in place. Prioritization of the different operating rules is important, and this may be set by the operator or be chosen by the optimization tool. This component requires information from the government policy sub-component ([G] in Fig. 1), specifically in terms of water source licensing and environmental flow regulations. These policies would typically be regulated by local or state government departments or the environmental protection authority. Operational rules set in this sub-component will inform the simulation sub-component [S] as they will need to be represented in any simulation model(s) of the system.

6. DESIGN DECISIONS SUB-COMPONENT [D]

This sub-component incorporates all of the design decisions that are available to the designer for the entire water supply and distribution system, from source to user. The critical aspects of this sub-component (items in **bold** can be optimized), as shown in Fig. 1 and Table 1 are:

*[D1]*the **water sources selected** to be used including natural catchments, harvested stormwater, recycled wastewater, groundwater, imported water, domestic rainwater, desalination, domestic greywater and sewer mining; and the **layout and capacity of source infrastructure**.

*[D2]*the **types of treatment selected** including centralized treatment at plants such as mechanical filtration, chemical dosing, ultraviolet treatment and ozonation, and decentralized in situ treatments such as gross pollutant traps, wetlands and biofilters; and the **layout, capacity, dosing rates and retention times for treatment facilities**.

*[D3]*the **type and configuration of the delivery system** used including potable, non-potable (for example dual reticulation systems to deliver recycled water), centralized and decentralized, and the infrastructure design variables such as **system layout, pipe sizes, lengths and materials, pump sizing, valve sizing, and tank sizing**.

*[D4]*the **types of water users** that are supplied by the system including potable, irrigation, agriculture, industrial, non-potable domestic/commercial and firefighting, and the demand rate and pattern for water use (for example, scheduling of irrigation demands).

Regulations on fit-for-purpose water use from the government policy component [G] in Fig. 1 inform what water sources can be used for particular applications and these are likely to be specified by state or federal government departments or health authorities.

Generally, sources such as harvested stormwater and recycled wastewater cannot be used for potable supply and rather serve non-potable demands in dual-reticulation systems or are supplied to irrigation, agricultural and industrial users.

There may be some systems, however, in which necessary approvals have been obtained to use these sources for potable supply. The design decisions are inputs to the water system infrastructure sub-component [W] which describes the system elements and data to be modeled.

7. INFRASTRUCTURE COMPONENT [INF]

The purpose of this component is to describe the infrastructure that needs to be modeled in order to evaluate the objectives and constraints of the problem. There are two sub-components; the water system infrastructure sub-component [W] and the electrical energy infrastructure sub-component [P]. Water system infrastructure includes the specific aspects of the water supply and distribution system and the data required, including construction and maintenance costs. Electrical energy infrastructure includes the power source (fossil fuel types and renewable types) and the electricity price and GHG emission factor data needed.

7.1 Water system infrastructure Sub-component [W]

This sub-component includes the specific infrastructure aspects of the water system design and the relevant data that is needed in order to simulate it. Most systems and optimization problems will not require all of these factors to be considered or modeled; however, the purpose of this framework is to cover a large range of the possible requirements for an optimization and hence the scope is intentionally broad.

The water system infrastructure sub-component [W] as shown in Fig. 1 represents a system with one water source, one treatment plant, one storage tank and one demand node. In reality, many systems will have more than one of each of these components, particularly the treated storage [W11] and demand node [W15]. Pumping of water between storages may occur in multiple stages, particularly when there is a large difference in elevation. For typical centralized potable WDSs, all treatment will occur at one water treatment plant. In decentralized systems such as for harvested stormwater schemes, however, treatment may occur in multiple stages. For example, a gross pollutant trap may be located on an urban creek before the water is collected in a harvest pond, then the water may be pumped to be treated through a wetland, and then treated again in a treatment plant.

The critical aspects of this sub-component (items in **bold** can be optimized) as shown in Fig. 1 and Table 1 are:

*[W1]*the rainfall or inflow scenarios for the water source; for example rainfall or streamflow scenarios for natural catchments and stormwater sources, or a sewer system flow pattern for recycled wastewater. Sources such as desalination and, depending on the temporal scale of the optimization, groundwater, do not usually require an inflow scenario. Rainfall and streamflow scenarios may be a data series obtained from measurements at gauging stations or modeled in a hydrologic simulation program [S1]. Multiple inflow scenarios may be used, particularly for

systems with highly variable inflows. Losses such as evaporation and infiltration may also need to be taken into account for sources with large open storages such as reservoirs and natural water ways.

[W2] the **source type** as described in [D1] with input from [W1].

[W3] the raw water storage; this may be a reservoir (typical for a natural catchment), a harvest pond for a stormwater system, a tank (for example for a recycled wastewater system) or an aquifer for groundwater. Associated data including **capacity, a volume curve, elevation, height and diameter** is required.

[W4] characteristics of available pumps such as performance curves (head, efficiency, and power against flow), cost, rated speed and variable speed pump (VSP) information where applicable.

[W5] the **pump** transferring water from the raw water storage to a treatment facility, requiring data from [W4].

[W6] pipe size and material information such as available diameters, unit costs, pipe wall roughness, wall thickness and embodied energy. For new pipes, this information will be easily obtained from the pipe manufacturer. For existing systems, however, there may be some uncertainty in these parameters if detailed records of the 'as constructed' system and any pipe replacements have not been kept. In addition to this, the pipe wall roughness of existing pipes will generally be uncertain. Pipe wall roughness increase as pipes age, and pipe condition assessment may be needed to provide an estimate.

[W7] the **pipe system** transferring water from the raw water storage to the treatment facility, **pipe lengths and layouts** need to be known as well as information from [W6].

[W8] the **treatment facility** that will produce water of the required quality based on the source type and demand type. Characteristics of the individual treatment methods as described in [D2] need to be known.

[W9] the **pump** transferring water to a treated storage, requiring the same data as [W5].

[W10] the **pipe system** transferring water to a treated storage, with the same information as [W7] required.

[W11] the **treated storage**, for example, a tank or multiple tanks that are typically at a high elevation point of the network in order to supply demands by gravity. Data required includes the **elevation, height, diameter** and maximum and minimum allowable water levels.

[W12] the **pipe system** transferring water from the treated storage to consumers, which again requires information as in [W7]. This pipe system

is likely to be more complex than those in [W7] and [W10], particularly for systems with many different demand nodes. For systems with only one source of water, [W7] and [W10] are likely to be single pipelines. For decentralized systems with only one specific consumer, [W12] will also most likely be a single pipeline. Most systems, however, have much more than one demand point and as such distribution systems are often looped or branched systems that require more complex analysis than single pipelines.

*[W13]*demand scenarios that will be applied to the demand nodes, consisting of a pattern of demand multipliers over a day, week or year. There may be multiple demand scenarios required for a system, for example, if there are different types of demand nodes (such as domestic, commercial, industrial) or different seasonal demands.

*[W14]*the peak demand is the demand rate that is typically used to size the system components and so will affect the design of the system. The demand scenarios [W13] are more likely to affect the operation of the system as the demand varies over the simulation time. The peak day demand (average demand over the peak day), the peak hour demand (the average demand over the hour with maximum consumption in the peak day) and average demand rates may also be required. Fire loading demands and other emergency conditions will affect the design of the system, for example storage tanks should be sized to be able to provide demand in the case of fires, other emergencies and system failures (e.g. if the supply to the tank is cut off).

*[W15]*the demand nodes for the consumers, these may be different types of users as described in [D4] and require information from [W13] and [W14]. Different types of users will have different demand rates [W14] and demand patterns [W13]. When simulating the system, an average demand rate will often be used with the demand pattern, rather than the peak demand. Systems with multiple demand nodes may prioritize different types of demands over other, for example, irrigation systems using non-potable water may prioritize high profile sport fields over reserves with no formal usage.

Choices made in the optimization of the design decisions sub-component [D] in Fig. 1 will be inputs to the water system infrastructure sub-component. There may be other parameters that are not decision variables in the optimization (as differentiated in Table 1) though are still required by this sub-component in order to simulate the system. The construction and maintenance costs of each of the infrastructure components needs to be known in order to calculate the initial construction cost and ongoing costs as part of life-cycle economic costing. Information collected through this sub-component will be input to the simulation sub-component [S] depending on the types of simulation models used and to the evaluation sub-component [E] through the construction cost or other factors calculated for the specific objectives of a problem.

7.2 Electrical energy infrastructure Sub-component [P]

The electrical energy infrastructure sub-component includes any power infrastructure that affects the electricity prices and GHG emission factors. The critical aspects of this sub-component as shown in Fig. 1 and Table 1 are:

*[P1]*the breakdown of power sources including fossil fuel sources such as coal and oil, and renewable sources such as solar, wind and hydrothermal.

*[P2]*the electricity price tariff structure, which may be a peak and off-peak structure, or multi-part (more than two price levels) and could include a peak demand charge which applies to the peak electricity power usage in each month.

*[P3]*the GHG emission factor, which is based on the power source breakdown [P1] and may vary with time, either in the short-term (with sources that do not have storage such as solar panels and wind turbines) or the long-term (as fossil fuel sources tend to be phased out and renewable sources become more popular).

Climate and energy policy [G5] in the government policy component in Fig. 1 will affect the power source breakdown and electrical energy pricing now and into the future. This is likely to be regulated by a federal government department or body. Information from this sub-component is used to calculate electrical energy costs in order to evaluate life-cycle economic costs and also to calculate life-cycle GHG emissions in the evaluation sub-component [E].

8. GOVERNMENT POLICY COMPONENT [G]

The government policy component covers policies by regulating bodies at any level (local, state, federal) that may affect other aspects of the framework. These policies need to be considered over the operational life-span of the system, for example, climate and energy policy may affect future energy sources and therefore affect future GHG emissions. The critical aspects of this component as shown in Fig. 1 and Table 1 are:

*[G1]*fit-for-purpose water use, which may be regulated by state or federal governments or health agencies and affects which water sources [D1] and water uses [D4] can be combined in the design decisions sub-component. It

may also guide which design decisions (for example, treatment) are appropriate.

[G2] water source licenses, which may be regulated by local or state governments or the environmental protection agency, depending on the catchment size, and will affect the amount of water available from particular sources for allocation in long-term operations [O4].

[G3] environmental flows, which similarly to water source licenses may be regulated by local or state bodies depending on the size of the catchment and affect the amount of water available for allocations [O4].

[G4] the discount rate applied to operational costs and GHG emissions in life-cycle analysis [E1]. This is unlikely to be set by a government body and rather will be informed from outside the decision making team, generally by recommendations from economists.

[G5] climate and energy policy set by state and federal governments will affect the energy sources available now and in the future, therefore affecting GHG emission factors and any GHG objectives [P].

9. ANALYSIS COMPONENT [ANL]

The analysis component uses information from the options, infrastructure and government policy components to simulate the system and evaluate how it performs relative to the objectives and constraints. Within an optimization algorithm, the analysis component is used to assess multiple combinations of decision variables from the options component to determine how they perform. There are two sub-components within the analysis component; the simulation sub-component [S] and the evaluation sub-component [E]. The simulation sub-component includes the modeling aspects of the problem and the key variables that are required to be output from the models in order to evaluate the system. Optimization objectives and constraints are defined in the evaluation sub-component, which also provides information to the optimization algorithm as to which of the potential solutions perform best.

9.1 Simulation Sub-component [S]

The simulation sub-component considers the type of simulation model that is most applicable to the particular system and problem, and specifies the key variables that need to be calculated in the model(s). The critical aspects of this sub-component as shown in Fig. 1 and Table 1 are:

*[S1]*the hydrologic simulator, which is required if rainfall scenarios need to be transformed to streamflow, typically for systems using natural catchment water or harvested stormwater.

*[S2]*the mass balance model, which may be required for systems that have multiple water sources with long-term allocation decisions, particularly if there are different rainfall and evaporation scenarios to be considered for the storages.

*[S3]*the WDS hydraulic simulator, which is required to analyze pump and pipe systems that transfer water between different storages and treatments and to consumers.

*[S4]*information on constraints, such as yield from a hydrologic model, environmental releases and system reliability from a mass balance model, and nodal pressures, pipe velocities, pump switches and tank levels from a hydraulic model.

*[S5]*the water levels in storages, which are important particularly when considering operational decisions, such as trigger levels, and for constraints, such as system reliability.

*[S6]*the power usage from any pumps or treatment facilities, which are important in informing the ongoing electrical energy costs as part of life-cycle economic costing. Generally a WDS hydraulic simulator is required to model detailed pump operations and therefore accurately estimate the pump power usage.

Each of the three types of models will require different simplifications or assumptions depending on the particular system. For example, mass balance modeling will generally only consider one pump operating point so may not accurately calculate the pump power usage. When deciding which type of model to use for a particular problem, the user will need to consider the different simplifications, assumptions, advantages and disadvantages of each model. Trade-offs between accuracy of outputs and simulation run times need to be considered. For example, when optimizing both short- and long-term operations of a system, there is likely to be a trade-off between using a hydraulic simulator for detailed hydraulic information and using a mass balance model for shorter run times. Most problems may ideally use elements from more than one type of model; however, using multiple models will increase computational complexity and run times. Wherever possible, the most applicable type of model should be selected and augmented with the required elements from other types of models. Depending on the particular system and optimization problem, there may be other key variables that need to be calculated in the simulation models. For optimization of pumping operations, which is the focus of the case studies in this paper, storage water levels and pump power usage are the most important. Existing hydrologic, mass balance and hydraulic simulators, for example, MUSIC, WATHNET and EPANET, have often been used in conjunction with optimization algorithms and should be taken advantage of where possible rather than creating individual simulators for different problems.

Information from the operation decisions sub-component [O] will be input to the simulation sub-component as the overall operating strategy for the system ([O3] and [O4]) will need to be modeled. Short-term operations are likely to be considered in a hydraulic simulator and long-term operations, including allocations, in a mass balance model. Parameter data on the physical components of the system from the water system infrastructure sub-component [W] are also required as inputs for this sub-component. Constraint information is provided to the evaluation sub-component to compare the systems performance against specified requirements. Energy usage is used to calculate objective functions such as life-cycle economic costs and life-cycle GHG emissions. Simulating systems prior to optimization is an important step to help inform the formulation of the optimization problem and provide a check that results from the optimization are reasonable.

9.2 Evaluation Sub-component [E]

The purpose of the evaluation sub-component is to compare the performance of each of the potential solutions to the objectives and constraints of the problem. The critical aspects of this sub-component as shown in Fig. 1 and Table 1 are:

*[E1]*the specific objective(s) to be considered in the optimization; typically, minimizing life-cycle economic cost is a primary objective (or a component of that such as construction cost or operational cost individually). Other possible objectives include minimizing spills from reservoirs and other storages, minimizing life-cycle GHG emissions (or a component of that such as embodied energy from construction or operational emissions), minimizing supplemental potable water supply (in systems using non-potable sources), maximizing water quality, maximizing reliability and minimizing environmental impact.

*[E2]*the objective function(s) to be optimized; multiple objectives may be evaluated as individual functions in a multi-objective optimization algorithm or combined into a single function for use in a single objective optimization algorithm. It is important to consider how each objective should be formulated, for example, when optimizing short-term pump operations to minimize ongoing costs, the objective function may be evaluated in terms of cost per volume of water pumped, as this will take into account the amount of water delivered to consumers. Reliability of a system may be formulated in different ways, for example minimizing the

time spent with water restrictions applied or minimizing the time spent below a certain storage level. Some objectives may be more difficult to quantify, such as minimizing environmental impact, so more specific objectives may be required, for example, maximizing environmental flow or minimizing the change in a water body's natural hydrological regime. Simplifications and assumptions may be required to formulate some objectives as mathematical functions. When performing multi-objective optimization, trade-offs between the different objectives should be considered by the development of Pareto fronts, allowing the decision maker to determine which Pareto optimal solution best fits their needs (see examples in Wu et al. 2010a, b, 2012a, b, 2013).

*[E3]*the specific constraints to be considered as described in [S4].

*[E4]*the evaluation of the constraints compared to the limits set by the user; maximum and/or minimum values for each constraint need to be specified. Some constraints may be flexible, for example, if an environmental flow is set by a regulator, the operator could consider increasing the set environmental flow as a decision variable in the optimization. There are several different ways constraints can be incorporated into the optimization algorithm. Penalty functions are often used for single-objective problems. They add value (often a monetary amount) to the objective function in a minimization problem and remove value from the objective function in a maximization problem based on the magnitude of the constraint violation, therefore making solutions that violate constraints less desirable (Nicklow et al. 2010). Care must be taken when formulating penalty functions to keep solutions that only slightly violate constraints in consideration during the optimization process, while ensuring the feasibility of the final optimal solutions. For multi-objective problems, a constraint-handling technique that will ensure feasible solutions are retained in preference to infeasible solution is often employed. An example of this is the constraint tournament selection procedure introduced by Deb et al. (2002).

Information about the objectives is received from the simulation sub-component [S] and from the calculation of construction, maintenance and electrical energy costs based on the water system infrastructure sub-component [W] and simulation sub-component. A discount rate for costs or GHG emissions may be set in the government policy sub-component [G] which will impact the ongoing costs and emissions in a life-cycle analysis. The discount rate may be informed by economists, such as the Stern review which recommends low discount rates for projects that lead to the production of GHG emissions (Stern 2006). Information about the performance of each potential solution in relation to the objectives and constraints is provided to the optimization algorithm in order to find the best solutions.

10. OPTIMIZATION ALGORITHM [OA]

The optimization algorithm is used to determine which solution(s), out of many potential solutions to the problem, performs best in relation to the objective function(s). The procedure used to set up the optimization will depend on the type of algorithm chosen; however, the first step is generally to define the decision variables, objectives and constraints of the problem. This will then guide what aspects of the system need to be modeled and what data is required in order to take into account all of the decision variables and that will provide information for all of the objectives and constraints. Multiple potential solutions to the problem form the 'solution space' and the optimization algorithm guides the search of this solution space towards the global optimum. The size of the solution space depends on the number of decision variables and number of choices available for those decision variables. More complex problems are often described as having a more 'rugged' solution space, meaning the optimization algorithm is more likely to get trapped in local optima and will have more difficulty finding the global optimum. When a single objective optimization algorithm is used, one optimal solution will be found, while in multi-objective optimization, a Pareto front will be developed with multiple solutions representing different trade-offs between the objectives.

Most optimization algorithms have parameters that need to be defined by the user, such as the number of generations or iterations and the population size in evolutionary algorithms. Although the choice of these parameters does not influence the components shown in Fig. 1, they have an effect on the optimal solutions found by the algorithm. In general, the most effective set of parameter values to use will vary between different optimization problems and the size of the solution space can only give some indication of what parameter values to use. In fact, multiple parameter sets should be tested in order to find the most appropriate values for the specific problem. Ideally, the chosen parameter set should find the same optimal solution regardless of the starting point or initial solution(s) for the optimization. Dandy et al. (1996) presented an improved genetic algorithm formulation for optimization of WDS design. Five different parameter sets were trialed on both their improved genetic algorithm and a comparatively simple genetic algorithm. They acknowledged that parameter selection does require some judgement on the part of the user, however, they found their optimization results to be relatively insensitive to the parameter choice, particularly for the improved genetic algorithm. As well as the effect of various parameter values, different optimization algorithms will be more suited to different problems. This issue has been addressed by the development of hybrid algorithms, such as AMALGAM (a multi-algorithm, genetically adaptive multiobjective approach proposed by Vrugt and

Robinson (2007)), which combines several different optimization algorithms to improve search efficiency. These hybrid algorithms also have the benefit of requiring little to no parameter specification by the user.

11. SENSITIVITY ANALYSIS

As identified in Table 1, values of some input parameters (for example, describing the network or water demand loadings) are uncertain or subject to change in the future. Sensitivity analysis can be performed to account for a wide range of possible future conditions when optimizing and simulating systems. Variation of a particular parameter may result in different Pareto fronts (in multi-objective optimization) or different optimal solutions (in single objective optimization), as seen in Wu et al. (2010b) when they considered variations in discount rates. These various Pareto fronts or optimal solutions along with the various parameter values that produced them can then be provided to the decision maker. Sensitivity analysis will also help to identify if there are any uncertain parameters that do not affect the optimal results. Robustness of the optimized solutions can also be explored *a-posteriori*: in general, solutions that perform well for many different possible conditions are more desirable from the decision makers' point of view. Climate change provides an additional source of uncertainty for the parameters identified in Table 1 – detailed discussion of this is omitted from Sections 'Demand', 'Rainfall and streamflow', 'Electricity and GHG emissions' and 'Discount rate' as it is covered in Section 'Climate change'.

12. DEMAND

In some applications, such as irrigation and agriculture, the demand rate and pattern may be deterministic [O1], either the water supplier has control over the consumption, or may be able to work with those who do to determine an optimal demand schedule. For other applications, such as domestic, commercial and industrial, the demand rate and pattern depends on the consumption of water by multiple individual users [D4, W13, W14, W15], and therefore has greater uncertainty. Historical consumption can provide some level of assurance as to how water may be used in the future, at least on an aggregated scale. Diurnal, weekly and seasonal demand variations need to be considered. In the future, factors such as climate change, population growth and water saving initiatives will affect how water is consumed and therefore impact demand rates and patterns.

Emergency conditions and system failure are by their nature unpredictable and this should be taken into account when designing and operating WDSs.

An example of how demand uncertainty can be considered in the optimization of WDS design is the study by Basupi and Kapelan (2015). The demand at each time step was based on a normal distribution with a gradually increasing mean (based on deterministic demand forecasts) and an increasing standard deviation. Monte Carlo or Latin Hypercube simulation was included in their methodology to consider multiple demand scenarios. Each solution in the Pareto front was also further analyzed against three demand projections – average, optimistic (low overall demand) and pessimistic (high overall demand). Their results demonstrated the value of flexible WDS design over deterministic approaches when considering uncertainty.

13. RAINFALL AND STREAMFLOW

Rainfall and streamflow inputs [W1] may be required for systems using natural catchment water, harvested stormwater or imported water, and they are often treated with higher uncertainty than demands (Seifi and Hipel 2001; Reis et al. 2005). Within the current climate, there may be multiple realizations of possible rainfall and streamflow series (for example dry or wet years). Beh et al. (2015) considered rainfall, as well as population and temperature, as uncertain variables in their optimal sequencing methodology for water supply system augmentation. They considered both climate and hydrologic variability: seven possible future climate scenarios provided different forecasted rainfall reductions, and within each of these seven scenarios, 20 stochastic replicates of the rainfall sequence were produced. Different Pareto fronts were produced for each climate scenario, with the more severe scenarios finding solutions that required greater system augmentation and therefore had higher costs and GHG emissions. The robustness of each Pareto solution was calculated based on the average reliability and vulnerability of the solution over the 20 rainfall sequences for the particular climate scenario.

14. ELECTRICITY AND GHG EMISSIONS

Power source(s) [P1], electricity tariffs and costs [P2] and GHG emission factors [P3] will generally be known for the present time, however, it may not be clear how they will change in the future. The mix of power sources changes naturally over time, as different power plants are built or decommissioned. This change in power source types over time, as well as technical advancements will affect the cost and GHG emissions associated with electrical energy generation. The electricity market and economic factors will also affect the cost of electrical energy over time. Changes in electricity and GHG emissions can be an important factor to consider during an optimization problem, as shown in the following examples. Blinco et al. (2014) studied the optimization of pump operations in WDSs in relation to the minimization of GHG emissions and the use of different power source scenarios, showing that optimal tank trigger levels can be influenced by the variation in emission factors. Wu et al. (2012a) considered three different electricity tariff scenarios, which increased over time, and three different GHG emission factor scenarios, which decreased over time, in the optimization of WDS design. The different electricity tariff and emission factor scenarios affected the solutions found in the Pareto front and their overall costs and GHG emissions.

15. DISCOUNT RATE

A discount rate [G4] may be used in life-cycle analysis for both ongoing economic costs and ongoing GHG emissions. In practice, discount rates on economic costs vary significantly between different organizations, generally from 2 to 10% (Rambaud and Torrecillas 2005), while many water utilities use discount rates in the range of 6 to 8% (Wu et al. 2010a). When selecting discount rates, consideration should be given to whether both economic costs and GHG emissions are discounted, if they have the same discount rate, and if intergenerational equity is taken into account using social discount rates. Various social discount rates have been proposed for discounting ongoing costs; the Intergovernmental Panel on Climate Change (IPCC) adopted a zero discount rate over a period of 100 years, after which no consideration for future costs or benefits is given (Fearnside 2002), other suggestions include 1.4% (Stern 2006) for projects that are impacted by climate change, 2–4% (Weitzman 2007) and a time declining rate (Gollier and Weitzman 2010). Wu et al. (2010b) gave an example of a sensitivity analysis of discount rates in the optimization of WDS design for minimization of costs and GHG emissions. Discount rates of 0, 1.4, 2, 4, 6, 8% and a time declining rate were applied to the economic costs, with GHG emissions either not discounted at all, or discounted at the same rate as costs. They found that the different discount rate scenarios produced different Pareto fronts; when GHG emissions were discounted, the solutions tended to have lower capital costs and higher operating emissions.

16. CLIMATE CHANGE

Management of water resources in the developed world has been based on an assumption of stationarity – that is, 'that natural systems fluctuate within an unchanging envelope of variability' (Milly et al. 2008). The effects of human-induced climate change make this assumption no longer valid (Milly et al. 2008), and introduce additional sources of uncertainty for many parameters. Uncertainty introduced by climate change is twofold – firstly, the impacts of climate change increase the uncertainty of future weather conditions; and secondly, our response to the threat of climate change and the types of adaption methods that will be utilized in the future are uncertain. Climate change affects the magnitude and temporal and spatial distribution of rainfall, temperature and other environmental factors, thus the possible rainfall and streamflow series to consider for the future will likely be different to the present. Changes to temperature and other environmental factors will also affect the hydrology of natural and urban catchments and therefore change how rainfall will transform to runoff or streamflow. Climate change impacts will also affect how people consume water, for example, higher temperatures and lower rainfall may drive people to water their gardens more. In order to simulate future climate conditions, general circulation models (GCMs) are often used in conjunction with future emissions scenarios. According to Mpelasoka and Chiew (2009), 'GCMs are the best tools available for simulating global and regional climate systems', however, the information provided is generally too coarse for applications to catchment runoff, and therefore some kind of downscaling is required. The modeling uncertainty of both the GCMs and downscaling methods increases the uncertainty of future climate scenarios (Paton et al. 2013). In 2000, the IPCC introduced several emissions scenarios (termed SRES scenarios) projecting future global GHG emissions (IPCC 2000). The various scenarios are based on different assumptions of the mix of energy generating technologies (fossil fuel or non-fossil fuel dominant) and population, economic and technological growth (IPCC 2007).

The extent to which we can reduce our GHG emissions will affect the magnitude of climate change impacts on rainfall and temperature. With the growing concerns of climate change and sustainability, renewable sources such as solar and wind will become more prevalent and replace fossil fuel sources such as coal and gas. This may affect electricity pricing and GHG emissions from power generation. Multiple future power source scenarios assuming different levels of climate change mitigation may need to be considered. Other climate change adaption strategies include economic incentives such as carbon taxes and cap and trade systems, which may affect economic analysis of WDSs. As discussed in Section 'Discount rate', when climate change and intergenerational equity are considered, social discount rates of 0, 1.4, 2–4% and time declining rates have been proposed.

Paton et al. (2013) analyzed the sources of uncertainty relating to climate change and their impact on water supply security. They considered 19 different scenarios with different combinations of six SRES scenarios, seven GCMs and six demand projections, as well as 1000 stochastic rainfall replicates. They found that the impact of the different sources of uncertainty on the optimal solutions varied over the 40-year planning period, with some having a greater effect in the short-term and others a greater effect in the long-term. Roshani and Filion (2014) investigated the impact that different climate change abatement strategies have on water main rehabilitation. They consider six carbon abatement strategies with different combinations of two discount rates (1.4 and 8%) and three carbon tax scenarios (no tax, 'fast and deep', and 'slow and shallow'). Using a low discount rate and implementing a carbon tax encouraged the optimization algorithm to find solutions that invested in rehabilitation early, to reduce the cost of continuing leaks, pipe repair, energy use and GHG emissions.

17. CASE STUDIES

The utility of the framework described in the previous sections will now be explored by two different case studies that have different water sources and many variables that need to be considered. These case studies are provided as an example of how the framework could be applied to optimize system operations. The first case study is a managed aquifer recharge (MAR) system that harvests stormwater from an urban creek for irrigation of reserves and sporting fields. This case study demonstrates the importance of analyzing the system by simulation prior to optimization in order to formulate the optimization problem. The second case study is a water supply system in a rural town that supplies potable water from multiple alternative water sources. This system is optimized for minimization of energy use of the many pumps used to transfer water from the various sources.

18. RIDGE PARK MANAGED AQUIFER RECHARGE – CASE STUDY 1

Ridge Park is located in the Adelaide metropolitan area in South Australia, within the City of Unley local government area. The scheme supplies harvested stormwater to sports fields and recreational reserves in the local area for non-potable irrigation use. The scheme is designed to harvest up to 60 ML of stormwater per year, which occurs over the winter, while in summer the harvested water is used for irrigation. During winter, stormwater from Glen Osmond Creek, an urban waterway, is collected in the Harvest Pond created by a dam on the creek (Fig. 2). Water is then pumped to the Bioretention Basin which provides some treatment, and then pumped to a small treatment plant that includes UV and filtration. Once the water has been adequately treated, it is stored in an above ground tank next to the treatment plant and final pump station. From the Storage Tank, water is injected into an artesian, fractured rock aquifer for long term storage. In summer, when no water is being harvested, water is extracted from the Aquifer and to the Storage Tank, before being pumped or gravity-fed to irrigation points. The Ridge Park Reserve is irrigated by a pressurized irrigation line, as it is at higher elevation than the Storage Tank. Fraser Reserve is also connected to the pressurized system to ensure adequate pressures for irrigation. In total, the pressurized system supplies almost 15 ML of water per year. The remaining seven reserves are on a gravity-fed line which supplies a total demand of roughly 35 ML per year. The layout and details of the system are given in Fig. 3. For more detailed data on this case study, please see the Additional files 1, 2, 3 and 4.

Figure. 2 Ridge Park managed aquifer recharge system process schematic

For existing systems, simulation analysis of the current operation is an important first step in formulating the optimization problem. Results of current operational simulations can highlight areas for improvement that can then be focused on in the optimization. The operation of the Ridge Park stormwater harvesting system was split between winter and summer operations and both were simulated in EPANET to determine current pump operational costs. Trigger levels (related to volumes in the three storages as shown in Table 2) control when the pumps in the Winter Harvesting and Injection system turn on and off (Table 2). The Bore Pump is also controlled by trigger levels in the Storage Tank.

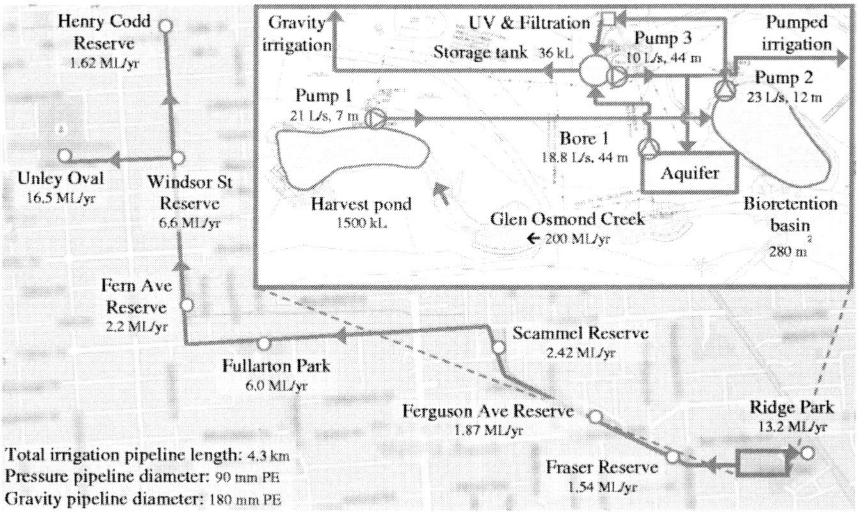

Figure. 3 Ridge Park managed aquifer recharge system layout and data

Table 2 Trigger levels for the Ridge Park system

Storage and trigger level type	**Current setpoint**		**Start pump**	**Stop pump**
	Volume (%)	**Level (m)**		
Harvest Pond High Level	80	1.6	1	-
Harvest Pond Low Level	50	1.0	-	1
Biofiltration Basin High Level	90	0.80	2	1
Biofiltration Basin Low Level	50	0.59	-	2
Storage Tank High Level	90	2.25	3	2, Bore
Storage Tank Low Level	70	1.75	Bore	3

Table 3 Irrigation demand schedule for the Ridge Park system

Reserve	Demand rate (L/s)	Duration/day (hr)	Start time	Irrigation days
Ridge Park 1	3.53	8.33	9:30 PM	Mon & Wed
Ridge Park 2	3.53	8.67	9:30 PM	Tues & Thurs
Fraser Reserve	1.41	5.83	9:30 PM	Mon & Wed
Ferguson Ave Reserve	2.00	5.00	9:30 PM	Tues & Thurs
Scammell Reserve	2.15	6.00	10:00 PM	Tues & Thurs
Fullarton Park 1	3.85	1.67	10:00 PM	Mon & Wed
Fullarton Park 2	3.85	6.67	10:00 PM	Tues & Thurs
Fern Ave Reserve	3.53	3.33	10:00 PM	Mon & Wed
Windsor St Reserve	2.20	8.00	8:30 PM	Tues & Thurs
Henry Codd Reserve	1.10	8.00	10:00 PM	Mon & Wed
Unley Oval	5.57	9.00	9:00 PM	Sun, Mon & Wed

During summer, Pump 3 is controlled by the irrigation demands, which are on a schedule so that different reserves are irrigated on different nights (Table 3). Pump 3 is a VSP and is operated at 80% of full speed for injection (such that the flow is less than the 7 L/s maximum for injection) and 75% of full speed for irrigation (such that the target pressure downstream of the pump is achieved at the expected demand rates).

Both systems were simulated for a period of 1 week in EPANET, with a 15 min hydraulic time step and 5 min reporting time step. Several week-long streamflow series for the available flow in Glen Osmond Creek at a daily resolution were used in the harvesting and injection model (Fig. 4). A peak/off-peak electricity price tariff applied to the entire system; a peak price of 25.53 c/kWh was applied from 7 am to 9 pm on weekdays, and an off-peak price of 15.26 c/kWh was applied over night and on weekends (tariff pattern and simulations started on a Sunday).

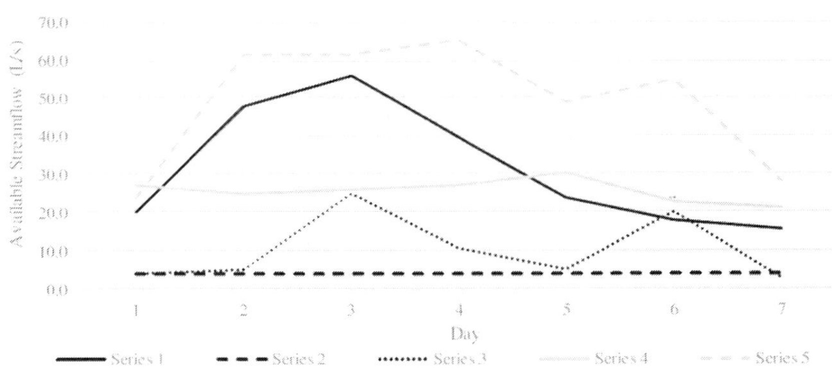

Figure. 4 Streamflow Series used for simulation of the winter harvesting and injection operation

19. WINTER HARVESTING AND INJECTION SYSTEM CURRENT PUMPING OPERATION RESULTS

When there was adequate water available, such as in Streamflow Series 1, 4 and 5, the volume of water injected into the aquifer (by Pump 3) was a little over 3 ML per week (Table 4). This was significantly less than the volume available, which reflects the limited flow rate of Pump 3 (7 L/s for injection to the aquifer), as well as the water that would be lost to overflow when the inflow rate is greater than the flow rate of Pump 1 (approximately 22 L/s). The total pump energy cost estimate for the harvesting and injection system ranged from $163 to $267 per week, with an average of $235 per week. Pump 1 was the most cost-effective to run, while Pump 3 was the most expensive. Pumps 1 and 2 operated at similar times throughout the day, however, Pump 2 has much lower efficiencies, which increased its energy use. Pump 3 operated at a lower flow rate but much higher head than Pumps 1 and 2, and was more likely to be switched on for the entire day, which contributed to its higher cost of operation. Pumps 1 and 2 turned on and off very frequently, and operated at a much higher flow rate than Pump 3 (Fig. 5). The flow rate of Pump 3 in Fig. 5(c) reduced over the week as the headloss through the bore increased from assumed clogging of the bore. As the storages are relatively small, in particular the storage tank, it did not take long for them to be filled and emptied (Fig. 6), which contributed to the frequent pump switches. The current trigger levels in the Storage Tank are very close together (70 and 90% volume) as a result of possible pump priming issues that occurred during the commissioning of the system. These close together trigger levels also contributed to the short fill and empty times.

Table 4 Current operation results for the winter harvest and injection system

Streamflow Series	**Available volume (ML/wk)**	**Cost (c/kL)**			**Volume injected (ML/wk)**	**Total cost ($/wk)**
		Pump 1	**Pump 2**	**Pump 3**		
1	19.0	0.64	2.28	5.49	3.14	267
2	2.29	0.68	2.32	6.19	1.76	163
3	6.19	0.69	2.23	5.87	2.44	222
4	15.4	0.64	2.24	5.46	3.18	258
5	29.7	0.63	2.25	5.47	3.16	264
Average	14.5	0.66	2.26	5.70	2.74	235

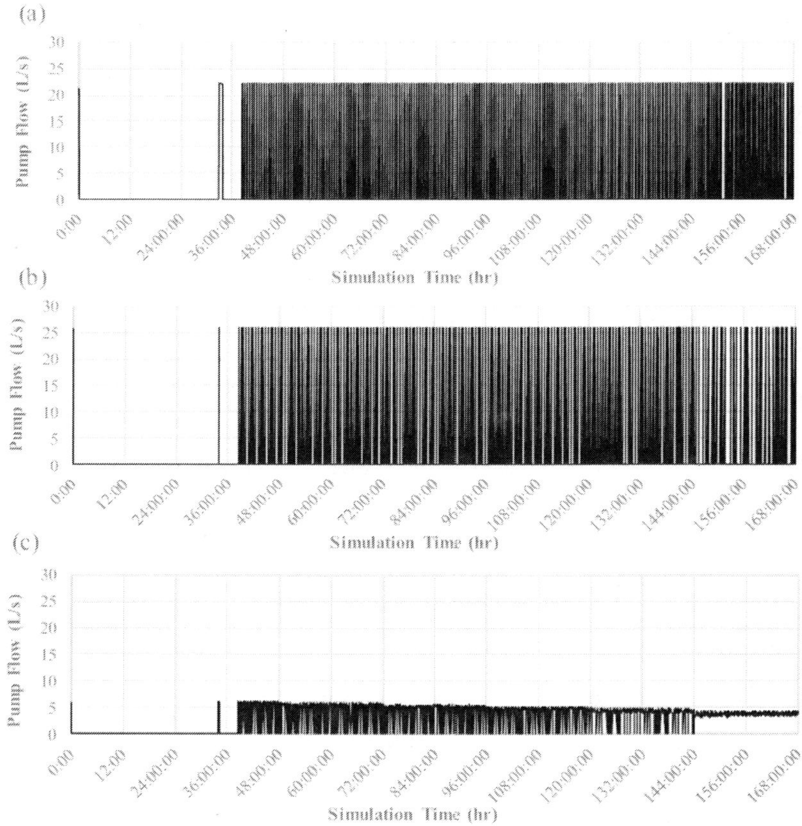

Figure. 5 Pump flows for Streamflow Series 2 for **a** Pump 1, **b** Pump 2 and **c** Pump 3 for a one week EPANET simulation

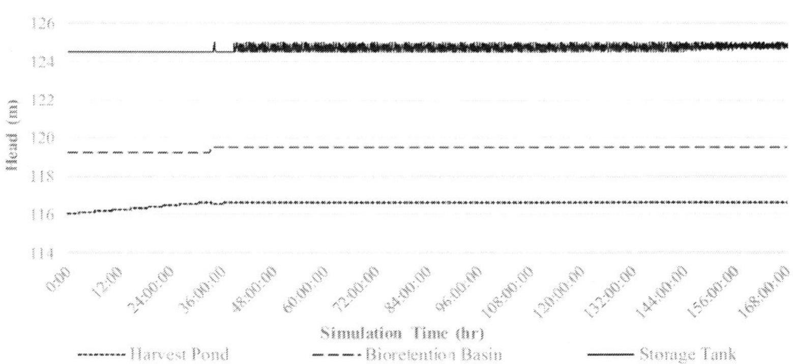

Figure. 6 Storage levels for Streamflow Series 2

20. SUMMER EXTRACTION AND INJECTION SYSTEM CURRENT PUMPING OPERATION RESULTS

Simulation of the irrigation system gave a total weekly pump energy cost of $90 (Table 5). The Bore Pump was more expensive overall, however, cost less per megaliter than Pump 3. This occurred because while the Bore Pump has a greater efficiency than Pump 3, it also has a higher flow and head, which increased the energy consumption. The higher volume pumped from the bore contributed to a lower cost rate than Pump 3.

Table 5 Current operation results for the summer extraction and irrigation system

Pump	Volume (ML/wk)	Cost (c/kL)
Bore Pump	1.93	3.52
Pump 3	0.57	3.97
Total		$90.3/week

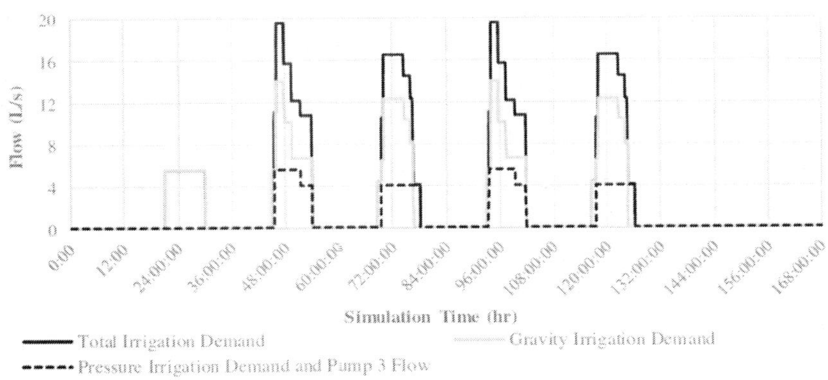

Figure. 7 Current demand rate and pump flows for the irrigation system

All of the pumping for this system occurred overnight (Fig. 7) when irrigation of all fields is allowed. The Bore Pump turned on and off very frequently when it was operating, again due to the small capacity of the Storage Tank which meant it did not take long for the pump to fill the operating volume (Fig. 8).

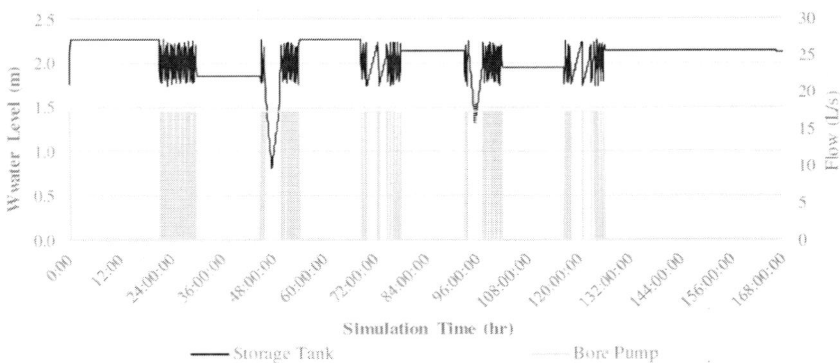

Figure. 8 Storage tank level and bore pump flow for the summer extraction and irrigation system

21. OPTIMIZATION FORMULATION

Initially, optimization of the Ridge Park system was considered to be an operational problem, however, results of the current operation simulation suggest that design decision variables need to be considered as well. Replacing Pumps 1 and 2 with models that would operate at much lower flow rates (to reduce the headlosses) and increasing the size of the Storage Tank will be considered along with operational decision variables (Table 6). These design decisions would aim to counter-act mismatched pump rates (Pumps 1 and 2 operating at a much higher rate than Pump 3) and small storage volumes that lead to frequent pump switches. Short-term operational decisions include trigger levels in the Harvest Pond, Bioretention Basin and Storage Tank that will govern when pumps are turned on and off, a schedule for irrigation (that is, which reserves will be irrigated at which times), and VSP multipliers for Pump 3. In the current operation, VSP multipliers for Pump 3 were selected to ensure the required flow rate (for injection) and pressure (for irrigation) were achieved. With different levels in the Storage Tank considered, the VSP multipliers for Pump 3 can be altered, especially if efficiency is improved. If the pump priming issues discussed earlier were to be resolved, trigger levels that utilize the full height of the Storage Tank (rather than the 20% range in water elevation that is currently used) would be considered in the optimization. There are also long-term decision variables deciding when to switch between summer and winter operation and vice versa (Table 6). As the scheme injects to and extracts from the aquifer through the same bore, it is not possible to frequently switch between injecting and extracting water, therefore there will be only two switch times per year; one going into winter operation and one going into summer operation. The decision variables presented in Table 6 may all be considered together in an optimization problem, however, they could also be analyzed prior to optimization in a simulation sensitivity analysis. Simulating the system initially with the different pump models and storage tank sizes could help to decide if these actions are worthwhile considering in an optimization formulation. Engineering judgement may be sufficient to determine which pump model(s) would be best to replace Pumps 1 and 2, and therefore reduce the size of the optimization problem.

Table 6 Possible decision variables for the Ridge Park MAR scheme

Short-term winter harvesting and injection operation	
Pump 1 Off	Harvest Pond Level Low
	Bioretention Basin Level High
Pump 1 On	Harvest Pond Level High
Pump 2 Off	Bioretention Basin Level Low
	Storage Tank Level High
Pump 2 On	Bioretention Basin Level High
Pump 3 Off	Storage Tank Level Low
Pump 3 On	Storage Tank Level High
Pump 3 Speed	Storage Tank Level
Short-term summer extraction and irrigation operation	
Bore Pump Off	Storage Tank Level High
Bore Pump On	Storage Tank Level Low
Irrigation Schedule	Days of Irrigation at each Reserve
	Start Time of Irrigation at each Reserve
Pump 3 Speed	Required Demand Rate

Long-term operations	
Day to Switch Between	Summer to Winter
Seasonal Operational Regimes	Winter to Summer
System Design	
Storage Tank Size	Doubled, 5 times, 10 times current size
Pumps 1 and 2	Selection of pump curves with lower operating rates

Constraints on the system include an environmental flow for Glen Osmond Creek, an extraction limit from the Aquifer and meeting the weekly irrigation volumes for each reserve in the summer (Table 7). If there was not enough water harvested over the winter to supply the summer demands, a potable back-up supply is available at a cost. The main objective for this case study is to minimize the pump energy cost; there is also a secondary objective of minimizing the number of pump switches. To create an incentive for the optimization to find solutions that harvest more water, the cost objective includes the energy cost for the harvesting and distribution operation as well as the cost of purchasing potable water if the harvested volume is not enough to supply demand. The objective function is formulated as the cost per volume of water harvested as another means to ensure enough water is harvested from the system during winter to supply summer irrigation. During the conceptualization and design of this scheme, regulations from the South Australian Environmental Protection Authority (EPA), the Department for Environment, Water and Natural Resources (DEWNR) and the Department of Health (DoH) were considered.

Table 7 Possible constraints for the Ridge Park MAR scheme

Constraint	**Value**
Glen Osmond Creek Environmental Flow	>2 L/s
Aquifer Extraction in Summer	<80% of Injection Volume
Pressurized System Demands	>15 ML/year
Gravity System Demands	>37 ML/year

A license to recharge water into the aquifer was required from the EPA, while the DEWNR regulates how much water can be extracted from MAR schemes. DoH regulations informed the level of treatment implemented and the irrigation practices, which must limit the risk of public exposure.

22. ORANGE INTEGRATED SUPPLY SYSTEM – CASE STUDY 2

Orange is a rural town roughly 250 km west of Sydney in the state of New South Wales, Australia. The water supply system serves a population of around 36,800 people with an average annual demand of approximately 5,400 ML. The majority of water supply is from the local surface water catchment, which culminates in the roughly 19,000 ML Suma Park reservoir (Fig. 9). Australia experienced severe drought between 2000 and 2010, and Orange was one of the hardest hit areas in New South Wales. Even with severe water restrictions almost halving the town's demand, Orange had less than two years of water supply heading into summer of 2009 and was relying only on surface water catchments (Montgomery Watson Harza, 2011). This prompted the Orange City Council to diversify their water supply, and they therefore developed two stormwater harvesting schemes and a long pipeline from an adjacent catchment, as well as re-opening several groundwater bores. Figure 2 shows a schematic process diagram of the system, which is described below, and Fig. 3 shows the layout (note that the 'Shearing Shed' Bore and 'Bore 5' in Fig. 2 are grouped as the 'Clifton Grove' Bores in Fig. 10). For more detailed data on this case study, please see the Additional files 1, 2, 3 and 4.

Water from the Ploughman's Creek Stormwater Scheme is treated through a series of wetlands, and then combined with water from the Blackman's Swamp Creek Stormwater Scheme. After treatment, this water can be used to top up Suma Park reservoir. Due to the severely low water supply levels during the drought, Emergency Authorization was initially given, and Council subsequently sought approval for use of the stormwater schemes on a permanent basis. Continuous water quality monitoring is undertaken to meet regulations of the Office for Water, the New South Wales Environmental Protection Authority and the Ministry of Health. To the authors' knowledge, this is the only system in Australia that has been approved to use harvested stormwater for potable supply. In order to use harvested stormwater for potable supply, the Council had to meet requirements of the Office for Water. The Macquarie pipeline transfers water from the adjacent Macquarie River catchment to Suma Park reservoir. It is 38 km long and requires more than 600 m of pumping head. Each of the three pumping stations has two pumps operating in parallel.

Water from the groundwater bores is pumped first to balancing storages and then to Suma Park reservoir, with a combined licensed volume of 462 ML per year. Water from all of the sources is combined in Suma Park reservoir and treated at a water treatment plant before being delivered to consumers.

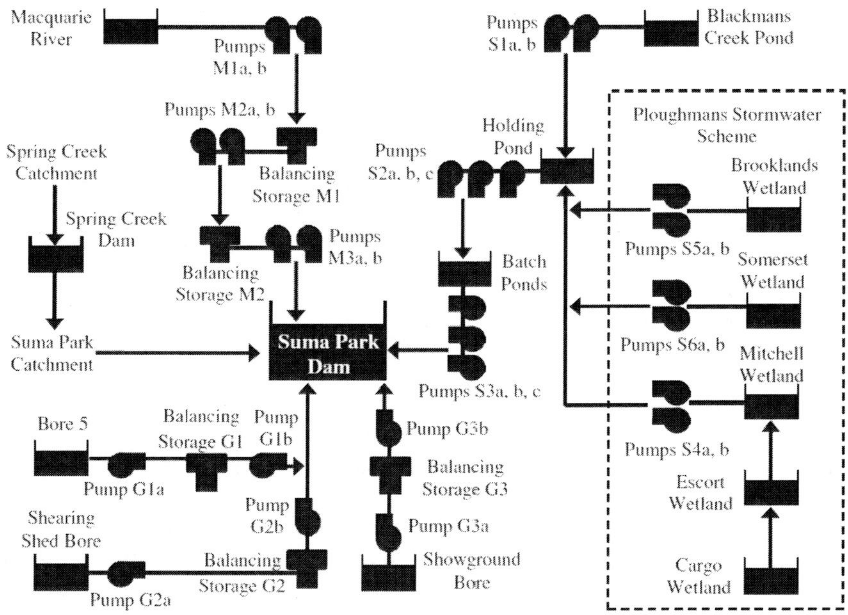

Figure. 9 Orange integrated supply system process schematic – inflow to Suma Park Dam

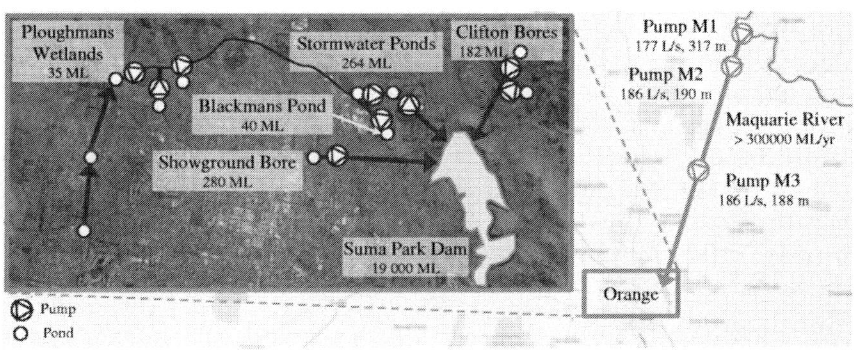

Figure. 10 Orange integrated supply system layout and data

The Orange City Council is interested in optimizing the operation of this while delivering a secure yield from Suma Park Dam. In addition to the primary objective of minimizing energy cost, there are objectives of minimization of spill from Suma Park reservoir, minimization of (perceived) environmental impact, maximization of (perceived) water quality, and minimization of energy use.

The Council has an explicit objective to minimize spill to ensure water and energy are not wasted by pumping from one of the three alternative sources to fill up Suma Park reservoir just before a rainfall event that would supply water from the natural catchment at no cost or energy output. As this system supplies potable demands, it is undesirable to apply water restrictions to consumers, thus minimizing time spent in restrictions is important. Objectives for the perceived environmental impact and water quality will be formulated as a preference ranking between the different sources based on community views of which sources are better for the environment and water quality.

The constraints of the problem include environmental flows for the Macquarie River (downstream of the pumping station offtake point) and stormwater schemes, a water source license for the Macquarie River and extraction limits on the groundwater bores (Table 8).

Table 8 Constraints for the Orange integrated supply system

Constraint	**Value**
Macquarie River Environmental Flow	>108 ML/day
Blackmans' Creek Environmental Flow	>20 ML/day
Ploughmans' Creek Environmental Flow from Pump S4	>0.4 ML/day
Ploughmans' Creek Environmental Flow from Pump S5	>2 ML/day
Ploughmans' Creek Environmental Flow from Pump S6	>2 ML/day
Clifton Grove (Shearing Shed and Bore 5) Aquifer Extraction	<182 ML/year
Showground Aquifer Extraction	<280 ML/year
Macquarie River Extraction License	<12 ML/day

23. ENERGY OPTIMIZATION FORMULATION

In this section, the developed framework is applied to the Orange case study to help set up the optimization procedure and identify the components and data to be modeled. Note that the model has been built taking into account all possible objectives of the system, however, the example of results presented here will focus on the minimization of energy consumption.

As all components of the system have already been constructed and considered sufficient for the operation of the system, there are no design decisions to consider, only operating decisions. For this case study, operating decisions consist of trigger levels in the various storages. These types of decision variables are chosen considering the control system available at each pump station (based on storage levels and not on time of the day) and the fact that the controls have to be defined for an operational horizon of 1 year or longer. As all of the pump stations have two or more pumps arranged in parallel, having different trigger level values may have a large impact on the operating point of the pumps and consequently their energy consumption. It is also likely that different trigger levels will be chosen for peak and off-peak electricity tariff periods when they are included in a cost optimization. For this system a peak/off-peak electricity tariff applies on weekdays, with weekends priced at the off-peak rate. A peak monthly electrical energy demand charge also applies to the Macquarie River pipeline pumping system. In order to assess the performance of different tank trigger levels, the infrastructure to be modeled includes the natural and urban catchments for the surface water and stormwater systems respectively, Suma Park reservoir, pipelines and pumps in the groundwater, Macquarie River and stormwater systems, and wetlands and storage ponds in the stormwater systems.

In general, the system could be modelled using hydrologic models, mass balance models, and/or hydraulic models. The choice of which model(s) will be used depends on the objectives and the processes to be modelled, on the available data and the computational times. In particular, hydrologic modeling is usually used to transform rainfall to runoff for the natural and urban catchments. For this case study, inflows inputs or approximate relationships between rain and flows were provided by previous studies by the Orange City Council. Hydraulic models are usually used for short term operations: pump energy costs can be computed accurately based on the hydraulic equations. Mass balance modeling is usually used for assessing the system in long term operations, as it can quickly compute the water available after evaporation and other losses in the system have occurred and after minimum environmental flows have been released. It cannot, however, take into account the non-linearity in the hydraulic equations and therefore assumptions need to be made in regard to the flow delivered by the pumps in the system. While hydraulic simulation would be most appropriate for the pumping stations in the system as they have multiple pumps and sometimes have connected pipelines, mass balance models would need to be used to compute the additional processes, such as evaporation and the release of minimum environmental flows that need to be taken into account given the long duration of the simulation. During an optimization process, simulating each potential solution using both a mass balance and a hydraulic model would increase considerably the computational time, particularly if data transfer between the two models was required. It is therefore suggested that the primary simulation tool should be a hydraulic solver. Rainfall-runoff modeling could be performed pre-optimization, and supplemental code added to a hydraulic model to account for functionality of a mass balance model. This would allow for consideration of the evaporation from and rainfall directly to reservoirs, changes to demands based on water restrictions and environmental flows that depend on the combined volume of two reservoirs (Spring Creek and Suma Park), infiltration losses when transferring water between reservoirs and peak power demand charges.

Another important issue to consider is what simulation time step should be used. Using a shorter time step will increase the accuracy of this hydraulic analysis and often results in feasible optimization times for storages that empty or fill in a day or two (as would likely be the case for the stormwater ponds and Macquarie pipeline balancing storages). Simulating the behavior of Suma Park dam is more challenging, however, as the variations in the water levels can have a period of several years. Thus, the computation times with a short time step become prohibitively long. A balance needs to be found between using a short enough time step for the detailed hydraulics and a long simulation time for the large storages without having a prohibitively large computational time. Given the data availability (there is 118 years of rainfall and inflow data available, with a daily time step) the time step chosen is 1 day.

Given that the time-step is automatically shortened by the hydraulic solver chosen (EPANET in this case), the model of the real system has been simplified in order to avoid excessive computational times. In particular, given that the levels in the balancing storages along the Macquarie pipeline vary rapidly, these storages were removed and the pipeline simulated with two parallel pumps, each representing the equivalent of the three stages of pumping (that is, the pump curves for Pumps M1a and b in Fig. 2 were adjusted such that they represented Pumps M2a, M3a and Pumps M2b, M3b as well). This simplification is considered acceptable as the pumps in series in the Macquarie pipeline will usually be operated at the same time, given that each pump will still be controlled also by the level of Suma Park Dam. Longer computational times were also caused by the small storages after the groundwater bores. The pumps used for extraction from the aquifers (Pumps G1a, G2a and G3a in Fig. 2) operate at relatively consistent rates, and as such they could be removed from the model and their energy use accounted for relative to the volume pumped from the second pump in each system (Pumps G1b, G2b and G3b respectively). To take into account the limited volume available from the groundwater bores, the storage tanks in the groundwater system each had a volume equivalent to a year's allocation for the respective bores. All of the stormwater pumps except for Pump S2c and Pump S3c, which are standby pumps and not in use, were included in the model. As well as the operating point of the pumps changing depending on the number of pumps used in parallel, there may be slight differences in efficiency and therefore energy use, and thus including all pumps here provided more accuracy.

All of the pumps included in the model were controlled using rule-based controls in EPANET, with conditions based on levels in one or more storages as well as time. Conditions based on downstream storages were considered as decision variables, while conditions based on upstream storages were fixed (Table 9). For the Macquarie pumps, there were also conditions based on the flow in the river to ensure that no water would be taken when there was not enough water available. There were four possible decision variables for each pump, a lower and upper trigger level in both the peak and off-peak time. For optimization of energy use, only two are required, as peak and off-peak tariffs are not considered. As the model was set up for other objectives including cost, which does use a peak and off-peak electricity tariff, the capability to choose different trigger levels in different periods was incorporated. A maximum of 15 pump switches per day per pump were allowed, and the end level of Suma Park Dam was constrained to 16 m (to be approximately the same as the start level). Based on license conditions, Macquarie River water can only be used when the Suma Park Dam level is below 90%, so choices for Pump M1a and M1b trigger levels in Suma Park Dam are more restricted than for other pumps.

Table 9 Decision variables and fixed controls for the Orange integrated supply system

Pump station action	**Storage(s) controlling operation**	**Decision variable or fixed**
Macquarie Pump M1a, M1b Off	Suma Park Dam Level High	Decision Variable
Macquarie Pump M1a, M1b On	Suma Park Dam Level Low	Decision Variable
Stormwater Pump S1a, S1b Off	Holding Pond Level High Blackmans Stormwater Pond Level	Decision Variable

	Low	Fixed
Stormwater Pump S1a, S1b On	Holding Pond Level Low Blackmans Stormwater Pond Level High	Decision Variable Fixed
Stormwater Pump S2a, S2b Off	Batch Ponds Level High Holding Pond Level Low	Decision Variable Fixed
Stormwater Pump S2a, S2b On	Batch Ponds Level Low Holding Pond Level High	Decision Variable Fixed
Stormwater Pump S3a, S3b Off	Suma Park Dam Level High Batch Ponds Level Low	Decision Variable Fixed
Stormwater Pump S3a, S3b On	Suma Park Dam Level Low Batch Ponds Level High	Decision Variable Fixed
Stormwater Pump S4a, S4b Off	Holding Pond Level High Mitchell Wetland Level Low	Decision Variable Fixed
Stormwater Pump S4a, S4b On	Holding Pond Level Low Mitchell Wetland Level High	Decision Variable Fixed
Stormwater Pump S5a, S5b Off	Holding Pond Level High Brooklands Wetland Level Low	Decision Variable Fixed
Stormwater Pump S5a, S5b On	Holding Pond Level Low Brooklands Wetland Level High	Decision Variable Fixed
Stormwater Pump S6a, S6b Off	Holding Pond Level High Somerset Wetland Level Low	Decision Variable Fixed
Stormwater Pump S6a, S6b On	Holding Pond Level Low Somerset Wetland Level High	Decision Variable Fixed
Groundwater Pump G1 Off	Suma Park Dam Level High	Decision Variable
Groundwater Pump G1 On	Suma Park Dam Level Low	Decision Variable
Groundwater Pump G2 Off	Suma Park Dam Level High	Decision Variable
Groundwater Pump G2 On	Suma Park Dam Level Low	Decision Variable
Groundwater Pump G3 Off	Suma Park Dam Level High	Decision Variable
Groundwater Pump G3 On	Suma Park Dam Level Low	Decision Variable

24. ENERGY OPTIMIZATION RESULTS

Minimization of pump energy use over the longer term is presented here as an example of optimization of this system. Note that the system is simulated over 1 year, at a daily time step in EPANET. Additional computer code was added to the EPANET hydraulic simulation to take into account other process such as rainfall to and evaporation from storages.

This code essentially adds a mass balance component to the hydraulic simulation. Historical rainfall for the catchments in the system was modelled in MUSIC hydrologic software to develop inflow series for the ponds and reservoirs. For this optimization the year with the closest to average rainfall was used, however, other years of rainfall were available and this optimization could be extended to consider other climate conditions.

NSGAII (Non-dominated Sorting Genetic Algorithm II) software was used for the optimization, with five random seeds, a population size of 50, 100 generations and probabilities of crossover and mutation of 0.8 and 0.02 respectively. In the best solution found, the system used a total of 793 MWh of energy over the entire year. Table 10 shows the volume of water pumped from each source to Suma Park Dam (and supplied from the local catchment) and the energy used by each of the pumps for the optimal solution. Pumping from the Macquarie is very energy intensive so this is only used at the very end of the simulation when the level in Suma Park Dam is very low, in order to achieve the end target level constraint (Figs. 11 and 12). Groundwater and stormwater sources are used initially to increase the level of Suma Park Dam to its maximum, and then not used again until around Day 160 when the level in the dam has dropped again. Only one of the Macquarie pumps is used, as, despite operating at a lower energy efficiency point, it uses less energy overall than operating two pumps in parallel.

In dryer years, both pumps may need to be utilized in order to ensure supply to Suma Park Dam. Nearly all of the available groundwater license is used; G1 and G2 have a combined license of 180 ML/year, and G3 280 ML/year. Groundwater is more energy intensive than stormwater, however, it can be used at any time throughout the year, while stormwater is reliant of inflow. Most of the stormwater provided to Suma Park Dam came from the Blackman's Creek scheme (S1) rather than the Ploughman's Creek scheme (S4, S5 and S6). While the storage capacity of the Blackman's Creek scheme is much lower, the pump capacity and energy efficiency is much greater than in the Ploughman's Creek scheme, so it provides more water.

Table 10 Volume of water pumped/supplied and energy used in the optimal energy solution

Source	Pump	Volume (ML)	Energy (MWh)	Energy Rate (MWh/ML)
Macquarie River	M1a	0	0	0
	M1b	74	150	2.02
	Total	74	150	2.02
Groundwater[a]	G1	24	11	0.46
	G2	146	79	0.54
	G3	235	106	0.45
	Total	405	196	0.48
Stormwater	S1a	258	39	0.15
	S1b	479	71	0.15
	S2a	828	65	0.08
	S2b	237	21	0.09
	S3a	1022	170	0.17
	S3b	22	5.5	0.25
	S4a	178	41	0.23
	S4b	12	3.1	0.27
	S5a	24	4.8	0.20
	S5b	56	11	0.19
	S6a	60	11	0.18
	S6b	26	5.0	0.19
	Total[b]	1044	447	0.43
Spring Creek and Suma Park Catchment	-	3865[c]	-	-

[a]The energy consumption for the groundwater pumps includes both the transfer and bore pumps, i.e. the energy for Pump G1 includes G1a (not modelling in EPANET, energy use estimated from volume) and G1b (modelled in EPANET)
[b]The total volume supplied by the stormwater schemes is measured as the combined volume supplied by Pumps S3a and S3b (which discharge to Suma Park Dam), while the total energy is the total of all pumps
[c]This is the volume supplied by the natural catchment for the town's consumption, the total inflow from the catchment is greater than this however some is used to provide environmental flows and some spills

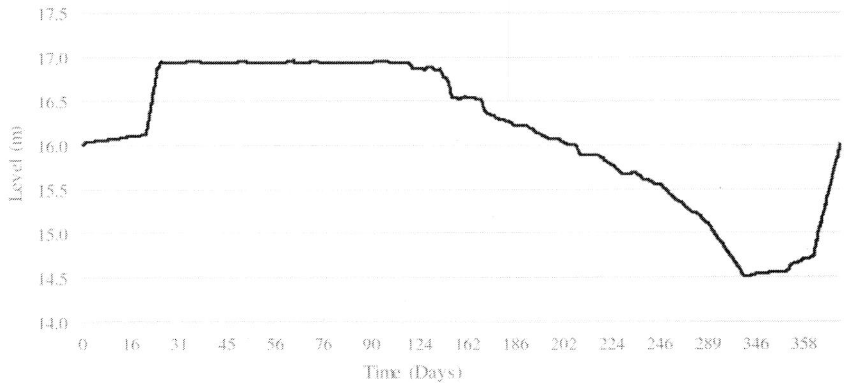

Figure. 11 Variation in Suma Park Dam level for the energy optimal solution

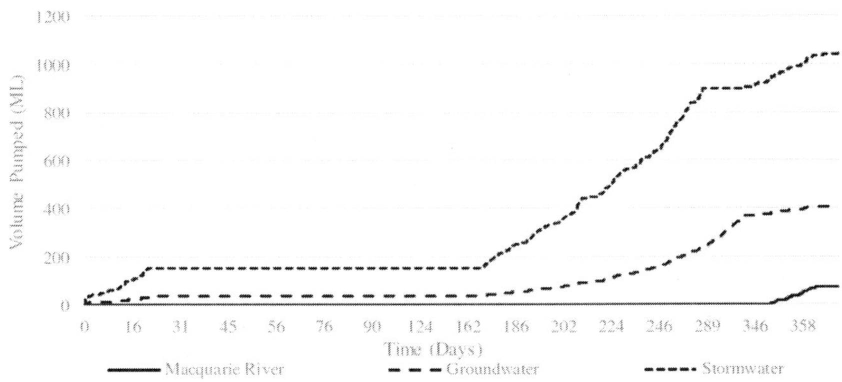

Figure. 12 Volume pumped from each source to Suma Park Dam for the energy optimal solution

25. CONCLUSIONS

A generalized framework for the optimization of the design and operation of water supply and distribution systems has been developed and two case study systems have been used as examples of how to apply it. The framework is comprised of several components; the options component describes the design and operational decision variables for the optimization, the infrastructure component covers the infrastructure aspects of the system that need to be modeled and their data requirements, the analysis component includes the simulation of the system and evaluation against the objectives and constraints, and finally the government policy component describes the regulations that may affect other aspects of the framework. These components fit within an optimization algorithm structure, which firstly generates potential solutions using the decision variables in the options component, models the system according to the infrastructure component and evaluates potential solutions using the analysis component. The evaluation of potential solutions then feeds into the solution space which informs how the decision variables are changed in the next set of potential solutions. Sensitivity analysis of parameters will significant uncertainty should be undertaken to ensure robust solutions. The framework also applies to simulation of systems prior to or without optimization.

The Ridge Park MAR Scheme Case Study harvests stormwater from an urban creek and stores it in an aquifer, to be extracted at a later time and used as non-potable supply for irrigation of sporting fields and reserves. For this case study, and similar ones, the simulation of the system may be simplified by splitting the system into two parts, one for the components of the system used in winter operation (harvesting and injection) and one for the components used in summer operation (extraction and irrigation). This system highlighted the importance of simulation and analysis prior to optimization, in order to focus the formulation of the optimization problem. The Orange Integrated Supply System Case Study uses multiple water sources; natural catchment water, harvested stormwater, imported water and groundwater to supply potable demands. For this case study, finding an appropriate combination of simulation models and time step and simulation duration is important in order to provide accuracy in representing both long- and short-term operations without excessive computational times. Optimization of pump energy use for this system indicated that the groundwater and stormwater supplies are more desirable to supplement natural inflows than the imported water from the Macquarie River, which required a lot of energy to transfer water over a long distance and against a high elevation head.

The framework is generalized, and so could be applied to other water supply and distribution systems, particularly those using non-traditional water sources, to optimize their design and operation. While the framework attempts to cover all aspects of water supply in a generalized manner, it does have some limitations. Along with the supply of water, there will always be a need to manage wastewater. Apart from considering recycled wastewater as a source, this framework does not cover wastewater systems in terms of collection, transport, treatment and potential discharge of wastewater into the environment. Treatment of raw water supplies is included in the framework, however, the details of such treatment and measurement of water quality throughout a water distribution system are not focused on as much as the design and operation of the systems. A difficulty of applying this framework will be the definition of the boundary of a system and which aspects should be analyzed. Currently, there does not exist commercial software that has all of the capabilities considered in the framework (i.e. both hydrologic and hydraulic simulation). This means that specialist simulation models may need to be developed for particular systems (as was done for the second case study). Future developments in simulation software may reduce the difficultly of combining hydrologic, mass balance and hydraulic considerations and remove the need for specialist tools built for individual systems. In the future, the framework should be tested with other case study systems to fully investigate its benefits.

ACKNOWLEDGEMENTS

The support of the Commonwealth of Australia through the Cooperative Research Centre program is acknowledged. This research was part of the CRC for Water Sensitive Cities Project C5.1 (Intelligent Urban Water Networks) and was supported by funding for post-doctoral research and a PhD top-up scholarship. The authors would like to thank Martin Haege (Geolyse), Chris Devitt (Orange City Council) and Matt Sanderson (The City of

REFERENCES

1. Ashbolt S, Maheepala S, Perera BJC (2014) A framework for short-term operational planning for water grids. Water Resour Manag 28:2367–2380. doi:10.1007/s11269-014-0620-4
2. Basupi I, Kapelan Z (2015) Flexible water distribution system design under future demand uncertainty. J Water Resour Plan Manag 141(4). doi:10.1061/(ASCE)WR.1943-5452.0000416
3. Becker N, Lavee D, Katz D (2010) Desalination and alternative water-shortage mitigation options in Israel: a comparative cost analysis. J Water Resour Prot 2:1042–1056. doi:10.4236/jwarp.2012.212124
4. Beh EHY, Dandy GC, Maier HR, Paton FL (2014) Optimal sequencing of water supply options at the regional scale incorporating alternative water supply sources and multiple objectives. Environ Model Softw 53:137–153. doi:10.1016/j.envsoft.2013.11.004
5. Beh EHY, Maier H, Dandy GC (2015) Adaptive, multiobjective optimal sequencing approach for urban water supply augmentation under deep uncertainty. Water Resour Res 51. doi:10.1002/2014WR016254
6. Blinco LJ, Simpson AR, Lambert MF, Auricht CA, Hurr NE, Tiggemann SM, Marchi A Genetic algorithm optimization of operational costs and greenhouse gas emissions for water distribution systems. In: 16th Conference on Water Distribution System Analysis, Bari, Italy, 14–17 July 2014. Procedia Engineering, pp 509–516. doi:10.1016/j.proeng.2014.11.246

7. Bogardi JJ, Dudgeon D, Lawford R, Flinkerbusch E, Meyn A, Pahl-Wostl C, Vielhauer K, Vörösmarty C (2012) Water security for a planet under pressure: interconnected challenges of a changing world call for sustainable solutions. Curr Opin Environ Sustain 4:35–43. doi:10.1016/j.cosust.2011.12.002
8. Cai X, Zeng R, Kang WH, Song J, Valocchi AJ (2015) Strategic planning for drought mitigation under climate change. J Water Resour Plan Manag 141(9). doi:10.1061/(ASCE)WR.1943-5452.0000510
9. Campisano A, Modica C (2012) Optimal sizing of storage tanks for domestic rainwater harvesting in Sicily. Resour Conserv Recycl 63:9–12. doi:10.1016/j.resconrec.2012.03.007
10. Chung G, Lansey K, Blowers P, Brooks P, Ela W, Stewart S, Wilson P (2008) A general water supply planning model: evaluation of decentralized treatment. Environ Model Softw 23:893–905. doi:10.1016/j.envsoft.2007.10.002
11. Cook C, Bakker K (2012) Water security: debating an emerging paradigm. Glob Environ Chang 22:94–102. doi:10.1061/j.gloenvcha.2011.10.011
12. Dandy GC, Simpson AR, Murphy LJ (1996) An improved genetic algorithm for pipe network optimization. J Water Resour Res 32(2):449–458. doi:10.1029/95WR02917
13. Dawadi S, Ahmad S (2013) Evaluating the impact of demand-side management on water resources under changing climate conditions and increasing population. J Environ Manag 114:261–275. doi:10.1016/j.jenvman.2012.10.015
14. Deb K, Pratap A, Agarwal S, Meyarivan T (2002) A fast and elitist multiobjective genetic algorithm: NSGA-II. Trans Evol Comput 6(2):182–197
15. Dobbie MF, Brown RR (2012) Risk perceptions and receptivity of Australian urban water practitioners to stormwater harvesting and treatment systems. Water Sci Technol Water Supply 12(6):888–894. doi:10.2166/ws.2012.069
16. Fearnside PM (2002) Time preference in global warming calculations: a proposal for a unified index. Ecol Econ 41(1):21–31. doi:10.1016/S0921-8009(02)00004-6
17. Fielding KS, Gardner J, Leviston Z, Price J (2015) Comparing public perceptions of alternative water sources for potable use: the case of rainwater, stormwater, desalinated water, and recycled water. Water Resour Manag 29:4501–4518. doi:10.1007/s11269-015-1072-1
18. Friedman K, Heaney JP, Morales M, Palenchar M (2014) Analytical optimization of demand management strategies across all urban water use sectors. Water Resour Res 50:5475–5491. doi:10.1002/2013WR014261
19. Gollier C, Weitzman ML (2010) How should the distant future be discounted when discount rates are uncertain? Econ Lett 107:350–353. doi:10.1016/j.econlet.2010.03.001
20. Hwang AH-S, Valeo C, Draper D (2006) Public perceptions and attitudes toward stormwater recycling for irrigation. Can Water Resour J 31(3):185–196. doi:10.4296/cwrj3103185
21. IPCC (2000) In: Nakicenovic N, Swart R (eds) Special Report on Emissions Scenarios. Cambridge
22. IPCC (2007) Synthesis Report: Contribution of Working Groups I, II and III to the Fourth Assessment Report of the Intergovernmental Panel on Climate change Climate Change 2007. IPCC, Geneva
23. Londoño Cadavid C, Ando AW (2013) Valuing preferences over stormwater management outcomes including improved hydrologic function. Water Resour Res 49:4114–4125. doi:10.1002/wrcr.20317
24. Makropoulos CK, Natsis K, Liu S, Mittas K, Butler D (2008) Decision support for sustainable option selection in integrated urban water management. Environ Model Softw 23:1148–1460. doi:10.1016/j.envsoft.2008.04.010
25. Marchi A, Dandy GC, Maier HR (2016) Integrated approach for optimizing the design of aquifer storage and recovery stormwater harvesting schemes accounting for externalities and climate change. J Water Resour Plan Manag 142(4). doi:10.1061/(ASCE)WR.1943-5452.0000628

26. McArdle P, Gleeson J, Hammond T, Heslop E, Holden R, Kuczera G (2011) Centralised urban stormwater harvesting for potable reuse. Water Sci Technol 63(1):16–24. doi:10.2166/wst/2011.003
27. Miller S, Shermer H, Semiat R (2015) Energy and environmental issues in desalination. Desalination 366:2–8. doi:10.1016/j.desal.2014.11.034
28. Milly PCD, Betancourt J, Falkenmark M, Hirsch RM, Kundzewicz ZW, Lettenmaier DP, Stouffer RJ (2008) Stationary is dead: whither water management? Science 319:573–574. doi:10.1126/science.1151915
29. Montgomery Watson Harza (2011). Orange drought relief connection concept investigation report, report prepared for the Orange City Council. http://watersecurity.orange.nsw.gov.au/client_images/1046362.pdf
30. Mpelasoka FS, Chiew FHS (2009) Influence of rainfall scenario construction methods on runoff projections. J Hydrometeorol 10(5):1168–1183. doi:10.1175/2009jhm1045.1
31. Muga HE, Mihelcic JR (2008) Sustainability of wastewater treatment technologies. J Environ Manag 99:437–447. doi:10.1016/j.jenvman.2007.03.008
32. Nagal R (2015) Making direct potable reuse a reality. J Am Water Works Assoc 107(7):76–82. doi:10.5942/jawwa.2015.107.0103
33. Naylor T, Moglia M, Grant AL, Sharma AK (2012) Self-reported judgements of management and governance issues in stormwater and greywater systems. J Clean Prod 29–30:144–150. doi:10.1016/j.jclepro.2012.02.003
34. Nicklow J, Reed P, Savic D, Dessalegne T, Harrell L, Chan-Hilton A, Karamouz M, Minsker B, Ostfeld A, Singh A, Zechman E (2010) State of the art for genetic algorithms and beyond in water resources planning and management. J Water Resour Plan Manag 136(4):412–432. doi:10.1061/(ASCE)WR.1943-5452.0000053
35. Oron G, Adel M, Agmon V, Friedler E, Halperin R, Lesham E, Weinberg D (2014) Greywater use in Israel and worldwide: standards and prospects. Water Res 58:92–101. doi:10.1016/j.watres.2014.03.032
36. Paton FL, Maier HR, Dandy GC (2013) Relative magnitudes of sources of uncertainty in assessing climate change impacts on water supply security for the southern Adelaide water supply system. Water Resour Res 49:1643–1667. doi:10.1002/wrcr.20153
37. Paton FL, Dandy GC, Maier HR (2014) Integrated framework for assessing urban water supply security of systems with non-traditional sources under climate change. Environ Model Softw 60:302–319. doi:10.1016/j.envsoft.2014.06.018
38. Rambaud SC, Torrecillas MJM (2005) Some considerations on the social discount rate. Environ Sci Policy 8:343–355. doi:10.1016/j.envsci.2005.04.003
39. Reis LFR, Walters GA, Savic D, Chaudhry FH (2005) Multi-reservoir operation planning using hybrid genetic algorithm and linear programming (GA-LP: an alternative stochastic approach. Water Resour Manag 19:831–848. doi:10.1007/s11269-005-6813-0
40. Rodriguez C, van Buynder P, Lugg R, Blair P, Devine B, Cook A, Weinstein P (2009) Indirect potable reuse: a sustainable water supply alternative. Int J Environ Res Public Health 6(3):1174–1209. doi:10.3390/ijerph6031174
41. Roshani E, Filion YR (2014) Water distribution system rehabilitation under climate change mitigation scenarios in Canada. J Water Resour Plan Manag 141(4). doi:10.1016/(ASCE).WR.1943-5452.0000450
42. Rozos E, Makropoulos CK (2013) Source to tap urban water cycle modelling. Environ Model Softw 41:139–150. doi:10.1016/j.envsoft.2012.11.015
43. Seifi A, Hipel K (2001) Interior-point method for reservoir operation with stochastic inflows. J Water Resour Plan Manag 127(1):48–57. doi:10.1061/(ASCE)0733-9496(2001)127:1(48)
44. Stern NH (2006) Stern review: the economics of climate change. Cambridge University Press, Cambridge
45. Stokes CS, Simpson AR, Maier HR (2014) The cost-greenhouse gas emission nexus for water distribution systems including the consideration of energy generating infrastructure: an integrated conceptual optimization framework and review of literature. Earth Perspectives 1(9). doi:10.1186/2194-6434-1-9

46. Umapathi S, Chong MN, Sharma AK (2013) Evaluation of plumbed rainwater tanks in households for sustainable water resources management: a real-time monitoring study. J Clean Prod 42:204–214. doi:10.1016/j.jclepro.2012.11.006
47. Vrugt JA, Robinson BA (2007) Improved evolutionary optimization from genetically adaptive multimethod search Proc Natl Acad Sci U S A 104(3):708–711. doi:10.1073/pnas.0610471104
48. Weitzman ML (2007) A review of the Stern review on the economics of climate change. J Econ Lit 45:703–724. doi:10.1257/jel.45.3.703
49. West C, Kenway S, Hassall M, Yuan Z (2016) Why do residential recycled water schemes fail? A comprehensive review of risk factors and impact on objectives. Water Res 102:271–281. doi:10.1016/j.watres.2016.06.044
50. Wu W, Maier HR, Simpson AR (2010a) Single-objective versus multiobjective optimization of water distribution systems accounting for greenhouse gas emissions by carbon pricing. J Water Resour Plan Manag 136(5):555–565. doi:10.1061/(ASCE)WR.1943.5452.0000072
51. Wu W, Simpson AR, Maier HR (2010b) Accounting for greenhouse gas emissions in multiobjective genetic algorithm optimization of water distribution systems. J Water Resour Plan Manag 136(2):146–155. doi:10.1061/(ASCE)WR.1943.5452.0000020
52. Wu W, Simpson AR, Maier HR (2012a) Sensitivity of optimal tradeoffs between cost and greenhouse gas emissions for water distribution systems to electricity tariff and generation. J Water Resour Plan Manag 138:182–186. doi:10.1061/(ASCE)WR.1943.5452.0000169
53. Wu W, Simpson AR, Maier HR, Marchi A (2012b) Incorporating variable-speed pumping in multi-objective genetic algorithm optimization of water distribution systems to reduce greenhouse gas emissions. J Water Resour Plan Manag 138(5):543–552. doi:10.1061/(ASCE)WR.1943.5452.0000195
54. Wu W, Maier HR, Simpson AR (2013) Multiobjective optimization of water distribution systems accounting for economic cost, hydraulic reliability, and greenhouse gas emissions. Water Resour Res 49(3):1211–1225. doi:10.1002/wrcr.20120

CHAPTER 6

Generalized storage–reliability–yield relationships for rainwater harvesting systems

L S Hanson[1,2] and R M Vogel[1]

[1] *Department of Civil and Environmental Engineering, Tufts University, Medford, Massachusetts 02155, USA*

[2] *Now at Institute for Public Research, CNA Corporation, Arlington, VA 22201, USA*

ABSTRACT

Sizing storage for rainwater harvesting (RWH) systems is often a difficult design consideration, as the system must be designed specifically for the local rainfall pattern We introduce a generally applicable method for estimating the required storage by using regional regression equations to account for climatic differences in the behavior of RWH systems across the entire continental United States. A series of simulations for 231 locations with continuous daily precipitation records enable the development of storage–reliability–yield (SRY) relations at four useful reliabilities, 0.8, 0.9, 0.95, and 0.98. Multivariate, log-linear regression results in storage equations that include demand, collection area and local precipitation statistics. The continental regression equations demonstrated excellent goodness-of-fit (R^2 0.96–0.99) using only two precipitation parameters, and fits improved when three geographic regions with more homogeneous rainfall characteristics were considered. The SRY models can be used to obtain a preliminary estimate of how large to build a storage tank almost anywhere in the United States based on desired yield and reliability, collection area, and local rainfall statistics. Our methodology could be extended to other regions of world, and the equations presented herein could be used to investigate how RWH systems would respond to changes in climatic variability. The resulting model may also prove useful in regional planning studies to evaluate the net benefits which result from the broad use of RWH to meet water supply requirements. We outline numerous other possible extensions to our work, which when taken together, illustrate the value of our initial generalized SRY model for RWH systems.

1. INTRODUCTION

In recent years, rainwater harvesting (RWH) has attracted increased attention for various reasons including its use as an alternative water supply and for urban water management, notably control of stormwater. RWH has maintained its importance as a water source for small scale agricultural needs and as a primary water source in remote locations in rural areas and on islands. In the past few decades, RWH has also become a popular supplemental (and generally non-potable) water source in urban and suburban areas in a wide range of climatic and socio-economic environments. The history of RWH, and evolution of system types, forms, and objectives has been well covered in the literature (see reviews by Boers and Ben-Asher 1982, Pandey *et al* 2003, Basinger *et al* 2010).

The elements of design of RWH systems vary according to the designer's goals for system performance. A designer interested in supplying potable water will value reliability, a suburban gardener may value the water saving efficiency, and a stormwater engineer may value capture efficiency. These disparate objectives and interests have led to the use of different metrics and constraints upon which to assess RWH system performance, though the physical components of the system-collection area, conveyance, and storage—remain largely consistent. In most cases, the most important and difficult design decision is how much storage capacity to build. Being unable to change the rainfall pattern in the location of the planned RWH system, designers focus on the system components and parameters they can control, namely, the collection area, storage volume, and demand level. But rainfall patterns have a strong relationship to the overall functioning of the system, and a better understanding of the effect of the rainfall pattern on system performance could inform system design, notably in sizing storage. While there is no substitute for a detailed engineering design study at a given project location, there would be great utility in a simpler and more generally applicable method of preliminary RWH storage capacity estimation.

1.1. RWH cistern sizing and simulation

RWH sizing approaches vary considerably in approach, methodology, the type of rainfall data used and the way in which those rainfall data are used. The simplest sizing methodologies are based on a simple water balance or a mass curve analysis (Handia *et al* 2002). Early versions of these analyses were performed with monthly rainfall records or average monthly rainfall (Watt 1978, Keller 1982). Algorithms for simulating the behavior of RWH systems using monthly rainfall records quickly replaced these methods (Schiller and Latham 1987). More recently, continuous simulation methods using daily rainfall records have been used to explore RWH system behavior (Fewkes 2000, Palla *et al* 2011), optimize performance relative to system cost (Liaw and Tsai 2004), or analyze potential potable water savings (Ghisi and Ferreira 2007). More detailed models investigating system performance under a complex demand pattern (i.e. multiple end uses being used at different times) have used hourly (Villarreal and Dixon 2005) or even 5 or 6 min rainfall data (Coombes and Barry 2007, Herrmann and Schmida 2000). Other RWH sizing and simulation approaches use semi-parametric (Cowden *et al* 2008) or nonparametric (Basinger *et al* 2010) stochastic precipitation generators to generate synthetic rainfall series to entirely replace or fill gaps in rainfall records. Guo and Baetz (2007) used an analytical approach from stormwater detention theory using the statistical characteristics of a continuous rainfall record (event totals, storm durations, and inter-event storm arrival times) as inputs to generate a complex series of equations enabling computation of necessary storage.

Ultimately, all of these approaches rely on local precipitation records, either as a hydrologic input or for computing parameters needed as an input to a stochastic precipitation model. The studies cited above have investigated how design parameters (storage volume, collection area, demand, etc) affect various measures of system performance (reliability, water saving efficiency, overflow volume), but relatively little effort has been devoted to determining how characteristics of a rainfall record, other than mean rainfall, affect overall RWH system performance. Furthermore, previous modeling efforts have focused on individual systems in a particular location, but little effort has been given to the generalization of approaches for sizing RWH systems across broad geographic regions that experience significant variations in climatology.

1.2. Using hydrologic statistics to study system behavior

The first step in being able to determine how characteristics of the rainfall pattern affect system performance is to describe the RWH system performance mathematically. Repeated simulations can generate empirical relationships between the design parameters and performance metrics to define local performance curves (Fewkes 1999). Regression approaches have been used to generate equations relating some system parameters. Liaw and Tsai (2004) used regressions to define isoreliability economic tradeoff curves between collection area and storage volume costs.

Lee *et al* (2000) developed a regression equation to relate the major system parameters (irrigation demand, collection area, storage) and performance (reliability) for a single agricultural RWH site in China. Since only a single site is considered, the character of the hydrologic input (rainfall) is effectively built into the regression model. Commenting on this study, Guo and Baetz (2007) remark, 'One series of such simulation modeling and regression analysis would have to be conducted for each geographical location of interest. The resulting regression equations would be applicable for only the locations studied.' Ironically, Guo and Baetz (2007) demonstrated analytically that differences in RWH system performance between locations *can* be explained by differences in the statistical parameters of the rainfall sequence.

Our primary hypothesis is that a multivariate regression framework similar to Lee *et al*'s might lead to RWH relationships with greater geographic generality by also including certain rainfall statistics in the regression. Multivariate regression has been used to generalize the behavior of reservoirs based on the statistical parameters of their inflow sequence. For example, McMahon *et al* (2007) use regional regression to estimate theoretical reservoir storage requirements for a particular streamflow station by using reservoir yield, reliability and various annual streamflow statistics as predictive variables (on a global geographic scale). Regional regression approaches have also proven useful in estimating other hydrologic parameters over broad geographic areas, most notably, streamflow (Vogel *et al* 1999).

The RWH system is conceptually similar to a surface water reservoir system fed by a stream, though there are important differences. Both have a fixed storage capacity, a characteristic volumetric demand, and a volumetric inflow determined by natural hydrologic processes. A surface water reservoir's inflow is determined by the volumetric flow of its influent streams that collect rainfall over the contributing watershed, which is analogous but more complex than precipitation falling on a RWH system's collection area.

The storage–reliability–yield (SRY) framework from reservoir analysis (Vogel and Stedinger 1987, Vogel and McMahon 1996, McMahon and Adeloye 2005) may be useful for RWH studies because it can handle a range water supply yield and reliability cases. The relationships of the storage capacity, reliability and demand (yield) variables have been explored for RWH systems (Schiller and Latham 1987, Fewkes 2000, Villarreal and

Dixon 2005), though not specifically in a SRY context. Given a daily rainfall record, a series of behavioural simulations can determine an empirical SRY relationship, which would provide a starting point for regression analysis.

Though some similar methodology is used, this study is significantly different from past regional hydrologic regression studies for a variety of reasons. Importantly, to our knowledge this is the first study which seeks to develop generalized SRY relationships for RWH systems that could be applied across an entire continent. Our approach, while analogous to the approach used by others for developing generalized SRY relationships for surface water supplies fed by rivers, differs in nearly every aspect. Unlike the early studies by Vogel and Stedinger (1987) and Vogel and McMahon (1996) and many subsequent studies summarized by McMahon and Adeloye (2005), which developed generalized SRY relationships based on hypothetical, theoretical inflow series, our approach employs actual rainfall series for 231 locations as inflows to the RWH systems. Similarly, many previous SRY relationships for surface water reservoirs incorporate critical assumptions regarding the probabilistic structure of both reservoir inflows and resulting storage capacities. Our study of RWH systems does not make any such probabilistic assumptions. Additionally, while the watershed area draining to a reservoir or stream gage location is essentially fixed, the collection area for a RWH system is an important design parameter, and any regression models developed should be normalized to collection area such that collection area can be easily changed.

Our primary goal is to establish a generally applicable, technically rigorous, and easily applied method for estimating the required storage for a planned RWH system with user-specified collection area, and a desired demand and reliability. Our approach is to use multivariate regression to generate regional equations relating storage capacity to demand level at given reliabilities, as well as certain parameters of the observed daily rainfall sequences to account for climatic differences across the entire geographic region of the continental United States.

2. METHODS

We simulate RWH systems that collect rainfall from a rooftop collection area with storage in a closed cistern and constant daily demand (yield), which may reasonably approximate a system designed to meet daily toilet flushing and/or other constant (likely indoor) daily needs. Long records of daily precipitation data for the 231 locations shown in figure 1 across the continental United States were used to simulate RWH systems and to generate SRY relationships. Simulations generated sets of points defining a fixed reliability storage–yield (SY) curve for several useful design reliabilities (80%, 90%, 95%, and 98%). Then, the SY data were used in a regional regression approach to obtain generalized equations for estimating storage for RWH systems using system parameters and daily rainfall statistics.

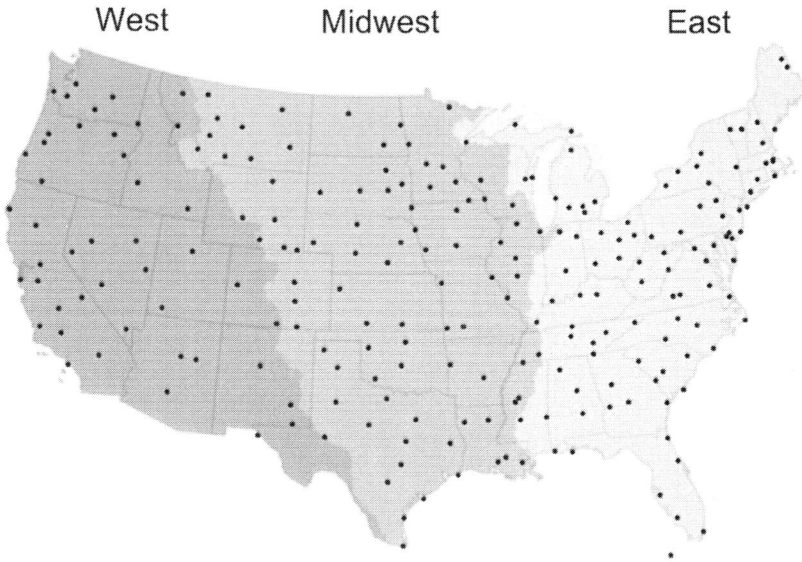

Figure 1. Map of regional divisions and locations of 231 first-order precipitation gages used in this study.

2.1. Simulation of RWH systems

A set of 231 first-order precipitation gages shown in figure 1 from the National Weather Service's Cooperative Station Network with high-quality daily records were selected as hydrologic inputs for simulation. The stations are well distributed across the continental United States and have a median record length of 59 years. The Yield-After-Spillage algorithm, widely used in the RWH literature (Jenkins *et al* 1978, Schiller and Latham 1987, Fewkes 1999, Mitchell 2007, Palla *et al* 2011), is selected for the simulations.

We assume the runoff coefficient of the collection area to be one, and no first flush is modeled. The precipitation gages report snow as liquid equivalent, so simulations are continuous even in sub-freezing conditions impractical for actual RWH operation.

Each simulation, run for a single combination of yield and storage, tracks daily storage and whether the daily yield is met each day. The number of days with failure is recorded, and used to compute reliability, q_t given by

$$q_t = 1 - \frac{d_f}{n}, \tag{1}$$

where q_t is reliability, d_f is the number of days with failure to meet full demand in the simulation, and n is the number of days simulated. All simulations are run for a unit rainfall collection area A_c of 1 m^2. Additionally, the yield is normalized by the mean daily rainfall of each station to generate the nondimensional yield ratio α, which describes the average fraction of available water used and is given by

$$\alpha = \frac{D}{A_c \mu}, \tag{2}$$

where D is the constant volumetric daily demand (L d^{-1}), μ is mean daily precipitation (mm d^{-1}), and A_c is collection area (m^2).

An algorithm iterated through values of storage until the desired reliability, q_t (e.g. 0.9), was reached for the given yield ratio α. A set of 20 simulations with α ranging from 0.05 to 0.99 defines a constant reliability curve. The reported storage volume is normalized to collection area to generate the physical storage ratio, S_r (3). Though S_r may have any unit of length (e.g. meter), S_r is expressed here in units of centimeters, but is more meaningfully understood as m^3 of storage per 100 m^2 of collection area.

$$S_r = 100 \times K/A_c \text{ (cm)}, \tag{3}$$

where S_r is the physical storage ratio, K is total storage capacity (m^3). Storage is expressed in this manner instead of the dimensionless storage ratio common to reservoir literature (McMahon and Adeloye 2005), which in this case would require further dividing by the mean annual rainfall, μ. This choice was made because collection area is an important system parameter, and further embedding μ into the definition of the storage variable would make calculating storage capacity more cumbersome. Furthermore, this form allows regressions to be fit to a variable (S_r) free of rainfall parameters, meaning rainfall parameters will only be on the right hand side of the regression, thus avoiding potential spurious correlation due to rainfall variables common to both sides of the equation.

The constant reliability curves can effectively describe the full SRY relationship when plotted. Figure 2 shows a set of simulated reliability contours generated using precipitation data from Los Angeles, CA. Figure 3 displays the 0.9 reliability contour from several locations to illustrate the wide variation in the shape and position of the SY curve between stations.

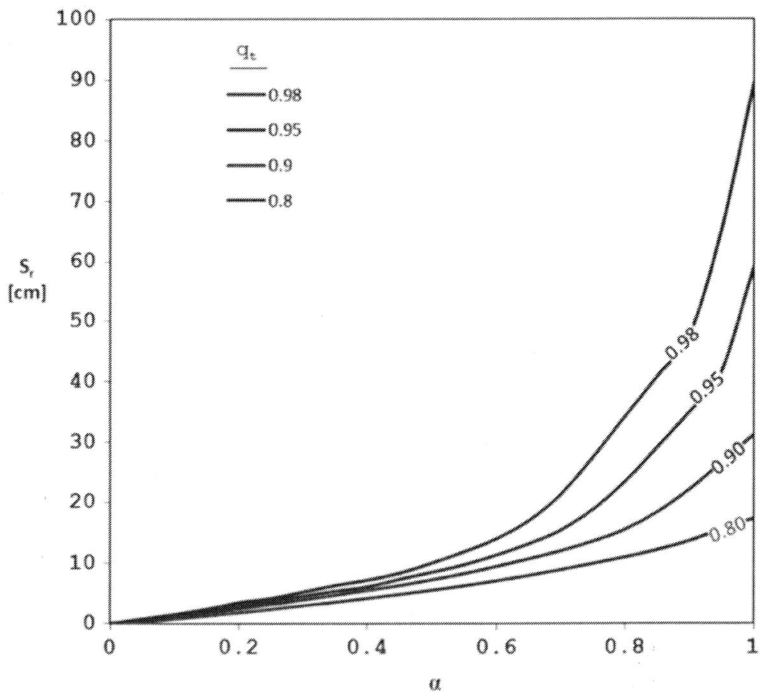

Figure 2. Storage–yield relationship for four reliability contours for a single station (Los Angeles, CA).

2.2. Regression modeling approach

We use multivariate ordinary least squares (OLS) regression as the method of generalizing the SY relationships across the entire US. The two goals of the regression modeling are (1) to determine which precipitation summary statistics most strongly influence differences in the SRY relationship between sites, and (2) to develop a generalized predictive model for estimating required system storage.

The regression equation is of the general form

$$S_r = e^{a} \cdot X_1^{\,b_1} \cdot X_2^{\,b_2} \cdot \ldots X_m^{\,b_m} \cdot Y_1^{\,c_1} \cdot Y_2^{\,c_2} \cdot \ldots Y_n^{\,c_n}, \tag{4}$$

where Sr denotes the physical storage ratio, Xi, $i = 1 \ldots m$, describe physical system parameters such as the yield ratio α, defined in (2) and Yj, $j = 1 \ldots n$, describe local climate statistics.

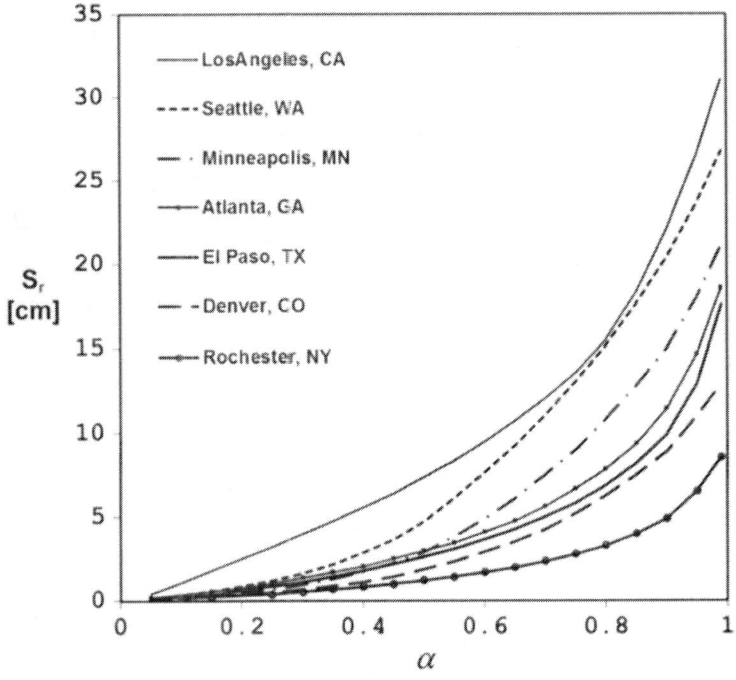

Figure 3. Variability of the Sr–α relationship across stations for the $qt = 0.9$ reliability curve.

The model in (4) is log-linear because taking the natural logarithm of (4) results in

$$\ln(S_r) = a + b_1 \ln(X_1) + b_2 \ln(X_2) \ldots + b_m \ln(X_m) + c_1 \ln(Y_1) + c_2 \ln(Y_2) + \ldots c_n \ln(Y_n). \tag{5}$$

The simulation of the RWH systems kept fixed all of the major system parameters except storage and demand on the system. Since storage is the dependent variable of the regression, the only system parameter ($X i$) that needs to be included in the regression is a representation of demand. Several expressions of average demand or yield of the system were investigated, but the yield ratio, α, from (2) was selected as simplest and most general, and has the benefit of having a constant domain across stations.

One difficulty with using yield ratio, α, is that the isoreliability SY curves on which the regressions are based are highly nonlinear, especially as reliability approaches 1. We sought to investigate if the $S r - \alpha$ relationship could be linearized to enable application of OLS regression methods. Even when both variables were logarithmically (base e) transformed, nonlinearity persisted. A single site regression analysis was performed for all stations to obtain the form of the regression model describing the relationship between $S r$ and α for a constant reliability curve. The model in (6) resulted in a fit with R^2-adjusted above 0.99 for all stations and reliabilities

$$\ln(S_r) = b_0 + b_1[-\ln(\alpha)]^d \quad (6)$$

While the model in (6) is no longer log-linear, the transformation caused by the d parameter can still be incorporated into a log-linear model. Replacing $\ln(X_1)$ in (5) with $[-\ln(\alpha)] d$, and exponentiation of both sides leads to

$$S_r = e^a \cdot e^{b_1 \cdot (-\ln(\alpha))^d} \cdot Y_1^{c_1} \cdot Y_2^{c_2} \cdot \ldots Y_n^{c_n} \quad (7)$$

Equation (7) relates $S r$, α, and some precipitation parameters that will be determined via regression analysis. The precipitation statistics investigated in the regressions include standard statistics calculated from the full record of daily precipitation depths, and also from the wet-day series (left censored to remove zero and trace values). Also included is the lag-1 serial correlation (autocorrelation) coefficient, ρ, along with a transformation $(1 + \rho)/(1 - \rho)$ familiar to reservoir analysis (see Phatarfod 1986, and Vogel and McMahon 1996 for two independent derivations of this transformation).

These parameters were added to the regressions using a staged regression analysis to determine how the addition of each parameter improved the overall goodness-of-fit. Regression models were developed for each reliability case using all stations in the study region. Then, the stations were broken down into three regions (divided along major river basin boundaries) shown in figure 1 to see how regressions might improve.

Table 1. Average precipitation characteristics of the three regions, and all US stations.

Region	Stations (n)	μ_y Mean (mm yr^{-1})	10th percent. (mm yr^{-1})	90th percent. (mm yr^{-1})	Cv	ρ	$P_{(x=0)}$	μ_w (mm d^{-1})	σ_w (mm)	Cv_w
East	97	1094	835	1382	0.229	0.114	0.648	8.71	12.2	1.50
Midwest	82	757	381	1327	0.280	0.145	0.740	7.91	11.9	1.36
West	52	417	159	984	0.268	0.225	0.780	4.94	6.7	1.42
All US	231	820	292	1329	0.256	0.150	0.710	7.58	10.8	1.40

Notes: μ_y denotes the mean annual rainfall of the stations, shown with the 10th and 90th percentiles of the stations in each region. Cv denotes the coefficient of variation, ρ denotes lag-1 correlation of daily rainfall depth, and $P_{(x=0)}$ denotes the daily probability of zero rainfall. μ_w, σ_w, and Cv_w denote the mean, standard deviation and coefficient of variation, respectively, of the daily wet-day rainfall series.

Table 1 displays the average precipitation characteristics of the regions, and all stations.

3. RESULTS

An effort was made to make the resulting models as simple as possible, by adding relatively few precipitation parameters. An initial search was performed on the precipitation variables to determine which had the most explanatory power in a single parameter regression. Using the data from all stations for the 90% reliability case, the transformation in (6) was applied, with d = 0.667 resulting in the highest goodness-of-fit. In each case shown in table 2, the following model containing the transformed α and a single precipitation parameter, Y, was fit to the data:

$$S_r = e^a \cdot e^{b\cdot(-\ln(\alpha))^{0.667}} \cdot Y^c \tag{8}$$

Table 2. Goodness-of-Fit of regressions (adjusted R^2), and T-ratio for a single precipitation parameter model (8) with regression coefficients for the 90% reliability case.

				Coefficients		
Parameter, Y	Description	R^2 adj.	T-ratio for Y	a	b	c
σ_w	Standard deviation, wet-day	91.3	57.42	0.4056	−2.307	0.7867
μ_w	Mean, wet-day	91.1	55.47	0.6987	−2.307	0.7801
$x_{0.5,\ w}$	Median, wet-day	89.8	46.15	1.434	−2.307	0.6706
σ	Standard deviation	89.0	40.46	2.581	−2.306	0.5395
$(1+\rho)/(1-\rho)$	Transformation of ρ	87.7	31.33	1.693	−2.306	1.713
ρ	Serial correlation coeff., lag-1	87.2	27.14	3.312	−2.306	0.5599
μ	Mean	87.0	25.56	2.021	−2.305	0.2936
Cv	Coefficient of variation	85.5	10.76	1.763	−2.305	0.3898
$P_{(x=0)}$	Dry day probability	85.5	10.56	2.447	−2.305	0.6727
γ	Skewness	85.3	6.65	1.818	−2.305	0.2171
Cv_w	Coeff. of variation, wet-day	85.3	7.07	1.986	−2.305	0.6356
κ	Kurtosis	85.2	5.6	1.869	−2.305	0.0840
γ_w	Skewness, wet-day	85.2	−3.92	2.415	−2.305	−0.1653
κ_w	Kurtosis, wet-day	85.1	−3.15	2.367	−2.305	−0.0521

The resulting regression coefficients and adjusted R^2 values are also reported in table 2, along with the T-ratio for the precipitation parameter Y. The parameters described with a subscript 'w' indicate the parameter is calculated from the wet-day record. All others are calculated from the full record daily rainfall series (depth expressed in mm).

Table 2 orders precipitation parameters from highest adjusted R^2 to lowest, indicating that of the precipitation statistics considered, $\sigma\ w$ and $\mu\ w$ have the most explanatory power. We note that μ is embedded in the definition of α (see equation (2)), so using it in regression models might introduce concerns over multicollinearity. However, the recent experiments performed by Kroll and Song (2013) document that in situations when the model form is unknown, as is the case here, it is reasonable to employ the multivariate OLS regression methods used here, even in the presence of multicollinearity, and especially considering the large samples used to develop these models. In any case, several parameters outperform μ.

Table 3 illustrates the best regressions with two precipitation statistics included, for all four reliability cases, when all stations are included. All of the regressions in table 3 take the form of (9), with $\sigma\ w$ and $(1 + \rho)/(1 - \rho)$ as the two precipitation parameters

$$S_r = e^a \cdot e^{b_1 \cdot (-\ln(\alpha))^d} \cdot \sigma_W^{c_1} \cdot \left(\frac{1+\rho}{1-\rho}\right)^{c_2} \quad (9)$$

Table 3. Regression coefficients for equation (9) for the four reliabilities, and measures of regression performance including: variance inflation factor (VIF), standard error (SE), adjusted R^2 (R^2 adj.), as well as T-ratios for the parameters.

	Regression coefficients							
Reliability	a	b	c_1	c_2	d	Max VIF	SE	R^2 adj.
0.80 *coeff.*	−0.778	−2.309	0.971	2.514	0.667	1.064	0.247	0.966
T	*−31.1*	*−340.5*	*199.7*	*85.1*	—	—	—	—
0.90 *coeff.*	0	−2.566	0.958	2.599	0.642	1.064	0.254	0.977
T	—	*−360.5*	*238.1*	*102.2*	—	—	—	—
0.95 *coeff.*	0.762	−2.934	0.952	2.433	0.577	1.064	0.247	0.971
T	*30.1*	*−371.2*	*195.6*	*80.7*	—	—	—	—
0.98 *coeff.*	1.768	−3.552	0.938	2.177	0.483	1.063	0.266	0.969
T	64.1	−361.1	97.3	66.8	—	—	—	—

Regression models using three precipitation parameters were investigated, but the small improvement in goodness-of-fit did not justify the added complexity. Additionally, the model coefficient for the third precipitation variable was often less significant than the regression constant, a, as measured by its T-ratio. Figure 4 compares the predicted versus simulated values of $S\,r$ for the four models summarized in table 3.

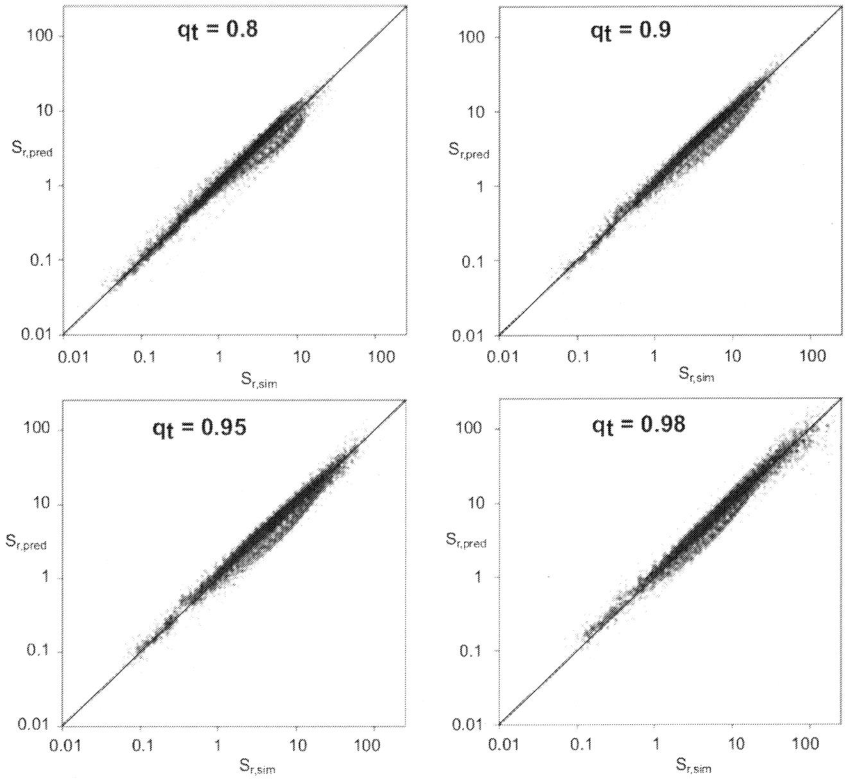

Figure 4. Predicted RWH Storage capacity ($Sr_{,\text{pred}}$) versus simulated ($Sr_{,\text{sim}}$) values for the four reliability models summarized by equation (9) and table 3 based on all 231 precipitation stations.

Table 4 summarizes the regression coefficients and fits for the three regions for each reliability case. The East, Midwest, and West regions use regression equations (9)–(11), respectively

$$S_r = e^{a} \cdot e^{b_1 \cdot (-\ln(\alpha))^{d}} \cdot \sigma_w^{\ c_1} \cdot C_v^{\ c_2}, \tag{10}$$

$$S_r = e^{a} \cdot e^{b_1 \cdot (-\ln(\alpha))^{d}} \cdot \mu_w^{c_1} \cdot \left(\frac{1+\rho}{1-\rho}\right)^{c_2} \tag{11}$$

Table 4. Regressions and fits for the three regional regression models, each for four reliabilities.

Region	Reliability	Regression coefficients					Max VIF	SE	R^2 adj.
		a	b_1	c_1	c_2	d			
East	0.80	−1.106	−2.168	1.054	2.253	0.675	1.286	0.129	0.989
East	0.90	−0.289	−2.694	1.062	2.818	0.589	1.288	0.151	0.988
East	0.95	0.567	−3.221	1.047	2.934	0.516	1.290	0.165	0.987
East	0.98	1.640	−3.735	0.932	3.176	0.468	1.248	0.222	0.979
Midwest	0.80	−0.539	−2.499	0.771	0.967	0.655	1.016	0.183	0.982
Midwest	0.90	0.330	−2.618	0.714	0.931	0.669	1.006	0.189	0.983
Midwest	0.95	0.885	−2.831	0.744	0.896	0.630	1.010	0.179	0.985
Midwest	0.98	1.749	−3.322	0.779	0.779	0.537	1.016	0.202	0.982
West	0.80	−0.831	−2.208	1.570	1.185	0.696	2.583	0.272	0.965
West	0.90	−0.124	−2.288	1.431	1.481	0.695	2.608	0.240	0.974
West	0.95	0.812	−2.631	1.054	2.068	0.613	1.463	0.243	0.973
West	0.98	1.814	−3.211	1.033	1.783	0.505	1.489	0.248	0.974

Note: Use (9) for East, (10) for Midwest, and (11) for West.

In general, the three regional regression models summarized in equations (9)–(11) and table 4 perform somewhat better than the regressions based on all 231 stations, especially the East and Midwest.

4. DISCUSSION

4.1. Significance of parameter selection

The high R^2 values (0.96–0.99) associated with these regression equations show that including a few relatively simple daily rainfall statistics can generate surprisingly accurate equations for estimating required storage for RWH systems across broad geographic and climatic regions. Even though the precipitation data set includes a wide range of climate types, the regression equations explain much of the difference in the $S\ r\ –\alpha$ relationships between sites with only a few rainfall statistics added to the model.

The most useful rainfall parameter appears to be the standard deviation of the wet-day series of daily rainfall depth. The coefficients for $\sigma\ w$ are positive in all of the regression equations presented, so required storage increases as $\sigma\ w$ increases. These results are in accordance with an intuitive understanding of how RWH systems respond to rainfall parameters. Given an average rainfall depth for a site, the minimum storage volume for a RWH system would be achieved if local rainfall delivered exactly the average rainfall depth on every day. As rain does not fall every day, the minimum storage would be achieved if the rainy days were evenly spaced, and all rainy days had equal rainfall depth. Increasing variability in either the wet-day precipitation depth (i.e. σ_w or perhaps more precisely Cv_w), or the relative length of dry and wet spells (i.e. increasing autocorrelation) increases the storage volume. Vogel and McMahon (1996) show that $(1 + \rho)/(1 - \rho)$ is effectively a factor which accounts for the inflation in the variance of a sample mean due to serial correlation of a series. Thus, this factor is closely related to the variance of the wet-day series, which may indicate why it was a helpful addition to the regional regressions.

When considering three sub-regions of the United States, the regression fits improved, presumably due to greater homogeneity in overall hydroclimatic characteristics, especially mean precipitation depth (see table 1), but parameter selection did vary between regions. The most powerful precipitation parameters in a regression will be those that best explain *differences* in the behavior of RWH systems between locations. For example, the *West* regression models showed a preference for μ_w over σ_w, which indicates that the wet-day mean rainfall was being used to explain differences in system behavior. This finding may be indicative of the wide climatic variation in the sample of West stations.

4.2. Use of regression equations and caveats

Some significant caveats should be considered when using these equations to estimate storage size of RWH system. Notably, the simulation methodology assumed constant daily demand, perfect collection efficiency, and no first flush device. Furthermore, the expression of time-based reliability indicated only the percentage of failure days within the entire record, but not the distribution of those failure days within the simulation period, or a particular period or season of interest. On the other hand, the time-based reliability is more conservative than volume-based reliability given constant daily demand. Extra caution should be exercised in using the results of the regression as α approaches 1 due to the extreme curvature of the SY curves.

Although the resulting equations were developed over very broad climatic and geographic regions of the United States, we do not recommend use of these equations outside the range of climatic characteristics summarized in table 1, or areas with climates considered extreme for the region (e.g. microclimates). It is always dangerous to extrapolate statistical models outside the range of variables used for their calibration. In that case, simulation methods to generate a set of local performance curves would be preferred. The regression equations presented *may* be potentially applicable in other geographic regions with similar latitudes or climates *if* climate characteristics are similar to those in table 1. Developing new regressions for other areas would be preferable, of course. Site selection for any future regional regressions should likewise take care not to include sites with microclimates or climates aberrant from the general character of the region.

In most cases, many of these limitations and caveats raised here may be remedied or at least addressed by choosing an appropriate factor of safety in selecting the storage volume.

5. VALIDATION AND SAMPLE CALCULATION

In order to validate the regression models, an example compares predicted SY curves to simulated SY curves for a station not included in the development of the model. The station selected is Spartanburg 3 SSE in Spartanburg, South Carolina, which has a 30 yr daily precipitation record, and mean rainfall of 3.37 mm d^{-1} (1232 mm yr^{-1}). The simulated RWH system has a collection area of 100 m^2. Based equation (2), with A_c = 100 m^2 and μ = 3.37 mm, the maximum supportable demand (α = 1) is 337.3 l d^{-1}. Additionally, ρ = 0.123, and σ_w = 15.5 mm. Figure 5 presents the simulated SY curves for the four reliabilities for this particular station as dashed lines for comparison with the results based on two different regressions. Using the values of μ, σ_w, ρ, A_c, the values of storage capacity, K, were computed from the full US and East (Spartanburg's region) regression models for a wide range of α values for each reliability case.

In general, there is strong agreement between the simulated and predicted SY curves, and the East regression model outperforms the US model in most cases. In this particular validation, the predicted values are often slightly higher than the simulated values, but if another station were used in a validation, this may not be the case.

Now, consider a specific design case. Suppose the proposed system is used to meet toilet flushing needs with constant daily demand, D = 150 l d^{-1}. Using equation (2) with mean daily rainfall of 3.37 mm d^{-1}, the yield ratio is

$$\alpha = \frac{150}{100 \times 3.37} = 0.44 \qquad (13)$$

With this value of α from (13) and the rainfall parameters ρ and σ_w, equation (9) can be used to calculate S_r using the coefficients in table 3 (US model) or 4 (East region model) for any of the four reliability cases. The regression equation output is expressed as S_r, which must be converted to storage capacity, K (m^3) using equation (3), though in this case with A_c =

100, $K = S\,r$. Alternatively, K can be calculated directly by substituting (3) into (9), resulting in:

$$K = \frac{A_c}{100} \cdot e^{a} \cdot e^{b_1 \cdot (-\ln(\alpha))^{d}} \cdot \sigma_W{}^{c_1} \cdot \left(\frac{1+\rho}{1-\rho}\right)^{c_2} \tag{14}$$

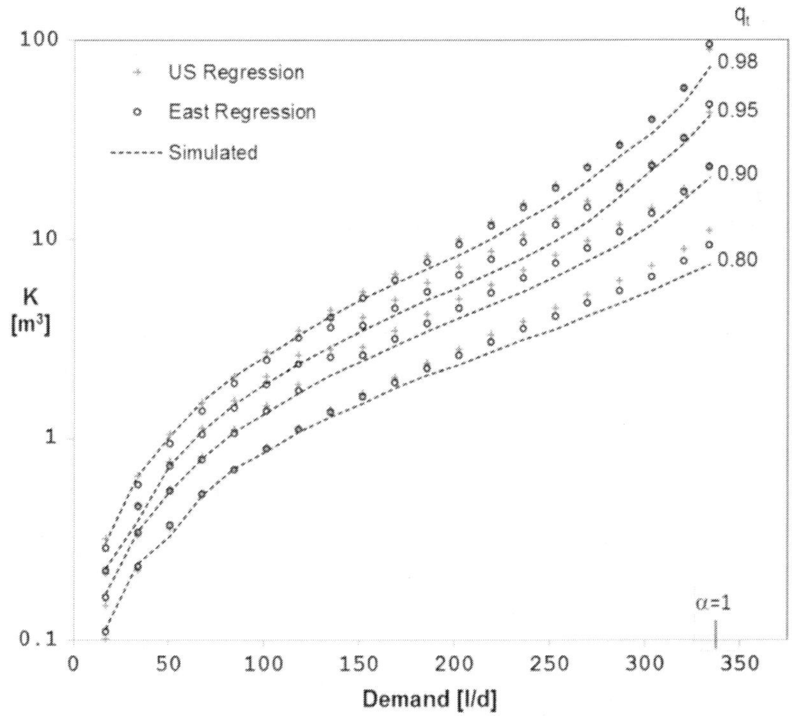

Figure 5. Simulated and regression model predicted SY curves for Spartanburg, SC, four reliabilities.

Suppose the designer desires a reliability between 90% and 95%. Using the given rainfall statistics, collection area, demand, and appropriate regression coefficients from tables 3 and 4, values K are shown in table 5 for both the simulation and the two regression models.

Given the results in table 5, the designer can roughly estimate required storage as between 2.6 and 4 m^3. For final system sizing, a more detailed engineering study is of course recommended.

Table 5. Simulated and calculated storage K (m^3) for sample location with A_c = 100 m^2, 90 and 95% reliability cases for demand of 150 l d^{-1}.

Reliability q_r	K simulation	K East model	K US model
0.90	2.43	2.55	2.78
0.95	3.46	3.56	3.94

6. SUMMARY: CONCLUSIONS, RECOMMENDATIONS AND EXTENSIONS

6.1. Conclusions

Determination of the required storage for a RWH system often presents a difficult design challenge. This study considered the problem of how to simplify the calculation of storage capacity by incorporating the information about RWH systems provided by a very wide range of simulation experiments across a very broad range of climatic conditions. Considering a RWH system as a storage reservoir, a daily RWH simulation model was used to develop generalized empirical SRY relationships at the 231 US locations shown in figure 1. To our knowledge, this is the first example of such a generalized RWH SRY relationship based on a continental dataset of precipitation measurements.

We hypothesized that differences in SRY relationships for RWH systems between stations could be explained by daily rainfall statistics, which make possible a generalized regression model capable of calculating required storage anywhere in the study area (continental United States). We show that by including daily rainfall statistics along with RWH system parameters into a multivariate log-linear regression framework can lead to a surprisingly powerful generalized SY equation at several useful reliabilities. Additionally, the regression analysis can highlight which characteristics of the rainfall distribution (other than the mean) most strongly affect RWH performance. Notably, standard statistics of the wet-day rainfall series had considerable explanatory power in the regressions, including its mean and standard deviation, μ_w and σ_w, respectively.

The regression models presented can be used to obtain a preliminary estimate of how large to build a storage tank based on desired yield and reliability, collection area, and local rainfall statistics. Since the climatic variables selected are relatively easily calculated with even incomplete daily precipitation records, the equations may be useful to wide range of users in estimating required storage. The rainfall statistics may be obtained from maps in regions with sparse or no rainfall data. Pooling of rainfall data from maps and localized short records to obtain blended estimators of rainfall statistics for use with our models may also prove useful. Mapped rainfall parameters would also allow using the equations to map storage requirements across a region.

Our framework could be extended to incorporate more RWH system design parameters, or test a broader set of precipitation statistics. One way would be to run many simulations with different values for runoff coefficients, first flush volume, and include those parameters in regression models. Incorporation of a broader set of daily rainfall statistics into the analysis could lead to improved regressions, but ease of calculation is an important consideration. Notably, measures of the variance of rainfall and measures of timing and/or seasonality may slightly improve regression fits, but substantial improvements are unlikely. Some initial tests indicate that the scale parameter of the Pearson-III distribution fitted to the full record series may slightly outperform σw (with which it is highly correlated).

6.2. Recommendations and extensions

There are a number of useful extensions to the generalized SRY relationships for RWH systems introduced here which we describe below. When taken together, the following extensions outline a research program with numerous benefits which would be difficult without the model developed here.

Within the context of water resource systems design and management, the derived RWH relationship could be used to optimize the operation, management and design of RWH systems. The regression equations could be used in planning studies which seek to examine cost-supply tradeoffs in urban settings, or evaluate the benefits of additional household water supply in rural areas or areas underserved by water infrastructure.

The expressions introduced here are useful for evaluating the impact of climate change on the performance (the reliability and/or yield) of existing or proposed RWH systems given expected changes to daily rainfall pattern. To the extent that downscaled climate models can predict changes in the daily precipitation distribution, regional regressions of this form could help predict changes in storage and/or RWH system yield under changing climate conditions. In that case, simpler models using only parameters that can be easily computed from climate model outputs may be attractive and effective.

Another natural extension would be to use the resulting SRY model to derive confidence intervals for estimates of the storage capacity or yield of

RWH systems, using the methodology outlined in the analogous companion paper by Kuria and Vogel (2014, this issue). Kuria and Vogel (2014, this issue) develop and apply generalized SRY relationships for surface water supply reservoirs fed by rivers, based on a global dataset of rivers, to derive the variance and confidence intervals of estimates of water supply yield and to quantify the length of streamflow record needed to secure a stable and secure water supply yield. Such analyses applied to RWH systems could prove useful for quantifying the value of rainfall records in data sparse regions.

This was an initial effort to develop generalized SRY relations for RWH systems, with many possible extensions. Whereas this study focused on determining the most explanatory precipitation statistics, future simulation approaches could model a RWH system with more realistic design parameters such as lower runoff coefficients, first flush devices, evaporative losses, delayed snowmelt, etc, to create more useful regressions for design. Investigating the effect of more complex demand patterns (e.g. for a RWH system designed for irrigation or stormwater control) on system performance could also be fruitful. Further work on generating a single regression equation capable of predicting storage at any desired reliability would be the final step toward a truly generalized SRY relationship, though there are significant difficulties in handling the nonlinearities of the SRY surface in a single regression model.

Perhaps the most natural and obvious extension to this initial study would be to extend our approach to other regions of the world, especially where RWH is an important source of water, either primary or supplemental, for both potable and non-potable uses. In summary, we highlight the singular nature of this study measured by both the value of the RWH model developed here combined with the numerous fruitful areas of future research which we have outlined.

ACKNOWLEDGEMENTS

Special thanks to Dr Paul Kirshen at the University of New Hampshire, and Dr Peter Shanahan at the Massachusetts Institute of Technology, both of whom served on Mr Hanson's MS committee at Tufts University, and offered valuable advice and suggestions on study design, methodology, and interpretation of results.

REFERENCES

1. Basinger M, Montalto F and Lall U 2010 A rainwater harvesting system reliability model based on nonparametric stochastic rainfall generator J. Hydrol. **392** 105–18
2. Boers T M and Ben-Asher J 1982 A review of rainwater harvesting Agric. Water Manage. **5** 145–58
3. Coombes P J and Barry M E 2007 The effect of selection of time steps and average assumptions on the continuous simulation of rainwater harvesting strategies Water Sci. Technol. **55** 125–33
4. Cowden J R, Watkins D W and Mihelcic J R 2008 Stochastic modeling in West Africa: parsimonius approaches for domestic rainwater harvesting assessment J. Hydrol. **361** 64–77
5. Fewkes A 1999 The use of rainwater for WC flushing: the field testing of a collection system Build. Environ. **34** 765–72
6. Fewkes A 2000 Modeling the performance of rainwater collection systems: towards a generalized approach Urban Water **1** 323–33
7. Ghisi E and Ferreira D F 2007 Potential for potable water savings by using rainwater and greywater in a multi-story residential building in southern Brazil Build. Environ. **42** 2512–22
8. Guo Y and Baetz B 2007 Sizing of rainwater storage units for green building applications J. Hydrologic Eng. **12** 197–205
9. Handia L, Tembo J M and Mwiindwa C 2002 Potential of rainwater harvesting in Urban Zambia. Proc. of the 3rd WaterNet/Warsfa Symp. 'Water Demand Management for Sustainable Development (Dar Es Salaam, Tanzania, 30–31st October 2002)
10. Herrmann T and Schmida U 2000 Rainwater utilisation in Germany: efficiency, dimensioning, hydraulic and environmental aspects Urban Water **1** 307–16
11. Jenkins D, Pearson F, Moore E, Sun J K and Valentine R 1978 Feasibility of Rainwater Collection Systems in California. Contribution No. 173, (California Water Resources Center) (Berkeley, CA: University of California)
12. Keller K 1982 Rainwater Harvesting for Domestic Water Supplies in Developing Countries: A Literature Survey (working paper 20) (Washington, DC: Water and Sanitation for Health (WASH), USAID)
13. Kroll C N and Song P 2013 Impact of multicollinearity on small sample hydrologic regression models Water Resour. Res. **49** 3756–69
14. Kuria F W and Vogel R M 2014 A global water supply planning model with uncertainty Environ. Res. Lett. under revision
15. Lee K T, Lee C-D, Yang M-S and Yu C-C 2000 Probabilistic design of storage capacity for rainwater cistern systems J. Agric. Eng. Res. **77** 343–8
16. Liaw C H and Tsai Y-L 2004 Optimum storage volume of rooftop rainwater harvesting systems for domestic use J. Am. Water Resour. Assoc. **40** 901–12
17. McMahon T A and Adeloye A J 2005 Water Resources Yield (Littleton, CO: Water Resources Publications)
18. McMahon T A, Vogel R M, Pegram G G S, Peel M C and Etkin D 2007 Global streamflows—part 2 reservoir storage-yield performance J. Hydrol. **347** 260–71
19. Mitchell V G 2007 How important is the selection of computational analysis method to the accuracy of rainwater tank behaviour modelling? Hydrol. Process. **21** 2850–61
20. Palla A, Gnecco I and Lanza L G 2011 Non-dimensional design parameters and performance assessment of rainwater harvesting systems J. Hydrol. **401** 65–76
21. Pandey D N, Gupta A K and Anderson D M 2003 Rainwater harvesting as an adaptation to climate change Curr. Sci. **85** 46–59
22. Phatarfod R M 1986 The effect of serial correlation on reservoir size Water Resour. Res. **22** 927–34
23. Schiller E J and Latham B G 1987 A comparison of commonly used hydrologic design methods for rainwater collectors Water Resour. Dev. **3** 165–70

24. Vogel R M and McMahon T A 1996 Approximate reliability and resilience indices of over-year reservoirs fed by AR(1) gamma flows Hydrol. Sci. J. **41** 75–96
25. Vogel R M, Wilson I and Daly C 1999 Regional regression models of annual streamflow for the United States J. Irrig. Drain. Eng.-ASCE **125** 148–57
26. Vogel R M and Stedinger J R 1987 Generalized reservoir storage-reliability-yield relationships J. Hydrol. **89** 303–27
27. Villarreal E L and Dixon A 2005 Analysis of a rainwater collection system for domestic water supply in Ringdansen, Norrköping, Sweden Build. Environ. **40** 1174–84
28. Watt S B 1978 Ferrocement Water Tanks And Their Construction (London: Intermediate Technology Publications)

CHAPTER 7

A Simple and Robust Method for Simultaneous Consideration of Overland and Underground Space in Urban Flood Modeling

Ah-Long Son [1], Byunghyun Kim [2,*] and Kun-Yeun Han [3]

[1] *National Disaster Management Research Institute, 365 Jongga-ro, Ulsan 44538, Korea*

[2] *Disaster Prevention Research Institute, Kyungpook National University, 80 Daehak-ro, Buk-gu Daegu 41566, Korea*

[3] *Department of Civil Engineering, Kyungpook National University, 80 Daehak-ro, Buk-gu Daegu 41566, Korea*

ABSTRACT

This study proposed two methods, boundary-type and pond-type, to link overland and underground space in urban flood modeling. The boundary-type treats the exit of underground space as an interface for inflow of floodwater by imposing open boundary condition and pond-type considers underground space as an underground pond by configuring pond terrain. The effect of underground space in urban flood inundation was examined by coupling one-dimensional (1D) stormwater management model (SWMM) and two-dimensional (2D) overland flood model. The models were applied to the Hyoja drainage basin, Seoul, Korea where urban flood occurred due to heavy rainfall in 21 September 2010. The conduit roughness coefficient of SWMM was calibrated to minimize the difference between observed and predicted water depth of pipe. In addition, the surface roughness coefficient of 2D model was calibrated by comparing observed and predicted flood extent. Then, urban flood analysis was performed on three different scenarios involving a case not considering underground (Case 1) and cases considering underground, boundary-type (Case 2) and pond-type (Case 3). The simulation results have shown that the boundary-type is simple but robust method with high computational efficiency to link overland and underground space in urban flood modeling.

Keywords: underground space; 1D SWMM; 2D flood model; urban flooding

1. INTRODUCTION

Rapid industrialization and urbanization has brought about an increase in population, industrial facilities, and transportation, which require building space utilization advancement and space layout planning for efficient utilization of overland space. Nonetheless, the lack of overland due to the expanding urbanization has led a variety of types of underground developments, such as underground roads, underground shops, subways, underground culverts, or substation for space utilization advancement, which has created highly complex combinations of overland and underground in urban areas. In recent years, the occurrence frequency of effective rainfall that exceeds design capacity of storm water drainage systems has increased due to the climate change and local heavy rainfall, resulting in flooding in overland and underground spaces.

1D/1D or 1D/2D coupled urban flood analysis models have emerged to assess the hydraulic performance of sewer network flow and overland flow simultaneously. 1D/1D models were developed and improved significantly [1,2], and studied for modeling approaches and real-time application about overland flooding modeling [3,4,5]. Mark et al. [1] and Leandro et al. [6] suggested the potentials and drawbacks of each model for urban flood modeling through extensive comparison of 1D/1D and 1D/2D models. Recently, 1D/2D models are becoming popular with the development of modeling techniques and the improvement of computer performance [7] despite the drawback of 2D overland model requiring considerable computational efforts including long simulation time. The resolution of topographic data is one of the most critical factors for accuracy of 2D overland model in urban inundation analysis [8,9]. That is, the accuracy of the urban inundation analysis is dependent on how well the real topography is considered. There have been numerous studies on the effect of urban topography including building. Shoji and Mikio [10] investigated the effect of building arrangement on the flood wave through the hydraulic model experiment. Mirei and Juichiro [11] investigated the behavior of 2D flood flows and the hydrodynamic force acting on structures through numerical model simulation and experiment. Alcrudo [12] studied mathematical modeling techniques to represent water depth and velocity in the vicinity and around buildings. In recent years, a number of studies of topographic data utilizing higher, more sophisticated resolution such as Light Detection And Ranging (LiDAR) and Digital Surface Model (DSM) have been conducted in order to calculate direction and pattern of flood flow in roads or around buildings [13,14,15,16,17,18]. However, 2D flood inundation modeling using a fine grid with high-resolution topography requires considerable computational time, thus the computational efficiency of 2D flood modeling is one of the challenges [19]. This problem has led to several methodologies to improve the performance of 2D urban flood modeling such as sub-grid method [20,21], 2D storage model with bilinear gridding technique [22], multi-layered coarse grid modeling [19], porosity shallow water model [20,21,22,23,24,25,26,27,28] and LiDAR filtering algorithm [29,30].

Most studies on urban flood modeling including the aforementioned studies did not consider floodwaters flowed into the underground space from the overland. That is, most studies on urban flood have been conducted by considering overland and underground space independently; few studies have been conducted on the analysis of the effect of the presence of underground space on urban flood considering the link between overland and underground spaces.

This study proposed two methods, boundary-type and pond-type, to link the overland and underground spaces in the urban flood modeling using the 2D flood model. The methods have been applied to examine the effect of underground spaces on flood characteristics including flood depth, area, and velocity of overland floodwaters in the urban flood modeling. To achieve this, the 1D SWMM and 2D overland flood models have been applied to the Hyoja drainage basin of Seoul where urban flooding occurred in 21 September 2010. The manhole overflow predicted from the SWMM was employed as forcing data in the 2D overland flood model. The accuracy and computational efficiency of the proposed methods were examined by comparing measured flood extent and depth and predictions, and computation time between models.

2. METHODS

2D flood inundation model was coupled with 1D SWMM to account for rainfall–runoff, pipe flow, surcharged overflow at manhole, flood flow entered into underground space, and overland flow on the street in urban area. The 1D SWMM simulates rainfall–runoff, sewer network and sewer overflow, and 2D flood inundation model accounts for overland flow caused by manholes and floodwater flowed into underground space.

2.1. Urban Flood Inundation Model

The causes of urban inundation are topographically low lands, poor drainage due to an increase of water level in the river, road surface drainage due to sewage backwater, lack of capacity or failure of pumps, and lack of drainage capacity of sewer system. In Korea, 25% of urban flooding is due to a lack of drainage capacity of sewer system [31]. In recent years, urban flooding has increased due to abnormal climate activity and local heavy rainfalls that exceed the design frequency of conveyance capacity in the urban drainage system.

In this study, the overflows at the manhole due to exceeding pipe capacity were calculated through rainfall–runoff analysis with SWMM which is a 1D dynamic urban storm sewer system analysis model. Then, an urban flood analysis was conducted by utilizing the manhole overflows as boundary conditions in the 2D overland flood model. The linkage process between 1D SWMM and 2D overland flood model applied in this study is shown in Figure 1.

Overflows at the manholes due to the surcharge in the urban drainage system cause overland flood inundation; some of them flow into underground, and thereby inundate the underground space. In this study, the 2D Godunov-type flood model [32] was applied to perform the linkage with overland flooding and inflow into underground space due to manhole overflows. The 2D flood model is based on shallow water equations and can be represented as a vector type as shown in Equation (1).

$$\mathbf{U}_t + \mathbf{F}(\mathbf{U})_x + \mathbf{G}(\mathbf{U})_y = \mathbf{S}(\mathbf{U}) + \mathbf{Q}(\mathbf{U}) \quad (1)$$

where U is a physical vector consisting of conservation variables; F(U) and G(U) are the fluxes in the *x* and *y* directions, respectively; S(U) is a source term; and Q(U) is a term that connects the SWMM and 2D overland flood model. In this study, urban flood inundation analysis was conducted by inputting the overflow calculated from SWMM to the 2D flood model through Q(U).

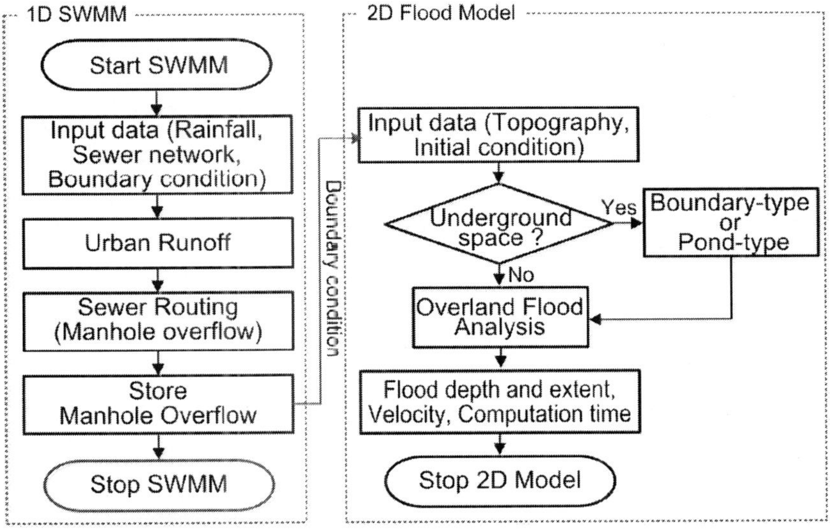

Figure 1. Flowchart of linkage process between SWMM (stormwater management model) and 2D flood model considering underground space.

In this study, the following two-step time frictional scheme was applied to calculate conservation variables U_i in the mesh center at n+1 time. In the first step (n and n^*), fluxes F(U) and G(U) are considered (Equation (2)), and in the second step (n^* and n+1), U^{n*}_i, S(U), and Q(U) are considered, thereby calculating conservation variables at the n+1 time (Equation (3)).

$$\mathbf{U}_i^{n*} = \mathbf{U}_i^{n} - \frac{\Delta t}{A_i}\left(\sum_{k=1}^{N_i} (\mathbf{F}\cdot\mathbf{n} - \mathbf{G}\cdot\mathbf{n})_k^n \, L_{i,k}\right) \quad (2)$$

$$\mathbf{U}_i^{n+1} = \mathbf{U}_i^{n*} - \frac{\Delta t}{A_i}(\mathbf{S}+\mathbf{Q})_i^{n*} \quad (3)$$

where A_i and N_i are area of mesh (i) and the number of interfaces in the mesh, respectively, and $L_{i,k}$is a length of the k-th interface in the mesh (i).

2.2. Methods to Link Overland and Underground Spaces

Subway stations accounted for the largest area in the underground spaces and flood inundation occurred there frequently in Seoul Metropolitan City, Korea. Accordingly, a subway station was taken into consideration as representative underground space to determine the effect of the underground space on the urban inundation modelings, a subway station was taken into consideration as representative underground space.

This study proposed two methods, boundary-type and pond-type, to link the overland and underground spaces in the 2D flood model. In the first method (boundary-type), the underground space in the 2D flood model was set as an interface, and inflow of floodwater to underground space was only possible at the exit (entrance) of subway station (Figure 2a). That is, a mesh was configured to set the subway station as a boundary condition, and the exit was imposed as the open boundary condition [32], thereby flow any floodwater above the threshold in the exit into the subway station. The other three sides were imposed as closed boundary conditions, so that inflow of floodwater did not occur as they were blocked by the wall.

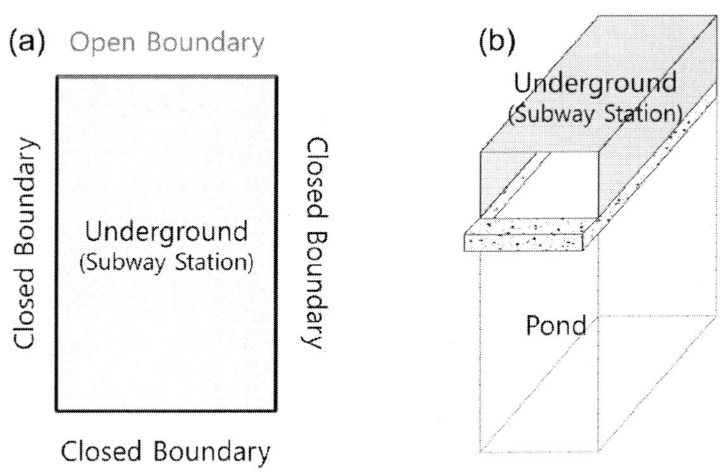

Figure 2. Schematic design to link overland and underground space: (**a**) boundary-type; and (**b**) pond-type.

In the second method (pond-type), the subway station was considered as an underground pond (Figure 2b). That is, the subway station at the 2D flood model was configured as a pond terrain. Here, size of the pond was set appropriately so that the water stored in the underground was not allowed to back flow to the ground. The floodwater over exit threshold only flowed into subway station through exit, and the three sides other than the exit side blocked by having the real structure elevation in the terrain mesh.

3. APPLICATIONS AND RESULTS

3.1. Study Area

Seoul Metropolitan City is divided into 239 drainage basins according to area size [33]. In this study, Hyoja drainage basin is selected as the study area because it was one of the most flood-prone areas among the 239 drainage basins. The area of the Hyoja drainage basin is 5.4 km^2, and the upstream of the basin is a steep mountainous terrain, which gradually becomes flat in the downstream. In the mid-to-downstream, most residential and commercial zones are densely populated, and there are many underground structures such as subway (Figure 3).

In the drainage system of study area, rainfall collects through sewage pipes flowing down to Baekundong Stream and Junghak Stream and finally joining at Cheonggye Stream. Baekundong Stream and Junghak Stream are urban streams that had been covered with cement. Figure 3 shows the geographical location and satellite image of the study area.

3.2. SWMM Calibration

In the study area, the 1D SWMM was applied to calculate an overflow that exceeded the conveyance capacity of the sewer pipes. As shown in Figure 4, the study area was divided into 165 sub-catchments, and input data of the SWMM including impervious ratio, slope, and CN of the sub-catchments were applied.

This study utilized the shapefile format data for specifications of the sewer pipes and manholes, and the level-2 land cover map and biotope map to calculate an impervious ratio and runoff curve number (CN) in the basin. These three data were provided by Urban Safety Division of Seoul Metropolitan City (https://safe.seoul.go.kr).

The 34 automatic weather stations (AWSs) are operating in Seoul Metropolitan City. Among them, AWSs in Jung-gu, Seongbuk-gu, and Seodaemun-gu are located around the study area and rainfalls measured at these three AWSs (http://www.kma.go.kr) were used.

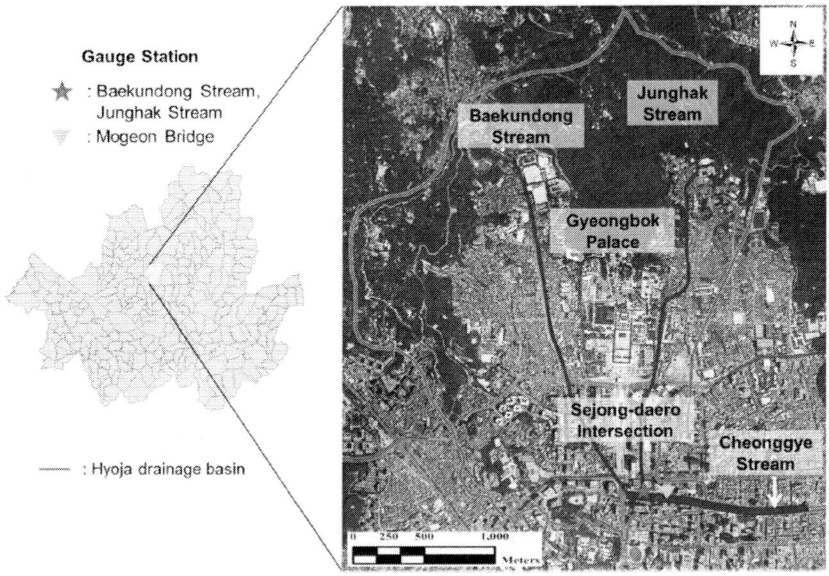

Figure 3. Study area including two covered urban streams with gauge station.

The pipes in the Sejong-daero crossroad were constructed as C-shape type (Figure 3), which was claimed to be one of the main causes of repeated flood due to the degradation of rainwater passage capability as a result of reduction in flow velocity and increase in friction [34]. Since the SWMM analyzed pipe flow assuming the sewage and rainwater pipes as a straight line, an energy loss coefficient was applied to take the C-shape type pipes in the Sejong-daero crossroad into consideration. For the value of the energy loss coefficient, 1.8, as proposed by Kim [34], was applied.

Lee et al. [35] reported that roughness coefficient of pipe and pipe slope were more sensitive than other parameters with regard to the concentration time of run-off at the urban area as rainfall scale increased. In addition, Kim et al. [36] highlighted Manning resistance parameter for sewer pipes is the greatest source of uncertainty relative to surcharge prediction and thus flood extent prediction. Thus, this study calibrated the Manning roughness parameter of conduit, which was one of the most sensitive parameters that affected the concentration time in the basin and overflow in the manholes. To do this, point rainfall data measured at the AWSs in Jung-gu, Seongbuk-gu, and Seodaemun-gu for two rainfall events that occurred at 12 June 2012 and 27 August 2012 were used. The point rainfall was converted into areal rainfall for every 10 min via the Thiessen's weighting method (Figure 5). For the downstream boundary condition of SWMM, a water depth observed at Mojeon Bridge in Cheonggye Stream was applied (Figure 5) (http://www.sisul.or.kr).

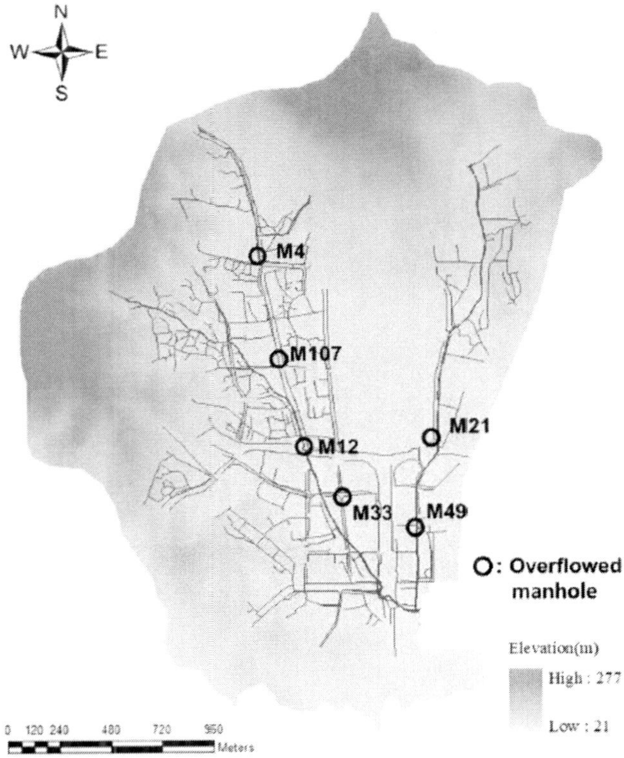

Figure 4. Digital Elevation Model (DEM) and sewer drainage network.

In the KWRA [37], the range of Manning roughness coefficient in concrete channels was proposed to 0.010–0.022 $m^{-1/3}s$. In this study, conduit flow analysis was conducted by changing the roughness coefficient by 0.001 $m^{-1/3}s$ increments within the proposed range, and the roughness coefficient was calibrated through comparison between calculated and observed water depth. A water depth at Baekundong Stream and Junghak Stream was observed via an ultrasonic gauge, which was installed at a location where two streams had just joined with the Cheonggye Stream (Figure 3) (http://www.sisul.or.kr). Baekundong Stream and Junghak Stream are urban streams that were covered up, and they were considered as combined pipes in this study.

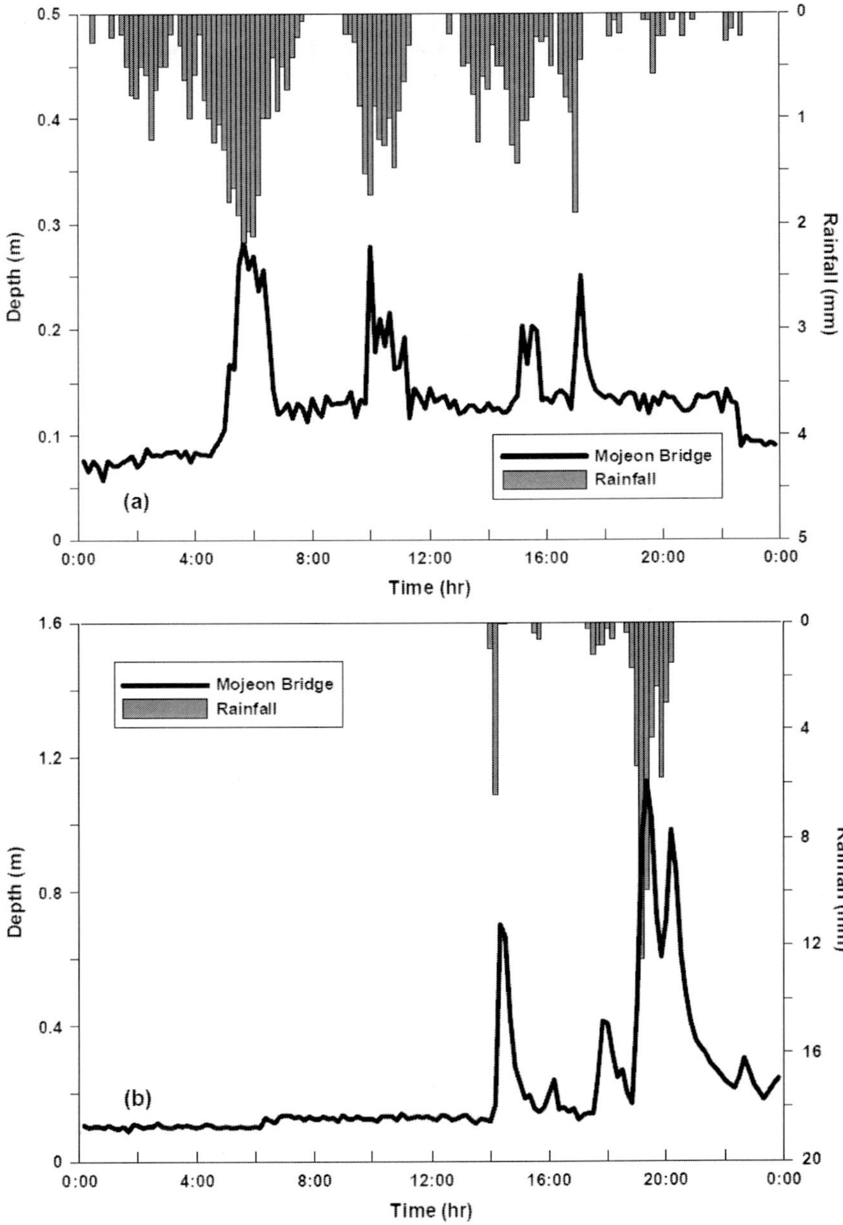

Figure 5. Rainfall hyetograph and boundary condition for urban run-off analysis: (**a**) 12 June 2010; and (**b**) 27 August 2010.

To calibrate the conduit roughness coefficient, the aforementioned two rainfall events and water depths in Baekundong Stream and Junghak Stream were employed. The conduit roughness coefficient was calibrated by adjusting it in SWMM and comparing the calculated and observed water depths. The root mean square error (RMSE, Equation (4)) and relative peak error (RPE, Equation (5)) were used as objective functions to minimize the difference of observations and calculations. In this study, nm = 0.020 $m^{-1/3}s$, where the least error between the calculated and observed values was set as the optimal roughness coefficient. Table 1 shows the quantitative errors including RMSE and RPE, and Figure 6 shows the comparison between observed and calculated water depth applying $n_m = 0.020\ m^{-1/3}s$ at Junghak Stream and Baekundong Stream.

$$\mathrm{RMSE} = \sqrt{\frac{\sum_{i=1}^{N} \left(I_{obs_i} - I_{com_i}\right)^2}{N}} \tag{4}$$

$$\mathrm{RPE} = \frac{\left|I_{obs_p} - I_{com_p}\right|}{I_{obs_p}} \times 100(\%) \tag{5}$$

where I_{obsi} and I_{comi} are the observed and calculated water depth at time (i), respectively; I_{obsp} and I_{comp} are observed and calculated peak water depth, respectively; and N is the number of data.

3.3. Urban Flood Analysis

3.3.1. Model Scenarios

Subway stations are the largest underground spaces in the study area and this study took Gwanghwamun subway station located at the Sejong-daero where urban flood occurred in 21 September 2010 (Figure 3 and Figure 7a). In this study, three scenarios were made to compare the flood influence according to whether underground space was present or not in the urban inundation modeling. In Case 1, the underground space was not considered, in Case 2, the underground space was considered as a boundary-type, and in Case 3, the underground space was considered as a pond-type. That is, in Case 2, an exit of the subway station was set to open boundary condition to have an inflow of the floodwater, and other sides were set to closed boundary condition to block the inflow of the flood into underground space. In Case 3, floodwater came into underground space through the exit of the subway station, and a real elevation of the subway station was taken for the other sides to block the flow.

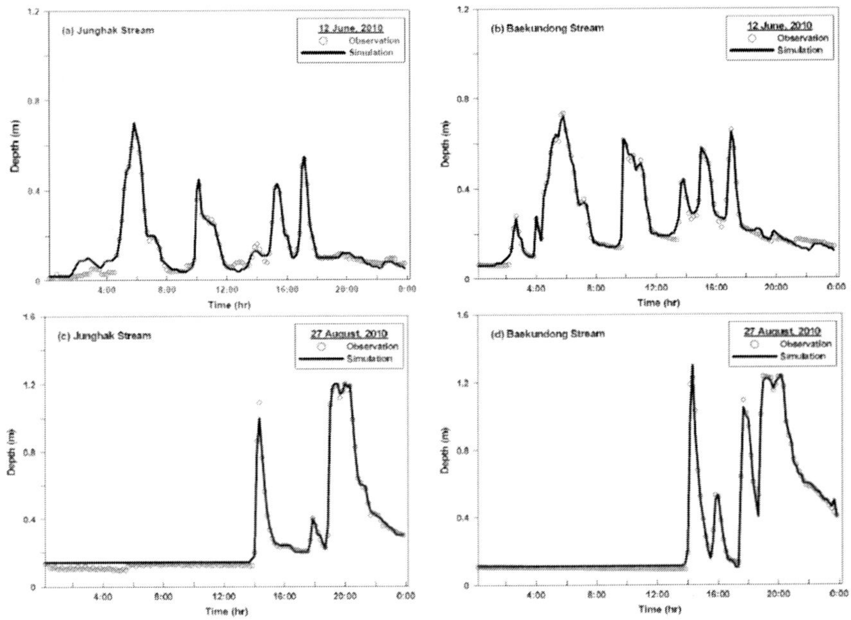

Figure 6. Comparison of observed and calculated water depths from SWMM at two covered urban streams: (**a**,**b**) 12 June 2012; and (**c**,**d**) 27 August 2010.

Table 1. Errors between observed and calculated water depth at two covered urban streams with gauge station.

Error	12 June 2010		27 August 2010	
	Junghak	Baekundong	Junghak	Baekundon
RMSE (cm)	2.29	1.88	2.27	2.87
RPE (%)	6.38	1.75	2.04	5.52

To apply 2D flooding model, unstructured mesh was generated using SMS Surface Water Modeling System (SMS) (Aquaveo, Provo, UT, USA). The mesh has a resolution of approximately from 2 m around subway station to 6 m. Figure 7b,c shows the generated meshes for Case 2 and Case 3, respectively. In Case 2, a mesh was configured to have an exit of the subway station set to open boundary condition and the other sides set to closed boundary condition (Figure 7b). In Case 3, a mesh was configured to consider the elevation of the underground space (Figure 7c). In Cases 2 and 3, am overland floodwater was flowed into the subway station, when flood depth was higher than the exit threshold elevation of the subway station. The elevations of the exit were measured in the site, shown in Table 2. The floodwater flowed into underground space was able to flow out to overland in Case 3 considering topography of underground space, whereas the water flowed outside of computational domain by open boundary condition was not considered in Case 2.

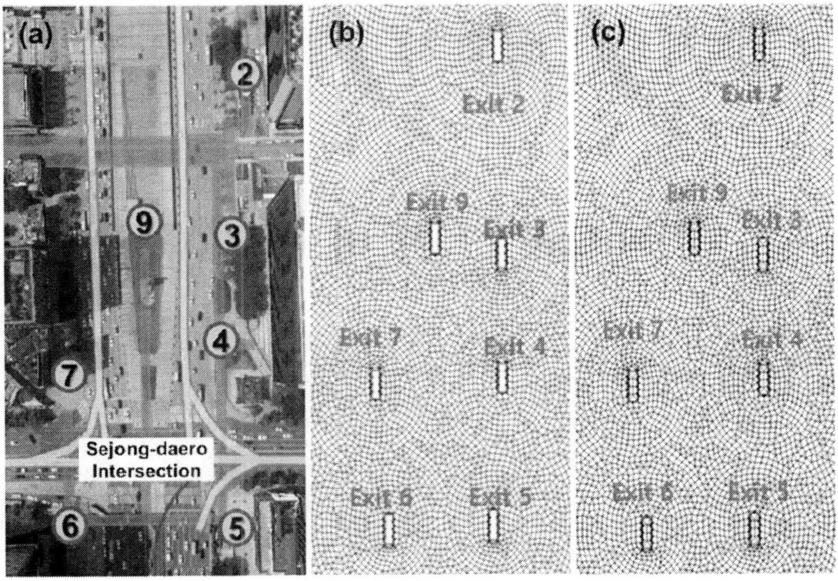

Figure 7. Mesh generation to link overland and underground spaces: (**a**) exit locations of Gwanghwamun subway station; (**b**) boundary-type; and (**c**) pond-type.

Table 2. Height of exit threshold in Gwanghwamun subway station.

Exit No.	2	3	4	5	6	7	9
Height (m)	0.22	0.10	0.33	0.49	0.32	0.29	0.0

The elevation was assigned to nodes of mesh from topographic data combining buildings of digital contour maps with 1:1000 scale (http://www.nsic.go.kr) and DEM with 2 m resolution. Yi et al. [38] built the DEM with 2 and 5 m resolution, respectively, from Airborne LiDAR with average point density of 2.5 pt/m^2 (http://www.ngii.go.kr) and KOMPSAT-2 (Korea Multi-Purpose Satellite-2) images with 1–4 m resolution (http://www.kari.re.kr). This study used this DEM with 2 m resolution and building process of topographic data involving LiDAR filtering algorithms was described in detail in Yi et al. [38].

3.3.2. 2D Flood Analysis Linking the Overland and Underground Spaces

In the study area, rainfall intensity of up to 94.4 mm/h and total rainfall of 243 mm for six hours were observed in 21 September 2010. This heavy rainfall exceeded the capacity of the sewer drainage system in the study area, thereby inducing an overflow in the manholes and causing urban inundation (Figure 8 and Figure 9a). Figure 8 presents the flood inundation around Sejoing-daero intersection and exit No. 7 of Gwanghwamun subway station at that time.

Figure 8. Flooding of study area on 12 September 2010 around: (**a**) Sejoing-daero intersection; and (**b**) Exit No. 7 of Gwanghwamun subway station (Photo by Yonhap News).

Figure 9a shows the converted areal rainfall from measured point rainfall at three AWSs and the observed water depth at Mojeon Bridge in 21 September 2010. These two data were used to calculate overflow at the manhole in the SWMM. Figure 9b,c shows a comparison of the observed and calculated water depth applying calibrated roughness coefficient in SWMM at two urban streams. In the urban run-off analysis, an overflow occurred at about 40 min after the heavy rainfall was started, and the overflow continued at the manhole from 13:00 to 17:00.

An overflow occurred at six manholes (Figure 4), and the calculated overflow is shown in Figure 10. The calculated overflow was used for the 2D flood inundation analysis linking the overland and underground spaces.

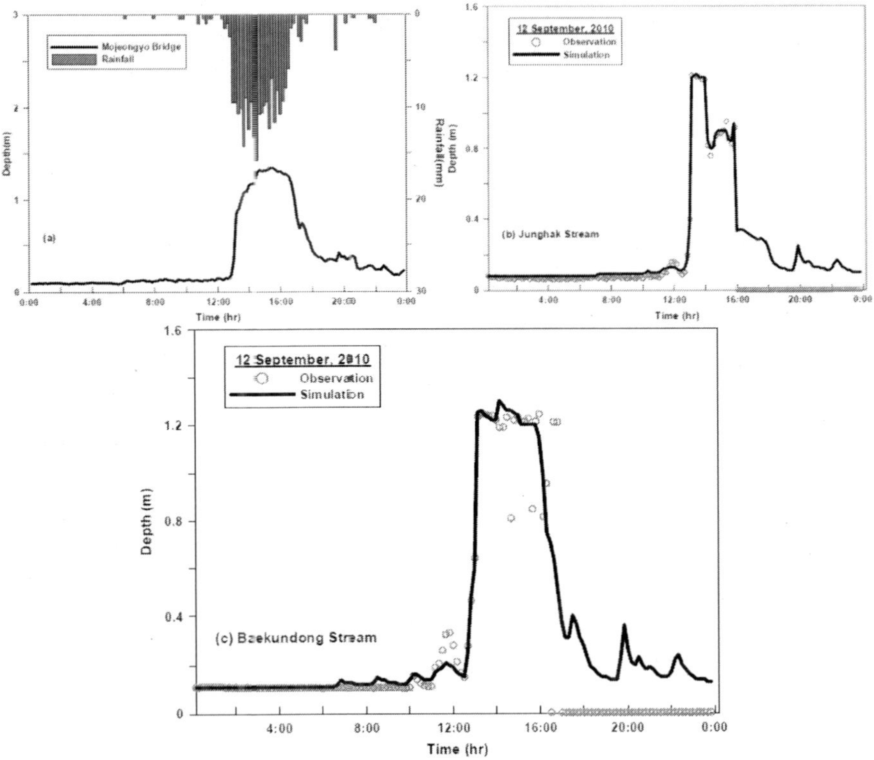

Figure 9. Flood event on 12 September 2010: (**a**) Rainfall hyetograph and boundary condition of SWMM; and (**b**,**c**) comparison of observed and calculated water depth at two covered urban streams.

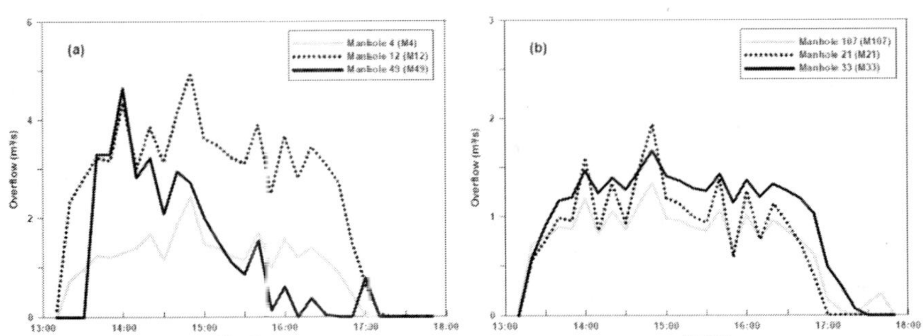

Figure 10. Calculated overflow at manhole: (**a**) M4, M12, and M49; and (**b**) M107, M21, and M33 where surcharging occurred (shown in Figure 4). These values are inputted to 2D flooding model as boundary conditions.

UFDMRC [22] proposed a manning roughness coefficient of the commercial zone in a range of 0.015–0.030 $m^{-1/3}s$. In this study, four roughness coefficients (0.015, 0.020, 0.025, and 0.030) were considered within a range proposed by UFDMRC [39] and increased by 0.005 to calibrate the surface roughness coefficient. The calculated flood extent through the application of each roughness coefficient and measured flood extent were compared, and a roughness coefficient that showed the greatest goodness of fit between them was selected as the optimal one. To do this, the manhole overflow (Figure 10) was applied to the 2D overland flood model (Case 1) in which the underground space was not considered. The measured flood extent was supplied from Seoul Metropolitan City (http://www.seoul.go.kr) and goodness of fit was calculated using Equation (6).

$$\text{Goodness of fit } (\%) = \frac{A_c \cap A_m}{A_c \cup A_m} \times 100(\%) \tag{6}$$

where A_c and A_m are the computed and measured flood extent area, respectively. The symbols ∩and ∪ represent the intersection and union of two domains, respectively. The value equals to 100 when two domains match perfectly, and 0 when no intersection area [40].

When nm = 0.015, 0.020, 0.025, and 0.030 $m^{-1/3}s$ were applied as roughness coefficient to 2D flooding model, the goodness of fit between measured and calculated flood extent was 39.52%, 39.88%, 40.22% and 40.11%, respectively (Table 3). This result showed low agreement compared with that of other studies presenting 54%–91% [40], 78%–92% [41] and 54%–95% [42]. Wadey et al. [43] divided the range of goodness of fit into three sections: good fit (>75%), moderate fit (50%~75%) and poor fit (<50%).

Table 3. Calibration of surface Manning's runoff coefficient.

Fit (%)	Surface Manning's Roughness Coefficient ($m^{-1/3}s$)			
	0.015	0.020	0.025	0.030
Goodness of fit	39.52	39.88	40.22	40.11
Modified goodness of fit	73.6	73.9	74.2	74.0

In order to find the reason for the low goodness of fit, an inundation region was divided into three sub-regions (A, B and C). Sub-region A refers to where the calculated flood extent and measured one were matched, sub-region B refers to where two extents were not matched within the measured flood extent, and sub-region C refers to where two extents were not matched outside the measured flood extent. When a proportion of each sub-region in entire region (A + B + C) was calculated, the area of sub-region C was relatively larger than the other two areas, and this was the cause of the low goodness of fit. This was because the measured flood extent had the following problems. First, the objective criterion depth to distinguish flooding was not established so the measured flood extent tends to be dependent on the subjectivity of post-flood field investigator. Second, inundation that occurred around the buildings or streets, which were far away from the main roads, was not taken into consideration clearly since the flood extent was examined based on the main roads only.

Thus, inundation regions based on flood damage reports for houses and shops provided by National Disaster Management Institute (http://www.ndmi.go.kr), media articles, and photo images showing real situations were collected in this study in order to overcome the inaccuracy of the measured flood extent. Figure 11 shows locations indicated by these new data marked with white inverted triangle. Most of the additionally investigated inundation areas were presented outside of the measured flood extent, which indicated that the real flood area was larger than the measured flood extent. Since the evidence for the accuracy of the area in sub-region C was not supportive, a modified goodness of fit equation was used in this study. The goodness of fit for each roughness coefficient calculated using Equation (7) is shown in Table 3, and n_m = 0.025 $m^{-1/3}s$ whose goodness of fit was the highest (74.2%) was selected as the optimal roughness coefficient.

$$\text{Modified goodness of fit } (\%) = \frac{\text{Area } (A \cap B)}{\text{Area } (A \cup B)} \times 100(\%) \tag{7}$$

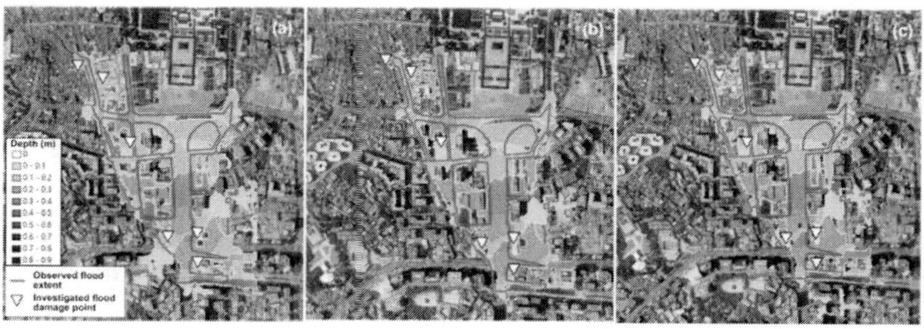

Figure 11. Comparison of observed and computed flood extent for: (**a**) Case 1; (**b**) Case 2; and (**c**) Case 3.

A manhole overflow (Figure 10) calculated using the SWMM was applied to the 2D overland flood model with $n_m = 0.025\ m^{-1/3}s$ to calculate a flood area, flood depth, and velocity for each scenario. The calculated results are shown in Figure 11 and Figure 12 and Table 4. Cases 2 and 3 showed a similar flood area, whereas Case 1 had a larger flood area than that of Cases 2 and 3 by 12% approximately (Table 4). At the west of Gyeongbok Palace and downstream of the study area, Case 1 was more inundated than in Cases 2 and 3 were (Figure 11). The goodness of fit between measured and calculated flood extent in Case 1 was 74.2% while those of Cases 2 and 3 were 82.7% and 84.2%, respectively, which revealed that a scenario in consideration of underground space had about 10% higher goodness of fit.

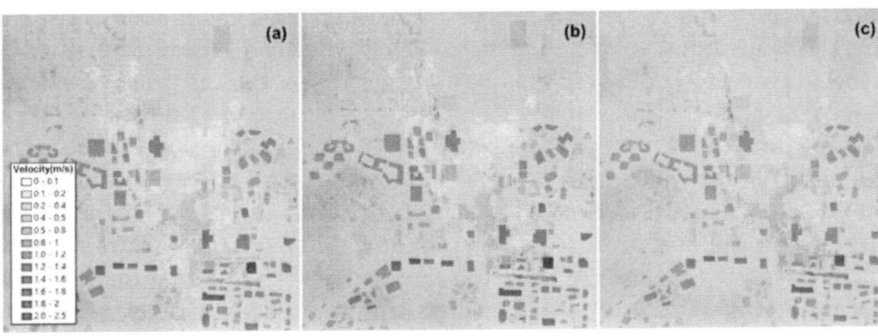

Figure 12. Computed flood velocity for: (**a**) Case 1; (**b**) Case 2; and (**c**) Case 3.

Table 4. Computed flood characteristics for each case.

Case	Case 1	Case 2	Case 3
Goodness of fit (%)	74.2	82.7	84.2
Avg. depth (m)	0.04	0.06	0.05
Max. depth (m)	1.01	0.96	0.98
Avg. velocity (m/s)	0.10	0.09	0.09
Max. velocity (m/s)	1.77	2.30	2.47
Execution time (min)	426	514	1865

Kim [34] presented a flood depth of two places (Exit No. 6 and 7 shown in Figure 7) through site survey. The calculated flood depth was compared with this site surveyed flood depth. A flood depth surveyed at the site of Exit No. 7 in Gwanghwamun station was 0.47 m (Figure 8b) while those for Case 1, Case 2, and Case 3 were 0.51, 0.48 and 0.46 m, respectively, indicating that Case 1 had 9% error and Cases 2 and 3 had 2% error approximately. A flood depth measured at the site of Exit No. 6 was 0.47 m while those for Case 1, Case 2, and Case 3 were 0.15, 0.16, and 0.15 m, respectively, indicating that Case 1 had 25% error and Cases 2 and 3 had 20% error approximately. Case 1 in the above two places showed 17% error on average and Cases 3 and 3 showed 12% error on average, indicating that more accurate results were obtained when underground space was considered in the case of a flood depth as the same as in the flood extent.

Although, there was not measured flood velocity at the time of flood event, this study examined the flood velocity for each case. A mean velocity in Cases 1, 2, and 3 was 0.10, 0.09, and 0.09 m/s, respectively, revealing no significant difference according to the consideration of underground space. However, the maximum velocity was 1.87, 2.30, and 2.47 m/s, resulting in a difference of 0.43–0.6 m/s between the cases (Table 4). In particular, Cases 2 and 3 showed a higher velocity distribution near the subway station than in Case 1.

Cases 2 and 3 showed more accurate results than Case 1 and two cases had similar accuracy in flood extent and depth. However, the execution time of Case 2 was almost four times faster than Case 3 under same situation (Table 4). which indicated that Case 2 had the highest computational efficiency among three cases.

A difference of flood depth and velocity was calculated using ArcGIS (ESRI, Redlands, CA, USA) to compare the difference of them for each case. Figure 13 and Figure 14 presented differences in maximum flood depths and velocities between each case (Cases 2 and 3) and the Case 1, respectively. For flood depth, no significant difference in mean flood depth was shown between cases but some region showed a large difference in flood depth, such as −0.71~0.60 m and −0.73~0.33 m (Table 5). As shown in Figure 13, a deeper flood depth was calculated in Case 2 and 3 than in Case 1 around the subway station, whereas a deeper flood depth was presented in Case 1 around Cheonggye Stream, which was a downstream of the study area.

The reason for this result was because floodwater that was not entered into the underground space was flowed to Cheonggye Stream, thereby increasing a flood depth and flood area around Cheonggye Stream. Furthermore, the reason for the deeper flood depth in Cases 2 and 3 around the subway station was due to the conveyance reduction of flood flow caused by the terrain mesh to present the topography of subway station.

For the velocity, a difference in velocity around Gwanghwamun station and Cheonggye Stream was revealed relatively large as similar as the flood depth (Table 5). The reason for this was the same as the cause of difference in flood depth for each scenario as explained above.

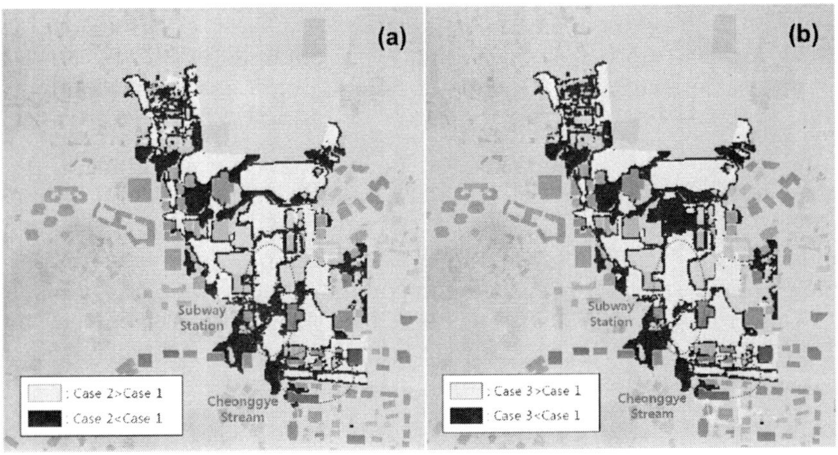

Figure 13. Difference of flood depth for: (**a**) Case 1–Case 2; and (**b**) Case 1–Case 3.

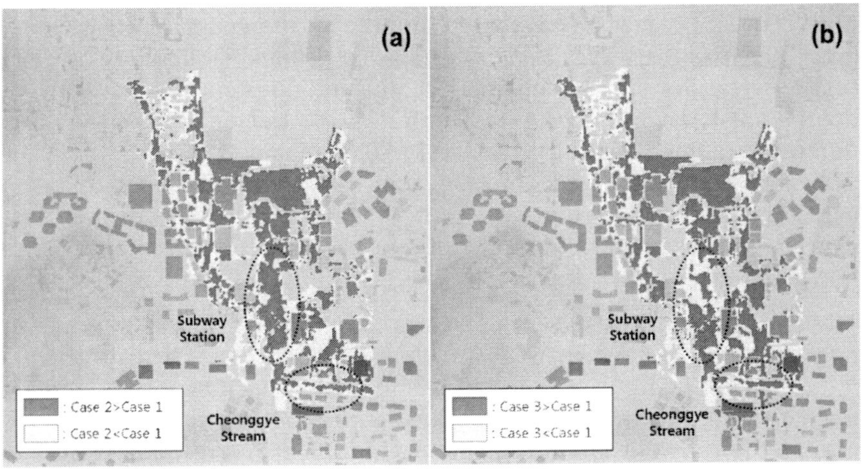

Figure 14. Difference of flood velocity for: (**a**) Case 1–Case 2; and (**b**) Case 1–Case 3.

Table 5. Difference of depth and velocity for each case.

Difference	Depth (m)			Velocity (m/s)		
	Min.	Max.	Mean	Min.	Max.	Mean
Case 1–Case 2	−0.71	0.60	0.006	−1.35	0.77	0.012
Case 1–Case 3	−0.73	0.33	0.004	−1.48	0.74	0.008

4. CONCLUSIONS

Although urban area is a complex topography composed of overland and underground spaces, few studies have been conducted on urban flood analysis in consideration of the two spaces simultaneously. This study proposed two methods of boundary-type and pond-type to link the overland and underground spaces in the 2D flood modeling, and examined the influence of underground space in urban flood inundation. An overflow at the manhole was calculated through sewer flow analysis using SWMM and then this overflow was applied to the 2D overland flood model as boundary condition. Model simulations were conducted in the Gwanghwamun area of the Hyoja drainage basin, Korea where flood inundation occurred on 21 September 2010. The predictions were compared with measured data including water depth in the covered urban streams considered as conduit in this study, and flood depth and extent in overland. The main study results are as follows:

1) The water depths measured at two covered streams were used for calibration of the conduit roughness coefficient in the SWMM. When nm = 0.02 m−1/3s was applied, a RMSE and RPE of the calculated and observed water depth at two streams were about 2.32 cm and 3.92%, respectively, showing the highest accuracy. In addition, surface roughness coefficient of 2D flood model was calibrated by minimizing the difference between measured and calculated flood extent, and nm = 0.025 m−1/3s presenting the highest goodness of fit (74.2%) was selected as an optimal roughness coefficient. In addition, in order to overcome the problem of the measured flood extent not being accurately investigated, flood damage report regions of residents, press media articles, and photos showing real situations as well as measured flood depth were additionally used.

2) The boundary-type (Case 2) and pond-type (Case 3) to link overland and underground spaces by considering floodwater flowed into the underground space were proposed. The predictions of Cases 2 and 3 were compared with those of Case 1 considering only overland flow to examine the effect of underground space in the urban flood

modeling. Regardless of the consideration of the underground space, mean flood depth, mean velocity, and maximum flood depth showed similar results, whereas flood extent and maximum velocity of Cases 2 and 3 were approximately 12% smaller and 30% faster, respectively, than those of Case 1.

3) Both of the boundary-type and pond-type presented more accurate in the predictions of flood extent and flood depth as compared with Case 1 not considering underground space. A model execution time of the boundary-type was similar to that of Case 1, whereas the pond-type took more execution time than the other cases. These results indicate that the boundary-type is simple but robust method with high computational efficiency for simultaneous consideration of overland and underground space in urban flood modeling. Thus, the boundary-type in urban flood modeling is expected to be usefully applied when accurate and fast flood information is required regionally such as urban flood responses, measures and planning or flood insurance.

ACKNOWLEDGMENTS

This research was supported by a grant (16AWMP-B079625-03) from Water Management Research Program funded by Ministry of Land, Infrastructure and Transport of Korean government.

REFERENCES

1. Djordjević, S.; Prodanović, D.; Maksimović, Č. An approach to simulation of dual drainage. *Water Sci. Technol.* **1999**, *39*, 95–103.
2. Mark, O.; Weesakul, S.; Apirumanekul, C.; Aroonnet, S.B.; Djordjević, S. Potential and limitations of 1D modelling of urban flooding. *J. Hydrol.* **2004**, *299*, 284–299.
3. Djordjević, S.; Prodanović, D.; Maksimović, C.; Ivetić, M.; Savić, D. SIPSON-Simulation of interaction between pipe flow and surface overland flow in networks. *Water Sci. Technol.* **2005**, *52*, 275–283.
4. Schmitt, T.G.; Thomas, M.; Ettrich, N. Assessment of Urban flooding by dual drainage simulation model RisUrSim. *Water Sci. Technol.* **2005**, *52*, 264–274.
5. Maksimović, Č.; Prodanović, D.; Boonya-Aroonnet, S.; Leitão, J.P.; Djordjević, S.; Allitt, R. Overland flow and pathway analysis for modelling of urban pluvial flooding. *J. Hydraul. Res.***2009**, *47*, 512–523.
6. Leandro, J.; Chen, A.; Djordjević, S.; Savić, D.D.A. Comparison of 1D/1D and 1D/2D coupled (Sewer/Surface) hydraulic models for urban flood simulation. *J. Hydraul. Eng.* **2009**, *135*, 495–504.
7. Pina, R.D.; Ochoa, S.; Simões, N.E.; Mijic, A. Semi- vs. Fully-distributed urban stormwater models: Model set up and comparison with two real case studies. *Water* **2016**, *8*, 58.

8. Casas, A.; Benito, G.; Thorndycraft, V.R.; Rico, M. The topographic data source of digital terrain models as a key element in the accuracy of hydraulic flood modelling. *Earth Surf. Process. Landf.* **2006**, *31*, 444–456.
9. Van de Sande, B.; Lansen, J.; Hoyng, C. Sensitivity of coastal flood risk assessments to digital elevation models. *Water* **2012**, *4*, 568–579.
10. Shoji, F.; Mikio, K. Prediction of flood-induced flows in urban residential areas and damage reduction. In Proceedings of the International Workshop on Floodplain Risk Management, Hiroshima, Japan, 11–13 November 1996.
11. Mirei, S.; Juichiro, A. Numerical and experimental study on 2-D flood flows with and without structures. *J. Hydraul. Eng.* **2003**, *129*, 817–821.
12. Alcrudo, F. *Mathematical Modelling Techniques for Flood Propagation in Urban Areas*; IMPACT Project Technical Report: Wallingford, UK, January 2004.
13. Marks, K.; Bates, P. Integration of high-resolution topographic data with floodplain flow models. *Hydrol. Process.* **2000**, *14*, 2109–2122.
14. Vojinovic, Z.; Seyoum, S.D.; Mwalwaka, J.M.; Price, R.K. Effects of model schematization, geometry and parameter values on urban flood modelling. *Water Sci. Technol.* **2011**, *63*, 462–467.
15. Fewtrell, T.J.; Duncan, A.; Sampson, C.C.; Neal, J.C.; Bates, P.D. Benchmarking urban flood models of varying complexity and scale using high resolution terrestrial LiDAR data. *Phys. Chem. Earth A/B/C* **2011**, *36*, 281–291.
16. Turner, A.B.; Colby, J.D.; Csontos, R.M.; Batten, M. Flood modeling using a synthesis of multi-platform LiDAR data. *Water* **2013**, *5*, 1533–1560.
17. Mason, D.C.; Giustarini, L.; Garcia-Pintado, J.; Cloke, H.L. Detection of flooded urban areas in high resolution synthetic aperture radar images using double scattering. *Int. J. Appl. Earth Obs. Geoinf.* **2014**, *28*, 150–159.
18. Leitão, J.P.; Moy de Vitry, M.; Scheidegger, A.; Rieckermann, J. Assessing the quality of digital elevation models obtained from mini unmanned aerial vehicles for overland flow modelling in urban areas. *Hydrol Earth Syst. Sci.* **2016**, *20*, 1637–1653.
19. Chen, A.S.; Evans, B.; Djordjević, S.; Savić, D.A. Multi-layered coarse grid modelling in 2D urban flood simulations. *J. Hydrol.* **2012**, *470–471*, 1–11.
20. Yu, D.; Lane, S.N. Urban fluvial flood modelling using a two-dimensional diffusion-wave treatment, part 2: Development of a sub-grid-scale treatment. *Hydrol. Process.* **2006**, *20*, 1567–1583.
21. Yu, D.; Lane, S.N. Interactions between subgrid-scale resolution, feature representation and grid-scale resolution in flood inundation modelling. *Hydrol. Process.* **2011**, *25*, 36–53.
22. Fewtrell, T.J.; Bates, P.D.; Horitt, M.; Hunter, N.M. Evaluating the effect of scale in flood inundation modelling in urban environments. *Hydrol. Process.* **2008**, *22*, 5107–5118.
23. McMillan, H.K.; Brasington, J. Reduced complexity strategies for modelling urban floodplain inundation. *Geomorphology* **2007**, *90*, 226–243.
24. Soares-Frazao, S.; Lhomme, J.; Guinot, V.; Zech, Y. Two-dimensional shallow-water model with porosity for urban flood modelling. *J. Hydraul. Res.* **2008**, *46*, 45–64.
25. Cea, L.; Vázquez-Cendón, M.E Unstructured finite volume discretization of two-dimensional depth-averaged shallow water equations with porosity. *Int. J. Numer. Methods Fluids* **2009**, *63*, 903–930.
26. Guinot, V. Multiple porosity shallow water models for macroscopic modelling of urban loods. *Adv. Water Res.* **2012**, *37*, 40–72.
27. Kim, B.; Sanders, B.F.; Famiglietti, J.S.; Guinot, V. Urban flood modeling with porous shallow-water equations: A case study of model errors in the presence of anisotropic porosity. *J. Hydrol.* **2015**, *523*, 680–692.
28. Özgen, I.; Zhao, J.; Liang, D.; Hinkelmann, R. Urban flood modeling using shallow water equations with depth-dependent anisotropic porosity. *J. Hydrol.* **2016**, *541*, 1165–1184.
29. Abdullah, A.F.; Vojinovic, Z.; Price, R.K.; Aziz, N.A.A. A methodology for processing raw LiDAR data to support urban flood modelling framework. *J. Hydroinform.* **2012**, *14*, 75–92.

30. Abdullah, A.F.; Vojinovic, Z.; Price, R.K.; Aziz, N.A.A. Improved methodology for processing raw LiDAR data to support urban flood modelling—Accounting for elevated roads and bridges. *J. Hydroinform.* **2012**, *14*, 253–269.
31. Shon, T.S.; Kang, D.H.; Jang, J.K.; Shin, H.S. A study of assessment for internal inundation vulnerability in urban area using SWMM. *J. Korean Soc. Hazard. Mitig.* **2010**, *10*, 105–117.
32. Kim, B.; Kim, T.H.; Kim, J.H.; Han, K.Y. A well-balanced unsplit finite volume model with geometric flexibility. *J. Vibroeng.* **2014**, *16*, 1574–1589.
33. Seoul Metropolitan City (SMC). *Environment White Paper: Environment of Seoul*; Seoul Metropolitan City Press: Seoul, Korea, 2014.
34. Kim, H.S. A Study on the Decision of Design Magnitude for Flood Control of Urban Basin Based on Flooding Characteristic Values: A Case Study on the Hyoja Drainage Basin, Located in Cheonggyecheon Basin in Seoul. Master's Thesis, University of Yonsei, Seoul, Korea, 2012.
35. Lee, J.T.; Hur, S.C.; Kim, T.H. Sensitivity analysis of the SWMM model parameters based on design rainfall condition. *J. Korea Water Resour. Assoc.* **2005**, *38*, 213–222.
36. Kim, B.; Sanders, B.F.; Han, K.Y.; Kim, Y.J.; Famiglietti, J.S. Calibration of stormwater management model using flood extent. *Proc. ICE Water Manag.* **2014**, *167*, 17–29.
37. Korean Water Resources Association (KWRA). *River Design Criteria*; Korean Water Resources Association: Seoul, Korea, 2009.
38. Yi, C.Y.; An, S.M.; Kim, K.R.; Choi, Y.J.; Dieter, S. Improvement of air temperature analysis by precise spatial data on local-scale: A case study of Eunpyeong Newtown in Seoul. *J. Korean Assoc. Geogr. Inf. Stud.* **2012**, *15*, 144–158.
39. Urban Flood Disaster Management Research Center (UFDMRC). *A Guideline on Flood Estimation in Urban Area: FFC06-1*; Urban Flood Disaster Management Research Center: Seoul, Korea, 2008.
40. Bates, P.D.; Dawson, R.J.; Hall, J.W.; Horritt, M.S.; Nicholls, R.J.; Wicks, J.; Hassan, M.A.A.M. Simplified two-dimensional numerical modelling of coastal flooding and example applications. *Coast. Eng.* **2005**, *52*, 793–810.
41. Schubert, J.E.; Sanders, B.F. Building treatments for urban flood inundation models and implications for predictive skill and modeling efficiency. *Adv. Water Resour.* **2012**, *41*, 49–64.
42. Smith, R.A.E.; Bates, P.D.; Hayes, C. Evaluation of a coastal flood inundation model using hard and soft data. *Environ. Model. Softw.* **2012**, *30*, 35–46.
43. Wadey, M.P.; Nicholls, R.J.; Hutton, C. Coastal flooding in the Solent: An integrated analysis of defences and inundation. *Water* **2012**, *4*, 430–459.

CHAPTER 8

Landscape Engineering, Protecting Soil,and Runoff Storm Water

Mehmet Cetin

[1] *College of Engineering, Department of Civil and Environmental Engineering, Temple University, Philadelphia, Pennsylvania, USA*

1. INTRODUCTION

Landscape engineering thinks about the application of mathematics and science to the creation of convenient outdoor living areas. These outdoor living areas are a consequence of the design and the construction process made feasible by landscape architects through landscape contractors. Beside landscape engineer's interested in containing site grading and drainage, earthwork calculations, and watersheds [1,12,13].

Landscape engineers employ engineering knowledge when designing and building spaces. They demand to know how to interpret contour maps which shows elevations and surface configuration by means of contour line sand and also consider how to interpret 2-D images, compute angles and grading requirements for pavements, parking lots, bridges, roads and other structures. Beside they understand the amount of fill needed for specific areas and figure out how water runoff and flow should affect their designs.

Developing and improving for landscape engineering [1,3,13,14]

1) Stormwater management: building up bioswales, landscape materials used to gather and install runoff, rain gardens and porous asphalt.

2) Mitigation of the urban heat island effect: diminishing the quantity of paved surfaces, building up green roofs and green walls, and using building materials with low reflectivity.

3) Wildlife habitat: protecting habitat and building up green roofs and rain gardens.

4) Social spaces: areas for walking, biking, gathering and eating.

5) Transportation: growth expanding pedestrian accessibility, limiting vehicle speeds and encouraging the use of public transportation.

All above are using for the mixing plastic with asphalt on the pavement for the above reasons. One of them we are using porous plastic asphalt on the pavement for diminishing the sounds of road, and protecting the soil, and making runoff out the structure.

Porous plastic pavement which is a permeable pavement surface permits the action of storm water in order to infiltrate directly into the soil. It usually constructs with an underlying stone supply that temporarily stocks surface runoff before infiltrating into the subsoil. It takes the place of traditional pavement. There are various types of porous surfaces, including porous asphalt, permeable concrete and even grass or permeable pavers.

Porous plastic asphalt pavements suggest an alternative technology for stormwater management. It is varied from classical asphalt pavement designs so that the structure enables fluids to come frankly through it, decreasing or monitoring the amount of run-off from the surrounding area. By permitting precipitation and run-off to running out the structure, this pavement kind functions as an additional stormwater management technique. The mainly benefits of porous plastic asphalt pavements should contain both environmental and safety benefits including improved stormwater management, improved skid resistance, decreasing of spray to drivers and pedestrians, as well as a potential for noise decreasing. Porous plastic asphalt is applicable to many uses, including parking lots, driveways, sidewalks, bike paths, playgrounds and recreational courts. In addition, with proper maintenance, including regular vacuuming or pressure washing of the pavement surface to prevent clogging by sediments, porous asphalt can have a minimum service life of twenty years [15,16,17,18].

Bitumen is a valuable binder for road construction. Different grades of bitumen which is 30/40, 60/70 and 80/100 are available on the basis of their penetration values. The steady growth in high traffic intensity in terms of commercial vehicles, and the important variation in daily and weather temperature demand developed road characteristics.

Nowadays the waste plastics are to exist tremendous, as the plastic elements have been piece of daily life for using. They either blended with Municipal Solid Waste and/or dispose of landfill. If not recycled, their present disposal is either by land filling that the method has certain impact on the environment. Because of that an alternate use for the waste plastics is also the needed.

Thinner polythene carry bags are most adequately disposed of wastes, which do not draw attention the attending rag pickers to accumulate for forward recycling, for lesser value. These polythene bags are simply suitable with bitumen at specified conditions. The waste polymer bitumen mix may be prepared and a study of the properties can throw more light on their use for road laying.

Permeable pavements have been sufficient in behalf of environmental problems and corroborating sustainable green construction procedures. They are planned to support surfaces for parking lots and pedestrian roads that will permit some of the precipitation to filter into the ground, decreasing the bulk of stormwater runoff and revitalizing groundwater [15,16,17]. Permeable porous asphalt and concrete are the most ordinarily used elements in permeable pavements. Permeable asphalt uses practically the similar elements as conventional asphalt, order than the percent range of plastic added, and the size of the aggregate is used to stay thin, permitting for minimum particle packing [16,17]. Permeable asphalt and concrete has been successfully implemented for road, retaining walls, streets, driveways, sidewalks, parking lots, and slope protection for the past 30 years in many countries, including the United States [17]. None of any researchers, to the authors' knowledge, have inquired for mixing permeable plastic asphalt so far that implement plastic waste as a binder bonder for permeable pavements. This study presents an alternative sustainable technology, familiar as permeable plastic asphalt, which should be implemented for permeable pavements. Permeable plastic asphalt material is produced from plastic waste, porous aggregates, and asphalt. The research evaluated the mechanical and hydraulic characteristics of permeable plastic asphalt.

An asphaltic paving material involves 4%, 5% and 6% percent, of granular recycled plastic (LDPE with 1%, 3%, and 6%), which supplements the porous aggregate component (aggregate size of 3/8, 4, 8, 16) of the mixture. The material produces a structurally superior paving material and longer lived roadbed. The plastic can contain any and all residual classes of recyclable plastic. The material produces roadbeds of higher strength with less total asphalt thickness and having greater water permeability, and is most useful for all layers below the surface layer. A process of shredding or mechanically granulating preferably forms the paving product.

A porous plastic asphalt pavement has benefit for using road users and road side environments such as decreasing road noise, developing drainage function, and driving security situation. Furthermore, the worth of porous asphalt is to support skid resistance, particularly in the wet weather, that is distinctly better than that of dense traditional asphalt [19].

Conventional asphalt was not usually acceptable in the high temperature, humid, and traffic. As a result that some of issues were occurred like the pores were clogged, strength of drainage, rutting and scattering came off by traffic loading [20]. Large percentage of air voids of the mixture is adopted in order to maintain the drainage function. Nevertheless, in the way that the proportions of air void grow, strength feature of the mixture diminish. This is the main theorem of asphalt mixture. For solving this, landscape engineers think about the plastic for design and planning as binder.

The aim of this research is to assess strength, which means the performances of porous asphalt mixtures, for testing Resilient Modulus to find Indirect Tensile Strength (ITS). Furthermore, it measures the durability of porous asphalt mixture using Marshall Immersion test.

It is the aim of this chapter (book) to outline the elementary essential parts fundamentals associate with the design, plan, and construct of pavements and to mix with plastic techniques that will allow a Landscape Engineer to plan and design a pavement to suit a variety of situations. After experiments next books chapters will be considered to help environment and nature same as protect soil and recycling. Next study will keep going to research the better experiments about permeable surfaces for using different plastic materials. As a result it gains economic, environmental, and practical design and plan for Landscape Engineering.

1.1. Landscape engineering

Landscape Engineering defined that is the art of developing land for people use and entertainment in such a manner as to obtain the utmost utility including with the utmost of beauty. It is a mistake to take into consideration the subject as implicated mainly with planting trees or to visualize its main function to be the supply of some decorative camouflage for some unsightly utility. Conversely, it has to be comprehended as a most important fundamental to life, and all art, which no utility would be necessary camouflage, and that every kind of artistic proceedings, instead of being without deep, has to be natural structural. Landscape engineering verifies the fundamental utilities, not as a necessary bad, but as necessarily well. In lieu of these utilities existing in the way of the objects, which the landscape engineer is succeed in doing, they turn out a most sufficient part of her/his own initiative. Apparently, once this point of view is admitted the Landscape Engineer and the others are definitely suiting for the same thing. Landscape Engineering impresses the main ideas, concepts, and techniques that cope with the functional, visual, and ecological perspectives of grading and landform cultivation. Landscape Engineering points out introduction to the processes, principles, and techniques of site engineering [1,3,13, 21,22,23].

Students who pursue to study their career in Landscape Engineering programs learn about Landscape Design and Planning, Transportation Plan, Structural Landscape Design, Urban Planning, Site Planning, Ecological Design, Environmental Design, and Horticulture. Essential aims of all of subjects in Landscape Engineering pay attention to cover sustainability, and eco-friendly landscaping processes. They usually not only study with their specific area of studies like sustainable urbanism, environmental hazard management, historic preservation, ecological design but also focus in land development along with construction management, or a compound of land development and architecture [2,14].

Landscape engineer is that construction is the first step after designing and planning. Landscape engineer have to be continually consider to reach suitable adjustments between the operating cost, the construction cost, and the maintenance cost. For only during the construction period can the required savings be influenced at a minimum of expense. Landscape engineer think about completed without mention of subgrade conditions or specifications on material and method of placing while designing and planning pavement, parking lot, etc. Drains are often paid attention as releases for storm water during the catch basins although the drain is laid to accumulate the soil. With new experiment in planning is the topographical survey, usually worked out with care and precision during considers drain and protects soil.

Landscape engineer has to keep in mind when planning the pavement that it is a business proposition as well as a picture designing and planning. That the Landscape planned with an eye to operation and maintenance cost must in time, have the better financial situation to protect appearance. Meanwhile design and plans are begun; we have to never forget that budget is a factor in maintenance, whether the project is considered on the constant care principle. The complete design of the pavement is that while every element of pavement construction must be considered from the point of view of beauty and aesthetic value, the weight of construction, operation and maintenance must be found and recognized at the time of beginning [3,22].

A Landscape Engineer conducts and takes advantage of the strengths of nature for the benefit and satisfaction of man. For instance, if you were designing and planning a new landscape for your property and wanted to direct the winds to preserve your amusing areas, a Landscape Engineer would help you make a determination which trees to get and where to plant them; thus, the wind could avoid your patio or courtyard. Landscape engineering involves with the application of mathematics and science to the constitution of practical outdoor living areas. These areas are a conclusion of the design and the construction process made possible by landscape architects, along with landscape contractors. Other main contemplates of landscape engineer's include drainage and site grading, watersheds and earthwork calculations. It is the Landscape engineer's act to vigorously interest in the design of the landscape, hence, to significantly supervise the creation of the landscape design. Landscape engineering exemplifies the traditional engineering elements of planning, operation, management, design and construction, and assessment. Before planning and designing, Landscape engineer should consider three main areas that focus on [4,5,24]:

1) Landscape plan and design that Landscape Engineers interest planning the individual landforms so that they go along with the objectives arrange in the termination design phase. It is necessity that the Landscape engineer confers with the rest of the architectural team for carrying out a sufficient visual result once the construction process is finished.

2) Termination design and plan that it implies setting targets as well as supporting the Landscape design of the project at hand. Landscape engineers who works very closely with the contractors and sub-contractors does for working with Landscape designers as well as the owner of a home or property for deciding what the desired look is considered together and the essential steps to be taken in order to success it.

3) Functioning assessment that the reason of performance evolution of the Landscape engineer results in estimating liability and financial assurance of the landscape design and plan project. Thus, performance evaluation is vital to both the closure planning progress and performance evaluation.

1.2. Soil protection

As we preserve air and water, which we use for breathing and drink, the soil is very important as well. The quality of soil takes care of ecosystems. Soil quality, which would be linked to water quality, is a measure of soil productivity. As a determination soil quality which particularly, the bulk of soil to operate within ecosystem borders to endure keep up environmental quality, biological productivity, and support plant and animal health [25,26,27].

Soil structure and water infiltration is very important for soil quality as well as aim attention at characteristics like organic matter content overall soil biological activity, nutrient availability, and total organic matter levels [25,27]. Soil can be protected from erosion, landslide, and compaction by mixing plastic materials. Soil conveys very easy underground by water. Soil, which is ancient, rock broken into sand, silt and clay is the recycled and continuously transforming.

The one of ways for protect and sustainably improve of soils is mixed cover pavement with plastic materials for prevent erosions of soil. Soil tests should be recommended for mixing with compaction plastic. It is evaluated compliance with hydrologic characteristics like drainage from plastic compaction.

1.3. Stormwater runoff

Stormwater runoff comes off while precipitations from rain or snowmelt infiltrate the ground. Impermeable surfaces like pavement, driveways, parking lots, sidewalks, and streets restrain stormwater runoff from naturally permeate into the ground.

For understanding the impact of stormwater runoff firstly need to consider how important water cycles through the urban environment. The bulks of precipitation infiltrate forests flows slowly underground, are infiltrated in by natural progress, and eventually arrive at lakes, streams while is being its natural. When we design and plan to compress the land with roads, parking lots, and buildings, the natural progress of water infiltrate in the earth is diminished. The existing grasslands and forests are put in place of concrete, roofs, and asphalt that do not permit rain to infiltrate the earth. Rather than, the precipitations soak through as faster as directly into streams, storm drains, and all without the profit of filtration. Designing, planning and constructing with mixing plastic material help to infiltrate water on the driveway, parking lots, pavement in place of concrete or asphalt. Thus, they permit stormwater to quickly drain into the ground [28,29,30].

Stormwater is water that directly results from a rainfall event is not absorbed into soil and rapidly flows downstream, increasing the level of waterways. The flow of water results from precipitation and that occurs immediately following rainfall or as a result of snowmelt. Stormwater is the portion of rainfall that does not infiltrate into the soil. Rainwater and snowmelt that runs off impermeable surfaces rather than infiltrate into the soil through a drainage system of underground pipes, stormwater carries nutrients, fine soils, plant debris, drippings from vehicles, and other substances from the drainage basin most of lakes, ponds, and wetlands are connected to the stormwater system. Water collected in a system of pipes which drain roads and industrial or trade premises Stormwater may contain contaminants present on drained surfaces.

Stormwater is concerning about two issues that are the volume and timing of runoff water and prospective polluting. Stormwater is needed to flood control and water supplies. Also stormwater lead to be water pollution for conveying the water. As a stromwater, watermanagement on the pavements should probably do urban environments self-sustaining in terms of water. Stormwater is pollution because impermeable surfaces like parking lots, roads, buildings, compacted soil do not enable rain to drain into the ground. More runoff is constituted than in the unprogressive condition. Thus, it should consume waterways like flooding after stormwater collection system is overwhelmed by the additional flow. So the water is out of watershed way through little drainage the soil, the storm event [31,32].

Stormwater is a problem so that it could collect chemicals waste, mud, dirt, and other pollutants and infiltrate in the storm sewer system or directly to a river, lake, coastal water, stream, or wetland. Anything, which inserts a storm sewer system, is released untilled into the waterbodies we run for fishing, swimming, and supplying drinking water.

The results of pollution of stormwater runoff could have many unfavorable effects on animals, plants, people, and fish. Sediment that could blur the water and make it complicated or unimaginable for aquatic plants to grown could demolish aquatic habitats. The pollution of stormwater usually influences drinking water sources, which could transform to imitate human health and grow drinking water treatment costs.

The solution of pollution of stormwater that is permeable pavement consists of mixing plastic. Traditional concrete and asphalt don't tolerate water to infiltrate into the ground. Rather than these surfaces depend on storm drains to switch unwanted water, Permeable pavement systems permit rain and snowmelt to saturate with, diminishing stormwater runoff [28,32]. Stormwater runoff is without filtered water that arrives at oceans, streams, and lakes by means of streaming on impermeable surfaces that contain driveways, parking lots, roads, and roofs.

1.4. Correlation between impermeable pavement and permeable runoff

Impermeable pavement in a watershed occurs in growth permeable runoff. The barely 10 percent impermeable pavement in a watershed would occur in stream decline.

Stormwater is pollution so that impermeable surfaces like parking lots, compacted soil, roads, buildings, do not permit rain to flows off from the land in the streams, further runoff is caused to be than in the immature condition [28,31].

This further runoff would spoil streams and rivers as well as bring about flooding after the stormwater system is overflow by the extra flow so that the water is reveled out of the watershed through the storm event, barely drains the soil, fills groundwater, or stocks stream baseflow in dry weather. Contaminant inserting surface waters during rain that is lead to contaminate runoff. Daily people activities appear in sediment of pollutants on parking lots, farm fields, driveways, roads, lawns, roofs. As soon as precipitation begins, water flow off and eventually makes its way to a lake, ocean, and river.

A traditional city block creates more than five times more runoff than the forest because of impermeable pavement. The waste of penetration from city may come out with depth groundwater changes [28,31,32].

The present drainage systems which accumulate runoff from impermeable surfaces like roads, parking lots, roofs insure that water is effectively transported to ways of water during pipes. As a matter affects little storms water occur in growth ways of water flows. Stomwater lead to some of issue that shows below.

1) Impermeable Pavement
2) Roads, sidewalks, rooftops, overly compacted soils
3) Do not allow for natural infiltration of stormwater
4) Increase temperatures (Heat Island Effect)
5) Degradation of water quality and natural habitats
6) Flooding, erosion and may reduce groundwater levels

1.5. Class of plastic for using contemporarily

The simplest way for a user to label the class of plastic used in a product is to recognize the resin identification code, which is familiar with the material container code as well, is generally plotted, shaped or symbolized in or close to the middle of the bottom of the product. In accordance with the society of the plastics industry (SPI) resin identification code (or material code) is to systematize mutual plastic resins and their characteristics.

The identification of plastics made known in 1988 by The Society of the Plastics Industry Trade Association (SPI) that provides to build the companies for easy recycles make collect the consumer plastics during the common pathways for coming together recyclable stuffs from household waste. It is based on willing for plastic manufacturers, however, it has evolved into comparatively standard on plastic products sold in the U.S. and internationally. For example, the identification of plastics is in service and is affirmed by the Canadian Plastics Industry Association (CPIA) in Canada. It supports specifics on the identification by mean of its Environment, health and safety strategic unit and its Environment and Plastics Industry Council (EPIC) [7,33,34].

The aim of the identification provides to make it simpler for recycling to plastics. Furthermore, it determines consumers with an easy, handy method for identifying the class of plastic resin used to create a specific product. In conformity with SPI identifications, the number is intentionally located in an unnoticeable place on the product so that the company purpose is not to affect the consumer's buying determination, barely to assist the progress of recycling of the product.

According to SPI, there are seven different classes of plastics. Showing Figure 1, the identification numbers imprint on the bottom of plastic products that is a number inside of triangle represents to mean their identifications for recycling. Have you ever been curious about what the numbers inside the little recycling symbol mean on all of the plastic packaging and plastic products which we consume for using?

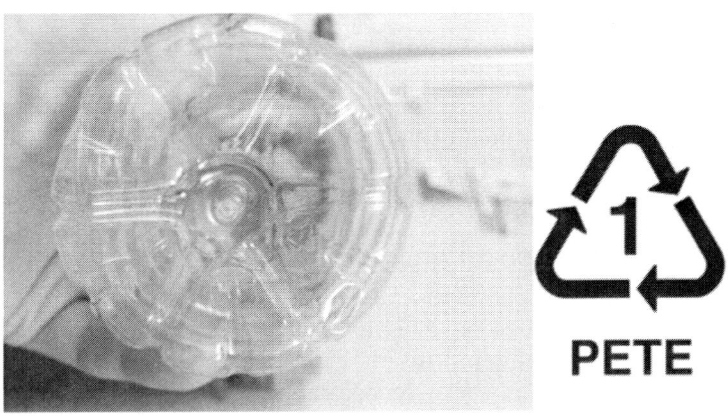

Figure 1. Samples of identification on the bottom of plastic water bottles

According to SPI the identification in 1988 reciprocates to the consider revising of the plenty recyclers side to side the countries. Here each class of plastics number and definition [7].

1) PETE, PET (Polyethylene Terephthalate) is clarity, strength or toughness, barrier to gas and moisture, resistance to heat. It uses for consuming plastic soft drink and water bottles, beer bottles,

mouthwash bottles, peanut butter and salad dressing containers, oven able film, oven able pre-prepared food trays.

2) HDPE (High Density Polyethylene) is stiffness, strength or toughness, resistance to chemicals and moisture, permeability to gas, ease of processing, and ease of forming. It uses to make plenty classes of bottles. The bottles are clear, have good limit qualities and stiffness, and are quite appropriated to packaging products with a short shelf life like milk so that HDPE has good chemical resistance; it is used for packaging many household and industrial chemicals such as detergents and bleach. It uses milk, water, juice, cosmetic, shampoo, dish and laundry detergent bottles, trash and retail bags, yogurt and margarine tubs, cereal box liners.

3) V, PVC (Vinyl, Polyvinyl Chloride) is versatility, ease of blending, strength or toughness, resistance to grease or oil, resistance to chemicals, clarity. It has well chemical resistance, weather ability, flow typical features and constant electrical qualities. Products made from Vinyl can be both flexible and rigid. It uses toys, clear food and non-food packaging, shampoo bottles, medical tubing, wire and cable insulation, film and sheet; construction products such as pipes, fittings, siding, flooring, carpet backing, window frames.

4) LDPE (Low-Density Polyethylene) is ease of processing, barrier to moisture, strength or toughness, flexibility, ease of sealing. It is used efficaciously in film uses because of its flexibility toughness, and approximate transparency, making it familiar for use in uses that heat sealing is essential. Furthermore, LDPE uses to procedure some flexible lids and bottles as well as in wire and cable uses. It uses squeezable bottles (honey, mustard), coatings for paper milk cartons and hot and cold beverage cups, container lids, toys, dry cleaning, bread, and frozen food bags.

5) PP (Polypropylene) is strength or toughness, resistance to chemicals, resistance to heat, barrier to moisture, versatility, and resistance to grease or oil. It has good chemical resistance, is strong, and has a high melting point making it well for hot-fill liquids. This resin is brought to light in rigid and flexible packaging, fibers, and large pattern parts for automotive and consumer products. It uses containers for yogurt, margarine, takeout meals and deli foods, medicine bottles, bottle caps and bottles for ketchup. Furthermore, for packaging, its plenty of uses are in fibers, appliances and consumer products, containing strong applications like automotive and carpeting.

6) PS (Polystyrene) is versatility, insulation, and clarity, easily foamed as known "styrofoam". It is clear, hard and brittle. Also, it has an approximately low melting point. General uses contain protective

packaging, food packaging, bottles, and food containers. It is usually connected with rubber to make high impact polystyrene (HIPS) that is used for packaging and constant uses necessity stiffness. It uses compact disc cases, food- service applications, grocery store meat trays, egg cartons, aspirin bottles, cups, plates, and cutlery.

7) Other is dependent on resin or combination of resins. Use of this number represents which a package is made with a plastic other than the six listed above, or is made of more than one plastic and used in a multi-layer combination. It uses usually shows the exits of polycarbonate which a hard, clear plastic used to make baby bottles, water pitchers, nalgene brand water bottles, three and five-gallon reusable water bottles, food containers, some citrus juice and ketchup bottles, compact discs, cell phones, automobile parts, computers, three and five-gallon reusable water bottles, some citrus juice and catsup bottles, oven-baking bags, barrier layers and custom packaging.

2. DEVELOPMENT OF DESIGN

2.1. Design factors

The area of pavement design is vigorous in which ideas are steadily changing as new data evolve into achievable. For mixing plastic pavement that are many majority of design and plan applicable, since alternatives relating to sustainability, suitability of designs and plans alter from area to area. Especially, supplies that are applicable for construction and foundation of pavements have a higher impress on design and plan. There are, nevertheless, fundamentals of design that are mutual to all problems irrespective of other uncontrollable situation [35,36,37,38,39].

The plan and design of pavement embrace with a work of soils and paving materials, their action under load, and the plan and design of pavement to convey which load under all hydraulic and weather situations. All pavements obtain their eventual support from the underlying subgrade. As a result, a comprehension of elementary plastic materials, pavement design and soil mechanics is necessary. Landscape Engineers are familiar that efficiency of pavement that are connected to a large volume upon the types of plastic, soils over that the pavement is designed and constructed, therefore, in relationships pavement efficiency between subgrade types are built. On the whole, the experiments of mixing plastic pavement demonstrated that pavement designed and constructed over plastic displayed higher degrees of distress than those designed and implemented over traditional pavement. Frost process and unfavorable drainage situations were observed early as two of the primary reasons of pavement lapse.

However, many landscape engineers are made use of standard cross components for most pavements. It means that a road, even though it crossed several mixing plastic and soil types, was designed and implemented using a constant thickness. The foreseeing was usually confirmed on the rest of economics. Beginning of 1980s, people recognized that the traditional pavement affects the environment and nature. All of thriving technological elements such as cars, bike for becoming a simple life causes to influence the pavements lapse [40,41,42,43].

2.2. Conventional pavements

AASHTO have been in charge for various tests, which roads designed and implemented in United States as well as some of state highway departments have implemented test pavements for the aim of assessment the influence of load and elements on pavement design.

There is a small suspicion that the outcomes of test results have had extreme effect on current design ideas. Furthermore, efficiency of example pavements in service has had important effect on design. This is not shocking, if one thinks that it is hard to do if not unreasonable to judge entirely design ideas in the laboratory. In addition, it has been familiar with for quite a while that reader belief in the final analysis imposes the competency in some of the design.

2.3. Definition pavement types

Basically, with regard to history of pavements have been classified two main types which are flexible and rigid. Flexible pavements comprise asphalt. Nowadays flexible pavement is very important with mixing plastic so that makes permeable areas.

Plan, design, and construction of permeable pavements have altered rather importantly in the last decades. On account of the current traditional plan, design, and construction pavements come up severe higher traffic levels, wheel loads, pavement lapse. A growth use of balanced is base and subbase. Balancers such as asphalt, plastic are repeatedly used to grow the structural strength of the pavement by growth rigorously. Because of the reason an extremely concentrated effort was made in the last several years to develop a more fundamentally based design analysis for asphalt.

2.4. Permeable plastic asphalt

Some of researchers assessed the impact of moisture susceptibility on porous asphalt samples [44]. Samples were based on wet and dry conditions and then tested for indirect tensile strength test (ITS). Results showed that ITS decreased noticeably when the sample was immersed in water.

2.5. Purpose of plastic asphalt

The aims of the study are need to do those. First of all; Disposal of waste plastic is a major problem, non-biodegradable, burning of these waste plastic bags causes environmental pollution. Secondly need is it mainly consists of low-density polyethylene, and to find its utility in bituminous mixes for road construction. Thirdly, Laboratory performance studies were conducted on bituminous mixes. Laboratory studies proved that waste plastic enhances the property of the mix, and improvement in properties of bituminous mix provides the solution for disposal in a useful way.

Waste plastics like polythene carry bags, etc. on heating usually at approximately 160°C. According to thermo gravimetric results has demonstrated that gas evolution isn't found during the temperature rank of 130 to 180°C. Furthermore the mellowed plastics have a binding feature. Therefore, the melted plastics can be implemented as a binder that they should be blended with binder like bitumen to improve their binding feature. As a result it should be a good modifier for the bitumen, implemented for road construction.

2.6. Function of the mix plastic asphalt

The growth of plastic city wastes has affected to wide and creative technologies that incorporate recycled plastics in miscellaneous uses. Scientists and Departments of Transportation have been interested in different researches regarding to the feasibility, economic and ecological impact and the complete efficiency of recycled plastic in connected landscape engineering projects. For instance, recycled thermoplastics like PET, HDPE, and LDPE, have been implemented in porous asphalt mixtures to put in place of aggregates with specified diameters [45,46,47,48,49,50]. Conclusion from these researches demonstrated improvement in strength, durability, and fatigue life. Nevertheless, the scale of improvement is a capacity of the plastic types and amount. The rest of researchers implemented recycled plastic strips to mechanically stabilize and aggregates by mixing plastic shreds with aggregates to compaction to defeat inadequacies in grading and diminishing the plasticity index

[51,52,53]. The scale of enhancement was affected by many factors like the class and volume of shreds, and aggregate classes. Furthermore, creative and innovative study has cause to the development of new mixed elements using recycled plastic waste for miscellaneous uses. The mixed is produced by heating and blending the absorption elements, recycled plastic, which flakes, shared, or unprocessed, granulates and by products to a highlighted temperature. The heated combination is then compacted into a particular mold to found a final product. The features of the mixed rely on the pressure, class of recycled plastic and granulates [54,55,56].

3. MATERIAL PROPERTIES

3.1. Recycled low-density polyethylene (ldpe)

The supplier for Recycled Low Density Polyethylene (LDPE) provided the test properties of the material with respect to density, tensile strength at break, elongation at break, impact strength, and melting point of the material as shown below Table 1 [8,10]. Recycled low-density polyethylene (LDPE), which is identification number four, was gathered together and implemented in this research. The cleaned LDPE has shredded as shown below Figure 2.

Recycled Low-density polyethylene (LDPE) material is used extensively to produce tote bags for domestic goods. These bags become solid waste after their use for short periods and cause serious waste disposal problems. To solve this environmental problem, and at the same time to improve the drain down and other related engineering properties of the porous asphalt mixture, reclaimed from LDPE bags was used in this investigation as additive in porous asphalt mixtures. LDPE material in shredded of used is as added ingredient.

Table 1. Recycled Low Density Polyethylene Features (LDPE)

Recycled Low Density Polyethylene Features (LDPE)	
Mechanical Properties	
Yield Strength	15-20 MPa
Elongation @ break	600-650 %
Bending Strength	10-40 MPa
Young's modulus (E)	200-400 MPa

Shear modulus	100-350 MPa
Tensile Strength (σ_t)	8-12 MPa
Physical Properties	
Density	910-928 kg/m^3
Thermal expansion	150-200 e-6/K
Water absorption	0.005-0.015 %
Melting Point	248 ^{o}F 120 ^{o}C
Thermal conductivity	0.3-0.335 W/m.K
Melting temperature	125-136 ^{o}C
Maximum Temperature	176 ^{o}F 80 ^{o}C
Minimum Temperature	58 ^{o}F 50 ^{o}C
Specific heat (c)	1800-3400 J/kg.K

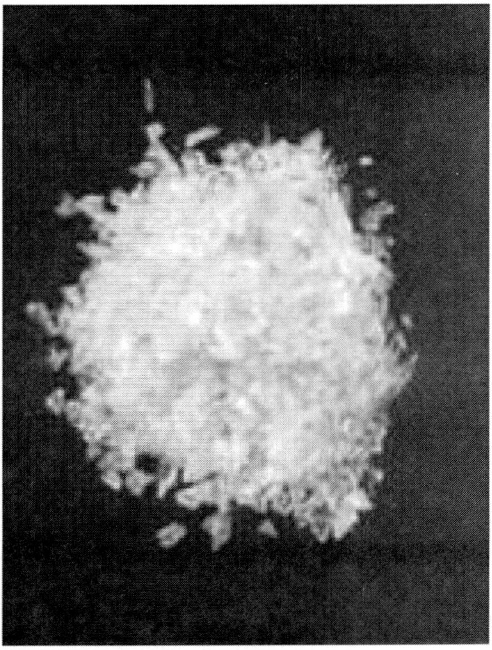

Figure 2.VShredded for recycled Low Density Polyethylene (LDPE)

3.2. Porous aggregate

Crushed limestone was chosen as the course aggregate for mixing LDPE. Bulk samples were sieved in conformity with the sieve sizes for AASHTO No. 8. According to Figure 3 demonstrates the gradation for aggregates that Porous aggregates confirming to the sizes 3/8 in., Nos. 4, 8, 16 (AASHTO No 8) were used for mixing with permeable plastic asphalt. Aggregates maintained on each sieve were washed, dried for 24 hours in 110°C in the oven and then located into their respective batches by sieve maintained. This procedure provided regenerate samples to meet AASHTO No. 8. Furthermore, it made to be better control over the gradation of each sample, so that gradation has important impact on the engineering and physical features of an aggregate mixed.

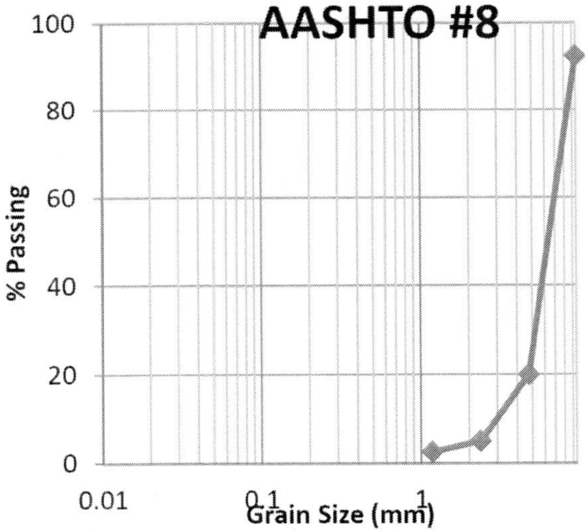

Figure 3. Grain size distribution of AASHTO No. 8

3.3. Bitumen

"PG 68-22" was used in the porous asphalt mixture [9]. Bitumen was mixed with Low-density polyethylene (LDPE) and porous aggregates. Mixes were prepared for three (4%, 5% and 6% bitumen) percentages of bitumen. Each % of bitumen has 1%, 3%, and 6% LDPE. Obviously, four types of mixtures with three different percent of bitumen of 4%, 5% and 6% were used at a mixing and compacting temperature of 160°C. These are: Without LDPE, 1% LDPE, with 3% LDPE, and with 6% LDPE.

4. SAMPLES PREPARATION, COMPACTION, AND TESTS

4.1. Mixing of shredded waste plastic (ldpe), aggregate and bitumen

The aggregate mix is heated to 160⁰C in oven, and similarly the bitumen is to be heated up to a maximum of 160⁰C. Plastic waste is shredding for mixing bitumen and aggregate to coat the plastics effectively. After that, put them in oven 160⁰C in order to mix and compact simultaneously. For protecting the moisture the spacemen, it compacted immediately after took out oven with approximately 160⁰C.

4.2. Mix design by marshall method marshall test

Use of the processed plastic bags is as an additive in bituminous concrete mixes. The processed plastic was used as an additive with heated bitumen in different proportions (ranging from 4 to 6 % by weight of bitumen) and mixed well by hand, to obtain the modified bitumen. The properties of the modified bitumen were compared with ordinary bitumen.

Varying percentages of waste plastic by weight of bitumen was added into the heated aggregates. Marshall sample with varying waste plastic content was tested for stability. Maximum value of stability was considered as criteria for optimum waste plastic content. The optimum modified binder content fulfilling the Marshall Mix design criteria was found to be 4, 5, and 6 % by weight of the mix, consisting of 1,3, and 6 % by weight of processed plastic added to the bitumen. In order to evaluate the ability of the mix prepared with the bitumen to withstand adverse soaking condition under water, Marshall Stability tests were conducted after soaking in water at 60°C for 24 hours.

4.3. Porous plastic asphalt for preparation and compaction

Shedder LDPE was simultaneously composed with binder and aggregates for heating and mixing approximately 160°C for two hours so that a uniform was achieved. Mixed plastic porous asphalt was poured and compacted into a mold, which is 4 inches diameter and 2.5 inches height, using a steel shovel. The Marshall test procedure was used for designing porous mix by compacting the sample with 50 blows on one face by Marshall hammer, at varying binder contents. The mixture design trials used asphalt content in the range of 4 – 6 %, by total weight of the mixture, excluding the weight of the fibers, with 1% increments. The LDPE fibers were added to the porous mixtures at a dosage rate of 1,3, and 6 % based on total mixture weight. The compacted samples were extracted from the mold when they had sufficiently cooled. After compaction, samples were be kept in the hot (conditioned) and cold (unconditioned) waters. Resilient Modulus and permameter test conducted. Samples were tested for hydraulic conductivity and indirect tensile strength. Figure 4 demonstrates the preparation of the Porous plastic asphalt samples.

4.4. Laboratory tests

Porous plastic asphalt samples were tested for hydraulic conductivity and indirect tensile strength. Hydraulic conductivity tests were operated using a falling head approach in conformity with a proceeding particularized in the researchers [57,58]. The indirect tensile tests were conducted in accordance with the ASTM C 6931-07 test methods.

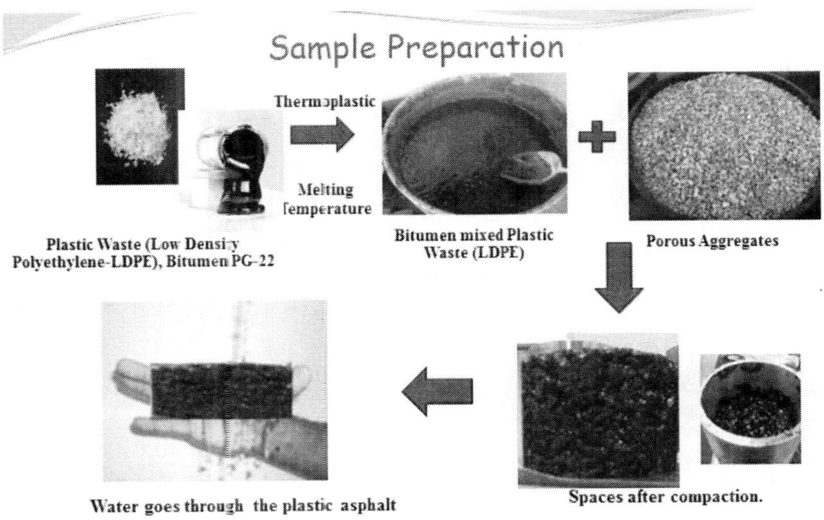

Figure 4. Sample preparation for permeable plastic asphalt that mixing of LDPE, porous aggregates, binder to create

From equation 1 calculate hydraulic conductivity for testing in the falling head permameater test. After used Permameters to measure k, use this following formula. Calculate the hydraulic conductivity of the sediment by using the following formula:

$$k = \frac{VxL}{(h_o - h)Axt} xln \frac{h_o}{h} \quad (1)$$

whereK = hydraulic conductivity of the sediment sample [L]/[T]

V = volume of water that passed through the sample $[L]^3$

L = sample length [L]

h0 = height of top mark above outflow port [L]

h = height of bottom mark above outflow port [L]

A = cross sectional area of sample. For the NEIU permameters, this is 31.65 cm^2. $[L]^2$

t = total time for discharge [T]

The indirect tensile test that is one class of tensile strength test implemented in order to stabilize elements. The test has been run on asphalt-stabilized elements [35,59,60,61]. The test has many advantages, the most obvious being simplicity of test procedure. From equation 2 calculate ITS for testing in the resilient modulus (MR) test. After used with the diametrical Mr test (repetitive indirect tensile modulus test) to calculate ITS, as using the tensile strength St of the material is given by:

$$S_t = \frac{2Pmax}{\pi td} \quad (2)$$

Where P= total applied load (lb)

t= sample thickness (in)

d= sample diameter (in)

5. TEST RESULTS

5.1. Permeability

According to Table 2 demonstrates the hydraulic conductivity (k) of the porous plastic asphalt samples. From Table 2, it has provided that the results of k diminished with the growth of porous plastic mixing. For instance, when the results of k mixed %4 binders with %1 LDPE were 0.204 in/s, the result of mixing %3 LDPE was 0.193 in/s. Obviously, while mixing with % of LDPE was increasing, the result of k was diminishing. Figure 5 represents the results of permeability of porous plastic asphalt with mixing %1 LDPE.

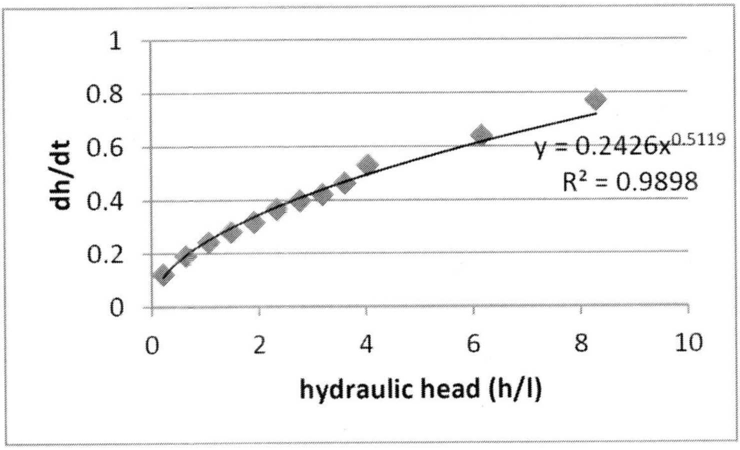

Figure 5. The results of k with mixing %1 LDPE

Table 2. The results of hydraulic conductivity

Binder	LDPE	k (in/s)
% 4	%0	0.217
	%1	0.204
	%3	0.193
	%6	0.178
% 5	%0	0.259
	%1	0.245
	%3	0.234
	%6	0.223
% 6	%0	0.301
	%1	0.294
	%3	0.288
	%6	0.272

5.2. Indirect tensile strength (its)

Use this test method to determine the tensile strength of compacted bituminous mixtures. The porous plastic mixed tested as a conditioned and unconditioned. After compaction, samples were kept in the hot water (conditioned situation) and cold water (unconditioned situation). The objective of this test was to measure the water resistance of the mixture after immersion for 24 hours at 60°C. After that, testing the resilient modulus provided the results of ITS. The purpose of results was to evaluate the resistance of porous plastic asphalt mix on plastic deformation. Furthermore, from equation 3 Tensile strength ratio (TSR) calculated that divided by conditioned to unconditioned situation. TSR < 70% considered Susceptible to Moisture. As a result it provides that the performance of strength of porous plastic asphalt. Moisture Susceptibility of Porous Plastic tested in accordance with ASTM C 6931-07. Samples were cured at room temperature, 100°C and 160°C for 24 hours. Samples dimension is Diameter=4", Height=2.5".

$$TSR = \frac{S1}{S2} \tag{2}$$

Where:

S1 = conditioned set (wet)

S2 = unconditioned set (dry)

According to Table 3 demonstrates that the results of unconditioned decreased with the growth of porous plastic mixing. For instance, when the results of mixed %5 binders with %3 LDPE were 57 psi, the result of mixing %6 LDPE was 54 psi. Obviously, while mixing with % of LDPE was increasing, the results of conditioned ITS was decreasing. On the contrary, the result of conditioned increased simultaneously with the growth of porous plastic mixing. Figure 6 represents that the result of conditioned ITS diminished while unconditioned ITS increased. According to the results of TSR is considered to susceptible to moisture. It provides that %3 LDPE and over is very strength.

The results of k and ITS of porous plastic asphalt mixtures were within the approximately predictable ranked come up in the literature for conventional asphalt showing Table 4. It is an indication that porous plastic asphalt should be used as a sustainable alternative for permeable pavements.

Table 3. The results of Indirect Tensile Strength test (ITS)

Binder	LDPE	Unconditioned (psi)	Conditioned (psi)	TSR (con/uncon)
% 4	%0	67	45	0.67
	%1	58	52	0.90
	%3	46	56	1.22
	%6	41	63	1.54
% 5	%0	71	54	0.76
	%1	62	59	0.95
	%3	57	62	1.09
	%6	54	68	1.26
% 6	%0	73	63	0.86
	%1	69	67	0.97
	%3	66	71	1.08
	%6	61	75	1.23

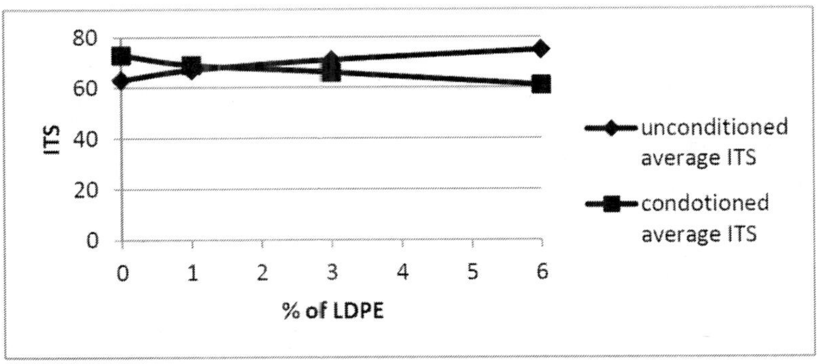

Figure 6. The results of ITS with mixing %6 LDPE

6. DISCUSSION AND CONCLUSION

Permeable Plastic Asphalt samples with different percentage of plastic to aggregate ratio were composed and then tested for hydraulic conductivity (k) and indirect tensile strength (ITS). Results represented which k and ITS results of samples were within the predictable results observed in the literature for porous pavements. In addition to it was come up with that the k, ITS values diminished as the percentage of plastic waste increased. The research results of Permeable Plastic Asphalt demonstrated that Permeable Plastic Asphalt should be a sufficient recycle and stormwater runoff with important reasonable economic and ecological associations.

Table 4. Summary of ITS porous asphalt from literature

References	ITS ranges (psi)
[62]	29.29-69.76
[63]	18.17-38.82
[64]	253.9-377.09
[65]	29-65.26
[66]	26.1-130.53
[19]	18.30-55.90

This research was taken in charge to evaluate the hydraulic and mechanical features of innovative porous plastic asphalt for implementing in permeable pavements. The test represented that the result of hydraulic conductivity diminished with the growth of porous plastic mixing. On the other hand, while the results of unconditioned decreased with the growth of porous plastic mixing, the result of conditioned increased simultaneously with the growth of porous plastic mixing. Furthermore, the results of TSR are the best result over 1 for susceptible to Moisture. The results of experiments were approximately come across that expected from literature. According to results that demonstrate porous plastic asphalt could be implement sustainable alternative pavements. It also provides also recycling.

This research concentrated on the hydraulic and mechanical properties of a permeable pavement like permeable plastic asphalt. Permeable Plastic Asphalt was composed of plastic waste, aggregates, and asphalt. Permeable Plastic Asphalt should provide a sufficient method for decreasing stormwater runoff, contributing a structural pavement sufficient for pedestrian and vehicular loadings. Furthermore, Permeable Plastic Asphalt should take the part of the currently ongoing recycling aims as a critical role that support to deflect a majority of plastic from landfills and incinerators.

The laboratory test indicates that aggregate, binder with mixing LDPE affected the results of k and ITS. Thus, permeability and strength of porous plastic asphalt is getting better with mixing LDPE. It proves that porous plastic asphalt help to diminish storm water runoff times diminish urban heat island effects. A new pavement increases for the sustainability of the nature that will be benefit users for many years. The design of plastic pavements contains developed pedestrian and public transportation as well as parking lot, driveway, bridges.

The results represents porous plastic asphalt should be implemented as a sustainable alternative for permeable pavements. Porous plastic asphalt is a peerless choice in that it undertakes two environmental problems that decreasing stormwater runoff and prevent to fill out with plastic waste at landfills. In the way that we keep up our way to green building and construction, porous plastic asphalt is new approach on the way to eco-friendly improvement. The innovative technology comes to grips with two environmental problems that are plastic waste and stormwater runoff. It provides to prevent a large quantity of plastic waste at landfills and incinerators, thus the plastic waste uses fro recycling. Also it decreases stormwater runoff and decreases the use of natural resources.

Permeable plastic pavement has whole with its permeability should be determined by valid void. Valid void should be directly implement to mixing ratio of permeable plastic pavement that both take control sufficiently the forming of run off and restrict urban flood.

All above results that Landscape Engineering considers that the progress of permeable plastic uses to efficiently integrate the mixing permeable plastic pavements with land use planning. Using recycling service with very powerful has approached to create pleasing environments in the world. According to test results that the advantages of plastic asphalt provide that is stronger road with increased plastic, better resistance during stormwater, without stripping and rutting, develop binder and better linking of the mix. Besides it support that is the strength of the road is increased. Using permeable plastic asphalt that is the cost of road construction diminish the maintenance cost of road gradually diminish as well. Obviously, as the plastic mix with pavement for using, the disposal of waste plastic will no longer be issue. As a result of that, using plastic helps to decrease in pores in aggregate save bitumen and help recycling.

Consecutive chapters of this research will count profoundly on the outcomes of the test pavements mentioned above, as well as efficiency data issued in the research. Detailed representing results of the several researches projects will be debated during the research.

7. FUTURE RESEARCH

Effectively managing the collection, separation and processing of plastic waste can limit the environmental damages limited by eliminating the waste from our streets. Thus, we can prevent to fill the landfill with plastic waste when we mix the plastic with the other elements with soil, asphalt and cement in order to use future studies.

Laboratory tests and real life implementation will study that the life expectancy of a plastic polymer road as compared to a conventional road. Future study will need to study for expanding of life expectancy for plastic. This study proved that investigates, summarizes preliminary results, and debates key properties to be considered for future alternative pavement. Future researchers will keep going to research the better permeable pavement for economic, environmental and nature.

A long-term monitoring project to document changes in performance, evaluation of different maintenance strategies, and lifecycle costs of permeable plastic asphalts is recommended for future research. Future experience is based on designs that provide to improve future properties such as increased new materials, and developed construction and maintenance activities.

ACKNOWLEDGEMENT

The author gratefully acknowledged Dr. Naji Khoury for his guidance and help.

REFERENCES

1. McKenna, GT. Sustainable Mine Reclamation and Landscape Engineering. PhD Thesisin Geotechnical Engineering, Department of Civil and Environmental Engineering,University of Alberta, Edmonton, 2002.
2. American Society of Landscape Architecture. ASLA.http://www.asla.org/ppn/ Article.aspx?id=1206&terms=landscape%20engineering(accessed 5 September 2012).
3. Zube E., Landscape Planning Education in America: Retrospect and Prospect.Landscape and Urban Planning, Volume 13, 1986, Pages 367–378, 1986.
4. The LA Group, Landscape Architecture and Engineering P.C.http://www.thelagroup.com (accessed 8 September 2012).
5. Valderrama A., Levine L., Yeh S., and Bloomgarden E., Financing Stormwater Retrofitsin Philadelphia and Beyond, Natural Resources Defense Council (NRDC), February 2012.http://www.nrdc.org/water/files/StormwaterFinancing-report.pdf (accessed 11Se ptember 2012).

6. The plastics industry trade association. SPI Resin Identification Code - Guide to CorrectUse. Washington.DC.http://www.plasticsindustry.org/AboutPlastics/content.cfm?ItemNumber=823&navItemNumber=1125 (accessed 2 November 2011).
7. LDPE, Husky, 2008 Polly America, 2000 West Marshall Drive, Grand Prairie, Texas75051, phone 800 527 3322, press 7654, www.poly-america.com (accessed 2 July 2011).
8. Asphalt 64-22 cas no 8052-42-4 PG-2010-321, NuStar Asphalt Refining, LLC, 4ParadiseRd. Paulsboro, NJ 08066 phone 856 224 7405.
9. Punith V. S., and Veeraragavan A., Characterization of Ogfc Mixtures ContainingReclaimed Polyethylene Fibers. Journal of Materials in Civil Engineering, ASCE, 341,March 2011.
10. Khan A., Gangadhar, Mohan M., and Raykar V., Effective Utilization of Waste Plasticsin Asphalting of Roads. Project Report Prepared Under The Guidance of R. Suresh andH. Kumar, Dept. of Chemical Engg., R.V. College of Engineering, Bangalore, 1999.
11. Zaro G., Nystrom K C., Bar A., Alvarez A. U., and Miranda A., Tierras Olvidadas:Chiribaya Landscape Engineering and Marginality in Southern Peru, The Society forAmerican Archaeology, Latin American Antiquity 21(4), pp. 355–374, 2010.
12. Corner J., Representation and Landscape: Drawing and Making in The LandscapeMedium, Word & Image 8, 246, July-Sept. 1992.
13. American Society of Landscape Architecture. ASLA Security Design SymposiumAbstracts. Chicago, IL, July 2004,http://www.asla.org/uploadedFiles/CMS/Resources/security designabstractfinal.pdf(accessed 2 July 2011).
14. Haselbach, L.M., Valavala, S., and Montes, F., Permeability Predictions for Sand-Clogged Portland Cement Pervious Concrete Pavement Systems, Journal ofEnvironmental Management, 81, 42-49, 2006.
15. Huang, B., Wu, H., Shu, X., and Burdette, E.G., Laboratory Evaluation of Permeabilityand Strength of Polymer-Modified Pervious Concrete, Construction and BuildingMaterials, 24, pp. 818-823, 2010.
16. Tennis, P.D., Leming, M.L., and Akers, D.J., Pervious Concrete Pavements. PorlandCement Association. 2004.Landscape Engineering, Protecting Soil, and Runoff Storm Water 719http://myscmap.sc.gov/marine/NERR/pdf/PerviousConcrete_pavements.pdf (accessed 2July 2011).
17. Montes, F., Valavaka, S., and Haselbach, L., A New Test Method for PorosityMeasurements of Portland Cement Pervious Concrete, Journal of ASTM International,2(1), 2005.
18. Subagio B., Kosasih D., Busnial, and Tenrilangi D., Development of Stiffness Modulusand Plastic Deformation Characteristics of Porous Asphalt Mixture Using TafpackSuper™, Proceedings of the Eastern Asia Society for Transportation Studies, Vol. 5, pp.803 - 812, 2005.
19. Nakanishi H., Kawanaka T., Ziqing L., and Baocun H., Study on Improvement inDurability and Function of Porous asphalt Pavement, Road Construction, Japan, 2001.
20. Daniels, S., and Repton H., Landscape Gardening and the Geography of GeorgianEngland, New Haven, CT: Yale University Press, 1999.
21. Harris D., and Hays D., On the Use and Misuse of Historic Landscape Views, inRepresenting Landscape Architecture, ed. Marc Treib., New York: Taylor and Francis,22-41, 2008.
22. Rieder K., Modeling, Physical and Virtual, in Representing Landscape Architecture, ed.Marc Treib., New York: Taylor and Francis, 2008.
23. Waugh F. A. Landscape Engineering in the National Forests. US. Department ofAgriculture Forest Service. 1918.http://archive.org/stream/landscapeenginee00waug#page/ n1/ mode/2up (accessed 5October 2012).
24. Doran, J.W., and T.B. Parkin. Defining and Assessing Soil Quality. In: J.W. Doran et al.(eds.) Defining Soil Quality for a Sustainable Environment. Soil Science Society ofAmerica, Madison, WI, Special Publication 35, pp. 3-22, 1994.
25. Mollison B., Increases in Porosity Enhance Infiltration and Thus Reduce Adverse Effectsof surface runoff, Permaculture: A Designer's Manual, Tagari Press, 1988.
26. Hubbard, A. T., Encyclopedia of Surface and Colloid Science Vol 3, Santa Barbara, California Science Project, Marcel Dekker, New York, 2004.

27. Beven K., and Robert E. Horton's Perceptual Model of Infiltration Processes,Hydrological Processes, Wiley Intersciences DOI 10:1002 hyp 5740, 2004. http:// earth. boisestate.edu/jmcnamara/files/2011/10/KBeven_HP2004.pdf (accessed 23August 2012).Stormwater Pennsylvania. http://www.stormwaterpa.org (accessed 25 August 2012).
28. Busco D., and Lindsey G., An Annotated Bibliography of Stormwater FinanceResources, Center for Urban Policy and the Environment School of Public andEnvironmental Affairs Indiana University- Purdue University Indianapolis, April 29,2002.http:// stormwaterfinance.urbancenter.iupui.edu/PDFs/Biblio%204%2029%2002.pdf(accessed 26 August 2012).
29. Lee J. H., and Bang K W., Characterization of Urban Stormwater runoff. WaterResearch. Volume 34, Issue 6, 1 April 2000, Pages 1773–1780
30. Dietza M. E., and Clausenb J. C., Stormwater Runoff and Export Changes withDevelopment in a Traditional and Low Impact Subdivision. Journal of EnvironmentalManagement, Microbial and Nutrient Contaminants of Fresh and Coastal Waters,Volume 87, Issue 4, June 2008, Pages 560–566http://www.sciencedirect.com/science/article/pii/S030 147970700103X (accessed 23August 2012).
31. Canadian plastics industry association.http://www.plastics.ca/home/index.php (accessed 2 November 2011).
32. Canadian plastics industry association. Environmental sustainability.http://www. plastics.ca/EnvironmentalSustainability/index.php/ (accessed 2 November2011).
33. Yoder E. J., and Witczak M. W. Principles of Pavement Design. Second Edition. JohnWily And Sons, Inc, New York. 1975.
34. Acum, W. E. A., and L. Fox, Computation of Load Stresses in a Three-layer ElasticSystem, Geotechnique, Vol. 2. pp. 293-300, 1951.
35. McCullough, B. F., A pavement Overlay Design System Considering Wheel Loads, Temperature Changes, and Performance, Graduate Report, ITTE, University ofCalifornia, Berkeley, 1970.
36. Hveem, F. N., Pavement Deflections and Fatigue Failures, Highway Research Board Bulletin 342, 1962.
37. Barber, E. S., Application of Triaxial Compression Test Results to the Calculation ofFlexible Pavement Thickness, Proceedings, Highway Research Board, 1946.
38. Fox, L., Computation of Traffic Stresses in a Simple Road Structure, Department ofScientific and Industrial Research, Road Research Technical Paper 9, 1948.
39. McLeod, and Norman W., An ultimate Strength Approach to Flexible Pavement Design,Proceedings, Association of Asphalt Paving Technologist, 1954.
40. Horner, and Raymond C., Effect of Base Course Quality on Load Transmission ThroughFlexible Pavements, Proceedings, Highway Research Board, 1955.
41. Palmer, L. A., and E. S. Barber, Soil Displacement under Circular Load Areas,Proceedings, Highway Research Board, 1940.
42. Poulikakis, L.D., and Partl, M.N., Evaluation of Moisture Susceptibility of PorousAsphalt Concrete using Water Submersion Fatigue Tests, Construction and BuildingMaterials, 2009.
43. Hassani, A., Ganjidoust, H., and Maghanaki, A., Use of Plastic Waste (Poly-ethyleneTerephthalate) in Asphalt Concrete Mixture as Aggregate Replacement, Waste ManageRes 2005: 23: 322–327, 2005.
44. Hinislioglu, S., and Agar, E. Use of Waste High Density Polyethylene as BitumenModifier in Asphalt Concrete Mix. Materials Letters, 58, 267-271, 2004.
45. Zoorob S.E., and Suparman, L.B, Laboratory Design and Investigation of Proportion ofBituminous Composite Containing Waste Recycled Plastics Aggregate Replacement Landscape Engineering, Protecting Soil, and Runoff Storm Water 721(Plastiphalt), CIB Symposiumon Construction and Environment Theory into Practice,Sao Paulo, Brazil, November, 2000.
46. Zoorob, S.E., Laboratory Design and Performance of Improved Bituminous Composites Utilising Plastics Packaging Waste", Conference on Technology Watch and Innovationin

Construction Industry, Belgium. Building Research Institute, Brussels, Belgium,April, 2000.

47. Conigliaro, A., and Watson, P., Determining the Best Formulation for a Unique AsphaltCold Patch Product Made with # 3-7 Rigid Plastic Aggregate, Chelsea Center forRecycling and Economic Development Technical Research Program, January 2000.
48. Mallick, R.B., and Teto, M., Evaluation of Use of Manufactured Waste Asphalt Shinglesin Hot Mix Asphalt, 2000, Chelsea Center for Recycling and Economic Development Technical Research Program, July 2000.
49. obhan, K., and Mashnad, M., Mechanical Stabilization of Cemented Soil–Fly AshMixtures with Recycled Plastic Strips, Journal of Environmental Engineering, ASCE,October 2003.
50. Sobhan, K., Stabilized Fiber-Reinforced Pavement Base Course With RecycledAggregate, Ph.D. Dissertation, Northwestern University, Evanston, Illinois, 303 pages,June, 1997.
51. Cavey J. K., Krizek, R. J., Sobhan, K., and Baker, W. H., Waste Fibers in Cement-Stabilized Recycled Aggregate Base Course Material. Transportation Research Record1486, Transportation Research Board, Washington D.C., 97–106, 1995.
52. Meyers, III, Swartz, J, Nathaniel, Kurczewski, N., and Kurczewski, M., RecyclableComposite Materials Articles of Manufacture and Structures and Method of usingComposite Materials, U.S. Patent No. 6,984,670, 2006.
53. Malloy, R. and Kashi, M., and Swan, C., Fly Ash/Mixed Plastic Aggregate and ProductsMade Therefrom, U.S. Patent No., 6,669,773, 2003.
54. Balkum, E , Aggregate Using Recycled Plastics, U.S. Patent No. 6,488,766., 2002.
55. Fwa, T.F., Tan, S.A., and Chuai, C.T., Permeability Measurement of Base Materialsusing Falling-Head Test Apparatus, Transportation Research Record, No. 1615, pp 94-99, 1998.
56. Khoury, N.N., Khoury, C.N., and Abousleiman, Y., Soil Fused with Recycled PlasticBottles for Various Geo-Engineering Applications. ASCE Conf. Proc. 309, 42, 2008.
57. Frocht, M. M., Photoelasticit, vol. 2 John Wiley and Sons, New York, 1957.
58. Hadley, W. O, Hudson W. R., and Kennedy T. W., Evaluation and Prediction of TheTensile Properties of Asphalt-treated Materials, Highway Research Board AnnualMeeting, Washington, D.C., 1971.
59. Kennedy, J. W., and Hudson W. R., Application of the Indirect Tensile Test to StabilizedMaterials, Highway Research Board Annual Meeting, Washington, D.C., 1968.
60. Suresha S. N., George V., and Shankar A.U. R., Effect of Aggregate Gradations onProperties of Porous Friction Course Mixes, Material and Structure, 2010.
61. Setyawan A., Design and Properties of Hot Mixture Porous Asphalt For Semi-FlexiblePavement Applications, Media Technic Sipil, July 2005.
62. Liu Q., Schlangen E., Garcia A., and Ven M. V.D., Induction Heating of Electrically Conductive Porous Asphalt Concrete. Construction and Building Materials, 2010.
63. Shen D., Wu C., and Du J., Laboratory Investigation of Basic Oxygen Furnace Slag forSubstitution of Aggregate in Porous Asphalt Mixture. Construction and Buildin Materials 23, 2009.
64. Xiao F., Zhao W., Gandhi T., and Amiskhanian A. N., Influence of AntistrippingAdditives on Moisture Susceptibility of Warm Mix Asphalt Mixture. Journal ofMaterials in Civil Engineering, October 2010.

CHAPTER 9

Stormwater Pollutant Process Analysis with Long-Term Online Monitoring Data at Micro-Scale Sites

Dominik Leutnant [1,*], Dirk Muschalla [2] and Mathias Uhl [1]

[1] *Institute for Water Resources Environment (IWARU), Muenster University of Applied Sciences, Corrensstr. 25, Muenster 48149, Germany*

[2] *Institute of Urban Water Management and Landscape Water Engineering, Graz University of Technology, Stremayrgasse 10/I, 8010 Graz, Austria*

ABSTRACT

Stormwater runoff quality was measured with online turbidity sensors at four common types of small urban subcatchments: (i) a flat roof; (ii) a parking lot; (iii) a residential catchment; and (iv) a high-traffic street. Samples were taken to estimate site-specific correlations between total suspended solids (TSS) and turbidity. Continuous TSS time series were derived from online turbidity measurements and were used to estimate event loads and event mean concentrations. Rainfall runoff event characteristics were subjected to correlation analysis to TSS loads. Significant correlations were found for rainfall intensities at sites with high imperviousness and decrease with increasing catchment size. Antecedent dry weather periods are only correlated at the parking lot site. Intra-event TSS load distributions were studied with *M* (*V*)-curves. *M* (*V*)-curves are grouped at runoff quantiles and statistically described with boxplots. All sites show, in general, a more pronounced first-flush effect. While wash-off of the flat roof tends to be source-limited, the parking lot and high-traffic street sites show a more transport-limited behavior. Wash-off process of the residential catchment appears to be influenced by a composition of different subcatchments.

Keywords: stormwater quality; online monitoring; stormwater pollutant processes; micro scale

1. INTRODUCTION

Stormwater runoff from urban environments is a significant source of pollutants which impacts the quality of receiving waters. Effective measures require realistic estimations of stormwater pollutants to adequately protect the receiving water. Usually, stormwater quality models are applied to support the implementation of urban drainage strategies. Current stormwater quality model concepts are based upon empirical equations or simple regression functions to replicate the complex nature of pollutant accumulation and wash-off. Although, these approaches offer a set of parameters for model calibration, quality models often show poor performance when simulating long-term conditions [1,2]. As a result, model outputs are highly uncertain. Improving quality models is therefore crucial to produce more reliable model results. In this respect, in-depth knowledge of processes is a key requirement which consequently demands measurement data. In recent years much effort has been spent to investigate the influence of meteorological influences and catchment characteristics on stormwater quality based on samples at small sites [3,4,5]. Additionally, online turbidity measurements have been successfully used for intra-event analyses in larger catchments [6,7].

With aiming towards new insights of stormwater quality processes this work combines both approaches by using online turbidity measurements at microscale sites to analyze stormwater pollutant processes. Results of a long-term monitoring campaign at four common types of urban subcatchments—(i) a flat roof; (ii) a parking lot; (iii) a residential catchment; and (iv) a high-traffic street—are presented. It is believed that monitoring at small urban environments is required to isolate relevant pollutant processes and to reduce interfering influences of catchment size and environment.

2. MATERIALS AND METHODS

2.1. Experimental Sites

Stormwater runoff and quality was continuously monitored at four microscale experimental sites: (i) a 50 m^2 flat roof FR; (ii) a parking lot PL with approx. 2350 m^2; (iii) a 9.4 ha residential catchment RC in a suburb of Muenster, Germany (separate sewer system); and (iv) a high-traffic HT street in the center of Muenster (2.5 ha, 30,000 vehicles per day). The 2% sloped roof is thoroughly covered with bitumen sheeting. Surfaces of the parking lot are asphalt (55%), porous pavement (40%, 8% thereof being joints) and small vegetated pervious areas (5%) which do not contribute to runoff. The impervious area has a slope of 2.5%. The residential catchment consists of streets (25%), flat and steep roofs (25%), and pervious area (50%). At site HT surfaces mainly consist of asphalt (60%), porous pavement (10%, 8% thereof being joints), flat and steep roofs (25%), and disconnected pervious area (5%).

2.2. Monitoring and Sampling Setup

Rainfall gauges are located at FR, PL, and RC (Pluvio2, OTT). Rainfall data from FR is also used for HT, being 2 km off FR. Runoff at **FR** runs from a downpipe into a horizontal measurement pipe (63 mm, PVC) in which an electromagnetic flowmeter (Promag50W25, Endress + Hauser) and quality sensors for turbidity, electrical conductivity, and pH (VisoTurb700IQ, TetraCon700IQ, and SensoLyt700IQ, WTW) are installed. Samples are taken from the measurement pipe with an automatic sampler (vacuum sampler ASP Station, Endress + Hauser). Sampling begins if runoff is above 0.03 L/s and repeats every 10 min. Each sample consists of five subsamples of about 200 mL. The capacity of the automatic sampler is 12 samples.

The control section at **PL** is a 300 mm circular concrete pipe (length 55 m, slope 1.8%). Runoff is calculated from measured water level by the Manning-Strickler-equation because of uniform flow conditions and no backwater effects. Manning's roughness coefficient n was experimentally determined with artificial inflows and ranges between 0.015 and 0.017. At **RC** and **HT** runoff is calculated from mean flow velocity and water level (POA, NIVUS). The control section at RC is a 900 mm circular concrete pipe (length 46 m, slope 1.8%). Manning's roughness coefficient n was also identified with artificial inflows and is about 0.0105. At HT, flow sensors are installed in a 500 mm circular concrete pipe (length 30 m, slope 0.6%).

In contrast to FR, quality sensors at PL, RC, and HT are integrated in a horizontal measurement pipe (63 mm, PVC, length: 1.5 m) of an external monitoring station (Figure 1). In case of an event stormwater is pumped to the measurement pipe by a peristaltic pump (Delasco 2Z3, PCM) through a hose (20 mm, PVC) whose orifice is fixed in the middle of the stormwater pipe 1.5 cm above the ground. The suction velocity in the hose is about 1.5 m/s with a corresponding flow of approx. 0.5 L/s. Stormwater flows with approx. 0.18 m/s through the measurement pipe and is later discharged to the sewer.

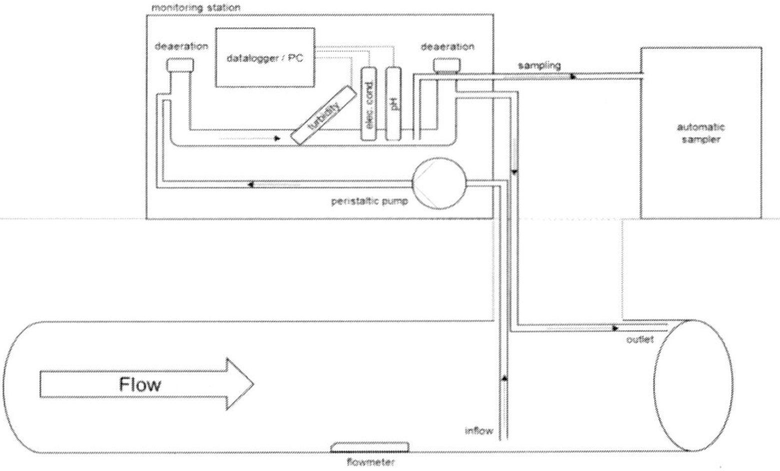

Figure 1. Schematic overview of the monitoring station.

At these sites, the sampling program starts if the water level in the stormwater pipe exceeds 1.5 cm. Samples of all sites are tested for total suspended solid (TSS) concentrations based on a standard method given in [8], and fine solids less than 63 μm (TSS63) according to the protocol given in [9].

An online measurement data management system was applied to supervise all monitoring stations and to reduce data loss [10].

2.3. Calculation of Continuous TSS Time Series

The turbidity of selected samples was measured to estimate the correlation to TSS. Initially, this has been conducted in the original sample bottle (PE, squared base, slightly transparent). Due to significant variance of the measured turbidity, a black cylindrical PE-HD bottle (diameter 10.8 cm, height 18.1 cm) has been used later. While measuring the turbidity, the sample is homogenized with a magnetic stirrer at 450 rpm. The five-minute mean of the turbidity is recorded. Calibration of turbidity probes was conducted with formazine primary standard solutions. TSS concentrations and the corresponding turbidity values were subjected to correlation analysis. The resulting linear regression equations are used to create continuous TSS time series from raw turbidity signals. Discussing the uncertainties through this conversion would exceed the scope of this paper and are therefore not presented here. The reader is referred to the literature [11,12,13].

2.4. Data and Analysis

All sensor signals were logged with a 1 min interval. High-resolution online runoff and quality data is available for approx. 2.5 years (FR and RC), 1.5 years (PL), and 0.5 years (HT), respectively. Rainfall runoff events were statistically analyzed using rainfall, runoff, and pollutant characteristics. Events with a minimum rainfall depth of $H > 2$ mm, maximum rainfall intensity in 60 min of Imax60 > 2.5 mm/h, and complete runoff/turbidity data are selected, only. Rainfall events below these criteria usually do not contribute to relevant runoff and are therefore excluded. Based on continuous runoff and TSS time series data, event volumes, event loads, and event mean concentrations are calculated according to Equations (1)–(3). Furthermore, event characteristics were subjected to correlation analysis with special emphasis to TSS loads.

$$\text{Event volume } (\text{m}^3): \; Vol = \sum_{i=1}^{n} Q_i t \tag{1}$$

$$\text{Event load (kg)}: \; Load = \sum_{i=1}^{n} Q_i C_i t \tag{2}$$

$$\text{Event mean concentration (mg/l)}: \; EMC = \frac{Load}{Vol} \tag{3}$$

where i = index of time series, n = number of data points of an event, Q_i = runoff at index i, Δt = time interval (i.e., 1 min), and C_i = TSS concentration at index i.

The intra-event distribution of TSS loads are examined by means of mass-volume-curves (*M* (*V*)-curves, [14,15]). *M* (*V*)-curves describe the proportion of transported mass at a given runoff volume proportion. This method is usually used to visualize transported mass proportions and to analyze the first-flush phenomenon. Knowledge of catchment-specific first-flush characteristics is crucial to designing cost-effective treatment or storage structures. However, *M* (*V*)-curves tend to be site-specific and vary greatly from event to event [6]. Aggregation of similar *M* (*V*)-curves is therefore required to extract relevant information. [16] for example, divide *M* (*V*)-curves in three different zones to classify similar events. Zone A contains curves with a dominant first-flush effect, while curves in Zone C tend to be more last-flush affected. Curves in Zone B are near the bisecting line and show a runoff-proportional mass transport.

In this paper, *M* (*V*)-curves are also used to characterize the two types of wash-off process, namely source-limited and transport-limited wash-off [17,18]. Source-limited runoff events have, in general, sufficient energy to wash off all available particles on the surface. This occurs if either masses on the surface are rather limited or the kinetic energy of rainfall/runoff is high enough. Transport-limited events are not able to completely remove available masses. Typically, these events occur either if the available masses are adequately high or the kinetic energy of runoff is insufficient.

M (*V*)-curves are calculated for the four study sites and compared. A seasonal differentiation is included. Instead of using zones, boxplots of the transported mass proportions are created at runoff volume quantiles to group *M* (*V*)-curves. With calculated and visualized interquartile ranges (IQR), the event variability and main wash-off trends can be observed and characterized.

3. RESULTS

3.1. TSS Sample Statistics

Table 1 summarizes TSS sample statistics at the four study sites. Statistics were also calculated for the dataset excluding outliers. Due to non-normality of the dataset, outliers are conservatively considered and defined as points beyond the mean ± four times the standard deviation. Mean and standard deviation are iteratively computed while potential outliers are excluded.

Table 1. Site-specific TSS sample statistics.

Site	Outlier Excluded	n	Events	TSS (mg/L)								
				Min	0.1-Perc.	0.25-Perc.	Median	0.75-Perc.	0.9-Perc.	Max	Mean	Sd
FR	no	193	40	0.6	2.2	3.7	6.7	14.2	45.5	674.1	22.4	60.2
	yes	182	39	0.6	2.2	3.7	6.5	11.7	27.8	85.1	11.8	15.5
RC	no	269	39	1.4	5.6	9.7	21.0	73.0	205.2	3645.7	114.3	339.7
	yes	256	39	1.4	5.5	9.4	18.7	63.4	132.7	569.1	55.6	91.5
PL	no	140	38	7.3	19.7	58.6	169.2	335.1	550.5	1842.0	248.3	278.1
	yes	139	38	7.3	19.6	58.5	168.4	333.8	546.8	1189.4	236.8	243.7
HT	no	92	17	2.9	26.1	53.3	77.4	98.6	129.2	237.1	79.1	41.2
	yes	no outliers detected										

At site **FR**, 193 samples were analyzed from 40 events. With the 0.75 percentile being 14.2 mg/L, the flat roof clearly shows low TSS potential and distributions are similar to other findings [19,20,21].

For site **RC**, 269 samples of 39 events were taken. The distribution of TSS concentration also reveals low TSS contribution. Compared to the results of [22], values are lower than TSS concentrations of a separated sewer system in Germany. The mean value of 114.3 mg/L and the standard deviation of 339 mg/L indicates high variation. However, these statistics are strongly influenced by the maximum value of 3645 mg/L. The 0.9 percentile being at 205 mg/L confirms this. 140 samples from 38 events were analyzed for site **PL**. TSS concentration ranges from 7.3 mg/L to 1842 mg/L, with the median at 170 mg/L. At **HT**, 92 samples of 17 events were collected. Compared to other studies at high-trafficked streets [23], the TSS statistics are significantly lower. For example, the median of 77.4 mg/L is less than half as the median in their study (175 mg/L).

3.2. Relationship between TSS and Turbidity

To create continuous TSS data from online turbidity data, correlation functions are determined. Due to the change of bottle type in which the turbidity was measured, correlation functions were established with only a subset of all samples presented in Table 2. The range of the sample subset is within the range of all samples with outliers being excluded. Only the maximum value at site RC is slightly higher (580.1 mg/L compared to 569.1 mg/L) and therefore still used for analysis. Both linear and non-linear relationships were tested. Since non-linear functions did not significantly outperform linear functions, only linear regression coefficients are listed in Table 3. The goodness-of-fit of the linear regression is visually verified and numerically expressed by r-squared. With the lowest r-squared being at 0.68, all linear regression models show a good fit of the underlying dataset.

Table 2. Site-specific TSS sample statistics of samples used for turbidity correlation.

Site	*n*	Events	TSS (mg/L)								
			Min	0.1-Perc.	0.25-Perc.	Median	0.75-Perc.	0.9-Perc.	Max	Mean	Sd
FR	36	4	1.9	2.8	3.7	5.2	9.4	22.6	43.3	9.0	9.7
RC	60	7	2.3	8.9	17.8	41.6	65.2	150.1	580.1	70.7	107.8
PL	96	33	2.3	6.5	18.2	40.9	97.5	137.2	459.9	67.4	77.3
HT	85	16	0.5	19.0	41.0	67.1	82.1	103.2	140.7	63.7	31.7

Table 3. Linear regression coefficients for correlation of TSS and turbidity (TSS = *f*(turbidity) = *a* + *b* * turbidity).

Site	*a*	*b*	R^2
FR	−3.52	1.89	0.835
RC	−20.9	3.69	0.823
PL	1.97	0.84	0.683
HT	7.93	0.97	0.681

3.3. Event Database

An overview of the event database with continuous measurement data is given in Table 4. It contains the number of total observed events and the number of events which are excluded from further analysis. Events are rejected if either selection criteria are violated or if measurement data is doubtful. In this respect, sites FR and RC show a high number of events with doubtful data. This is mainly caused by almost constantly low turbidity values (FNU < 15) in the course of an event. For site FR this can be justified with few particles in the runoff. At site RC, this is also caused by pumping difficulties. Gaps due to measurement failures of runoff and quality sensors are rarely present. Turbidity gaps are only observed if stormwater contained substances which caused intensive foaming in the measurement pipe. However in total, 65 events were analyzed at FR, 23 at site RC, 46 at PL, and 16 at HT. Descriptive statistics of selected event characteristics are given in Table 5.

Table 4. Description of event database with continuous monitoring data.

Site	Total Observed Events	Events Violating Selection Criteria	Events with Doubtful Data	Valid Events	Valid Events/Total Observed Events
FR	415	275	75	65	16%
RC	324	199	102	23	7%
PL	152	87	19	46	37%
HT	40	11	13	16	40%

3.4. Correlation Analysis

Table 6 shows Pearson correlation coefficients for TSS loads and selected variables at the four study sites. A strong correlation of rainfall intensities and mean/max runoff to TSS loads can be observed at site FR. This effect is also evident but less intense at sites PL and HT. However, the variable I_{mean} (mean rainfall intensity) has only a strong influence at FR (0.8). Rainfall depths seem to be strongly correlated to TSS loads at site HT, only.

Table 5. Descriptive statistic data (min, 0.1-, 0.25-, 0.5-, 0.75-, 0.9-percentiles, max, mean, standard deviation) of site-specific event characteristics; rainfall depth: H, max. rainfall intensity in 60 minutes (I_{max60}), max runoff (Q_{max}), runoff volume (Vol), TSS loads (Loads), and TSS event mean concentrations (EMC).

Site	Min	0.1-Perc.	0.25-Perc.	Median	0.75-Perc.	0.9-Perc.	Max	Mean	Sd
-	*H* (mm)								
FR	2.0	2.1	3.1	4.3	7.3	9.6	22.7	5.6	3.8
RC	2.2	3.8	4.3	7.1	13.4	18.9	29.1	9.8	7.2
PL	2.1	2.6	3.1	5.1	10.0	18.8	31.0	8.0	6.9
HT	2.3	3.2	3.6	6.2	8.7	17.5	21.8	8.0	5.9
-	*Vol* (m^3)								
FR	0.02	0.02	0.1	0.2	0.3	0.4	1.2	0.2	0.2
RC	18.2	45.5	65.2	116.3	233.5	428.5	715.8	192.0	191.0
PL	1.4	2.6	3.4	5.7	11.8	29.7	71.2	11.6	13.8
HT	81.2	105.5	121.5	166.3	336.1	581.4	784	268.5	211.4
-	Loads (kg/ha)								
FR	0.0	0.0	0.1	0.2	1.7	4.9	19.4	1.7	3.6
RC	0.1	0.3	0.6	0.9	3.5	7.4	9.4	2.6	2.9
PL	0.1	0.5	0.9	1.3	2.6	6.3	11.1	2.3	2.6
HT	1.6	3.1	3.6	8.0	13.6	29.2	47.5	12.6	12.7

Site	Min	0.1-Perc.	0.25-Perc.	Median	0.75-Perc.	0.9-Perc.	Max	Mean	Sd
-	I_{max60} (mm/h)								
FR	2.6	2.8	3.3	4.6	8.2	15.2	49.4	7.4	8.1
RC	2.5	2.7	3.0	3.7	5.2	6.6	10.4	4.4	2.1
PL	2.5	2.7	2.9	5.5	10.4	18.7	44.9	8.6	8.2
HT	1.1 *	1.3	1.8	2.3	2.6	4.7	7.5	2.7	1.7
-	Q_{max} (L/s)								
FR	0.01	0.01	0.1	0.1	0.3	0.4	1.6	0.2	0.3
RC	8.9	10.5	15.9	28.8	95.4	148.4	215.7	61.3	66.5
PL	1.0	2.3	4.0	7.8	13.1	23.3	54.9	11.1	11.3
HT	13.5	16.3	19.8	27.5	52.3	81.8	133.1	40.7	32.9
-	EMC (mg/L)								
FR	0	1	3	9	35	94	250	33	55
RC	4	11	18	50	92	152	364	77	94
PL	5	13	24	49	80	112	254	60	49
HT	27	38	54	120	172	242	297	125	84

Note: * Event is considered valid although the Imax60 criteria is violated.

The overall rainfall duration does not correlate with TSS loads at any site. Correlation of the variables runoff volume (Vol) and antecedent dry weather periods (ADWP) to TSS loads can be noticed only at site HT and PL, respectively.

Table 6. Site-specific Pearson correlation coefficients (FR: flat roof, RC: residential catchment, PL: parking lot, HT: high-traffic street) for TSS loads and selected variables: rainfall depth, duration, and intensities (H, D_p, I_{mean}:I_{max60}), runoff characteristics (Q_{mean}, Q_{max}, volume), and antecedent dry weather period (ADWP). Bold values indicate correlation coefficients >0.5.

Site		H	Dp	I_{mean}	I_{max1}	I_{max5}	I_{max60}	Q_{mean}	Q_{max}	Vol	ADWP
FR	Loads	0.39	−0.09	**0.80**	**0.68**	**0.80**	**0.82**	**0.90**	**0.88**	0.37	0.20
RC		0.19	−0.03	0.40	0.47	0.43	0.26	0.45	0.35	0.29	−0.06
PL		**0.50**	0.08	0.48	**0.69**	**0.69**	**0.63**	0.49	**0.64**	0.38	**0.56**
HT		**0.80**	0.30	0.41	**0.73**	**0.74**	**0.64**	**0.62**	0.59	**0.78**	−0.06

3.5. Intra-Event TSS Load Distributions

Intra-event distributions of TSS load are studied with site-specific *M* (*V*)-curves (Figure 2). Clearly, all sites show large variability of intra-event TSS load distribution which confirms findings of other studies [6,7] also for microscale sites. However, from the four study sites it can be observed that the more curves are taken into account the variability increases. Therefore, boxplots at runoff volume quantiles are used to depict the main tendency of wash-off behavior. This enables a visual comparison between site and season-specific *M* (*V*)-curves.

Figure 3 shows boxplots of *M* (*V*)-curve distributions at given runoff volume quantiles for each of the study sites. At site **FR**, in most cases a large portion of pollution loads tend to be washed-off in the first period of an event. In addition, distances between the first and third quartile (interquartile range, IQR) increases until 20% of runoff volume and decreases afterwards. This generally indicates a decreasing event variability. In this respect, after 60% of runoff volume, most pollutants are already washed off. With regard to site **PL** and **RC**, the IQR rises until 20% of runoff volume and almost constantly continues up to 60% of runoff volume. At site **HT**, the IQR is merely changing in the first 80% of runoff volume. Although, the number of events taken into account is likely to affect the interquartile ranges, *M* (*V*)-curves from site HT are noticeably closer to the bisecting line than *M* (*V*)-curves from site FR.

Similarly, *M* (*V*)-curves from PL are closer to the bisecting line compared to the *M* (*V*)-curves from site RC.

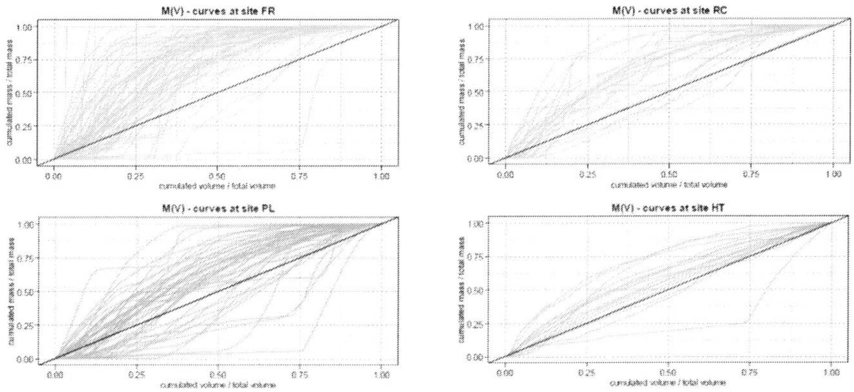

Figure 2. Site-specific *M* (*V*)-curves (FR: flat roof, RC: residential catchment, PL: parking lot, HT: high-traffic street).

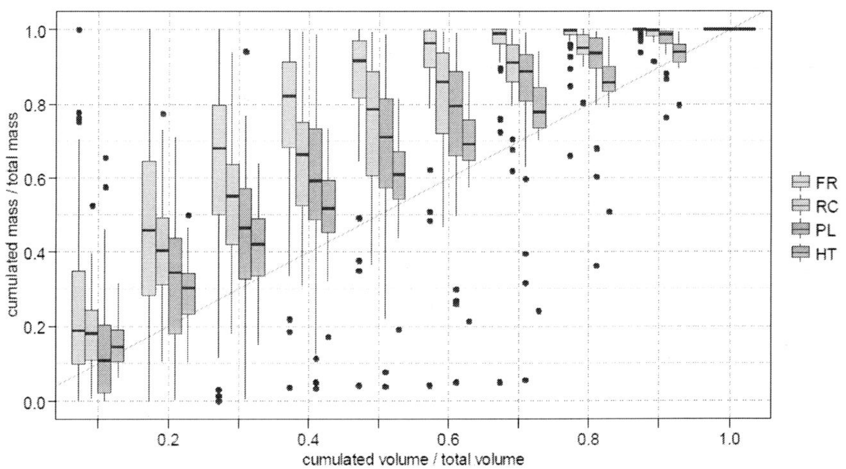

Figure 3. Site-specific boxplots of *M* (*V*)-curve distributions at runoff volume quantiles (FR: flat roof, RC: residential catchment, PL: parking lot, HT: high-traffic street). Box ranges correspond to the first and third quartiles. Median is indicated by a solid black horizontal line. Whiskers comprise lowest/highest value within 1.5 × inter-quartile range. Outliers exceed whiskers' ends and are indicated by solid black dots.

3.6. Seasonal Intra-Event TSS Load Distribution

Figure 4 shows *M* (*V*)-curve distributions for different seasons. At FR, the *M* (*V*)-curves start steeper in spring, summer, and autumn periods, which indicates a more pronounced first flush. Contrarily, in winter, the *M* (*V*)-curves show a less dominant wash-off behavior at the beginning of the events. At site PL, the variability is highest during spring and autumn periods. Events during summer months show similar wash-off behavior, which is indicated by relatively low IQR. The three events in the winter are characterized by a delayed wash-off, but cannot be statistically interpreted due to small number of events. *M* (*V*)-curve distributions at site RC are comparable to PL with the highest variability during spring and autumn months. Pollutants tend to be washed-off in the first periods of an event. For site HT, monitored events are available in the autumn and winter months, only. Both seasons show comparable wash-off behavior, which is characterized by runoff almost proportional to washed-off loads, low IQR, and close distance to the bisecting line.

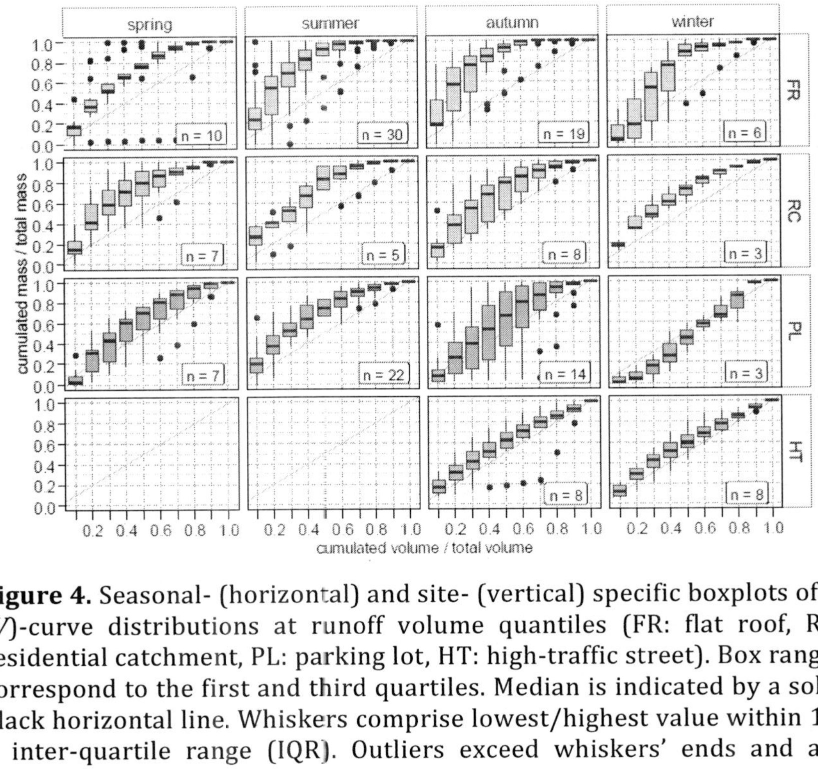

Figure 4. Seasonal- (horizontal) and site- (vertical) specific boxplots of *M* (*V*)-curve distributions at runoff volume quantiles (FR: flat roof, RC: residential catchment, PL: parking lot, HT: high-traffic street). Box ranges correspond to the first and third quartiles. Median is indicated by a solid black horizontal line. Whiskers comprise lowest/highest value within 1.5 × inter-quartile range (IQR). Outliers exceed whiskers' ends and are indicated by solid black dots.

4. DISCUSSION

From the correlation analysis it is stated, that firstly rainfall intensity (I_{max5}, I_{max60}) has a strong influence on TSS loads at small catchments with a high proportion of impervious surfaces (FR, PL, HT). Secondly, this effect decreases with increasing catchment size. Thirdly, in residential catchments which consist of multiple subcatchments (e.g., roofs, streets, parking lots) the correlation between rainfall event characteristics and TSS loads are strongly attenuated. The low correlation of the antecedent dry weather period suggests that this parameter is inappropriate to describe the pollutant build-up. However, in this study, the average antecedent dry weather period is about three days. This means pollutants are mostly accumulated shortly after an event and therefore exposed to other influential processes such as wind-driven processes.

Analysis of M (V)-curves suggests, that firstly, microscale sites show, in general, a more pronounced first-flush effect and only a few events with a delayed wash-off process. Secondly, the wash-off process at FR seems to be source limited because of the majority of particles are washed-off after 60% of runoff volume and the IQR is significantly low at the end of the events. Thirdly, in contrast, PL and HT show a more transport-limited wash-off because the IQR is closer to the bisecting line at the end of the events. Finally, it is assumed, that RC's wash-off processes are influenced by a composition of subcatchments, such as roofs, streets, and parking lots, which is explained by the intermediate position of RC in comparison to FR, HT, and PL. In fact, runoff from different surfaces is superposed and therefore pollution transport processes are mixed.

From seasonal M (V)-curves it can be observed, firstly, that M (V)-curve distributions at FR show the largest variability in the first 50% of runoff volume throughout the seasons except for spring. The delayed wash-off process during winter months can be caused by a low pollutant potential on surfaces, coarser particles with high densities, or by events with low rain intensities. Secondly, variability of M (V)-curve distribution, in general, is largest during autumn, especially for sites FR, RC, and PL. It can be assumed that this is mainly caused by high variability of rainfall intensities in conjunction with varying pollutant masses available at the surface. It must also be noted, that only few events were monitored during the winter months, which must be taken into account for further statistical analysis.

5. CONCLUSIONS

A long-term monitoring campaign was conducted to analyze stormwater pollutant processes at microscale sites with online sensors. Rainfall runoff events were statistically analyzed and intra-event TSS load distributions were site- and season-specifically examined by means of *M* (*V*)-curves. The correlation analysis reveals a strong relationship between rainfall intensity and event loads for small catchments with a high proportion of impervious surfaces, but not for the larger residential catchment. Furthermore, grouping *M* (*V*)-curves with boxplots at runoff volume quantiles enables the comparison of wash-off behaviors of different catchments. In general, the wash-off process at site FR (flat roof) tends to be source-limited. In contrast, sites PL (parking lot) and HT (high-traffic) show a transport-limited behavior. A seasonal analysis of *M* (*V*)-curve distributions demonstrated the large variability, especially during autumn.

With these results, this paper clearly highlights the need for a spatially more detailed assessment of stormwater quality runoff. This can be drawn from the subcatchment-specific wash-off behavior. Consequently, it can be recommended to use different wash-off models for different catchment types to adequately address transport-limited and source-limited catchments.

ACKNOWLEDGMENTS

The research work and software developments are part of the research project "Modelle für Stofftransport und -behandlung in der Siedlungshydrologie" (STBMOD) which has been supported by the German Federal Ministry of Education and Research (BMBF, FKZ 03FH033PX2).

REFERENCES

1. Dotto, C.B.S.; Kleidorfer, M.; Deletic, A.; Fletcher, T.D.; McCarthy, D.T.; Rauch, W. Stormwater quality models: Performance and sensitivity analysis. *Water Sci. Technol.* **2010**, *62*, 837–843.
2. Sage, J.; Bonhomme, C.; Al Ali, S.; Gromaire, M.-C. Performance assessment of a commonly used "accumulation and wash-off" model from long-term continuous road runoff turbidity measurements. *Water Res.* **2015**, *78*, 47–59.

3. Alias, N.; Liu, A.; Goonetilleke, A.; Egodawatta, P. Time as the critical factor in the investigation of the relationship between pollutant wash-off and rainfall characteristics. *Ecol. Eng.* **2014**, *64*, 301–305.
4. Egodawatta, P.; Miguntanna, N.S.; Goonetilleke, A. Impact of roof surface runoff on urban water quality. *Water Sci. Technol.* **2012**, *66*, 1527–1533.
5. Liu, A.; Egodawatta, P.; Guan, Y.; Goonetilleke, A. Influence of rainfall and catchment characteristics on urban stormwater quality. *Sci. Total Environ.* **2013**, *444*, 255–262.
6. Métadier, M.; Bertrand-Krajewski, J.-L. The use of long-term on-line turbidity measurements for the calculation of urban stormwater pollutant concentrations, loads, pollutographs and intra-event fluxes. *Water Res.* **2012**, *46*, 6836–6856.
7. Sun, S.; Barraud, S.; Castebrunet, H.; Aubin, J.-B.; Marmonier, P. Long-term stormwater quantity and quality analysis using continuous measurements in a French urban catchment. *Water Res.* **2015**, *85*, 432–442.
8. United States Environmental Protection Agency. *Methods for the Chemical Analysis of Water and Wastes*; Environmental Protection Agency: Cincinnati, OH, USA, 1979.
9. Dierschke, M.; Welker, A. Bestimmung von Feststoffen in Niederschlagsabflüssen. *GWF Wasser Abwasser* **2015**, *156*, 440–446.
10. Leutnant, D.; Henrichs, M.; Muschalla, D.; Uhl, M. OSCAR—An online supervisory control and urban drainage data acquisition system with R. In Proceedings of the 10th International Conference on Urban Drainage Modelling, Mont-Sainte-Anne, QC, Canada, 20–23 September 2015; pp. 135–138.
11. Bertrand-Krajewski, J.-L. TSS concentration in sewers estimated from turbidity measurements by means of linear regression accounting for uncertainties in both variables. *Water Sci. Technol.* **2004**, *50*, 81–88.
12. Hannouche, A.; Chebbo, G.; Ruban, G.; Tassin, B.; Lemaire, B.J.; Joannis, C. Relationship between turbidity and total suspended solids concentration within a combined sewer system. *Water Sci. Technol.* **2011**, *64*, 2445–2452.
13. Lepot, M.; Aubin, J.-B.; Bertrand-Krajewski, J.-L. Accuracy of different sensors for the estimation of pollutant concentrations (total suspended solids, total and dissolved chemical oxygen demand) in wastewater and stormwater. *Water Sci. Technol.* **2013**, *68*, 462–471.
14. Geiger, W.F. Flushing effects in combined sewer systems. In Proceedings of the 4th International Conference in Urban Storm Drainage, Lausanne, Switzerland, 31 August–4 September 1987; pp. 40–46.
15. Bertrand-Krajewski, J.L.; Chebbo, G.; Saget, A. Distribution of pollutant mass vs. volume in stormwater discharges and the first flush phenomenon. *Water Res.* **1998**, *32*, 2341–2356.
16. Lacour, C.; Joannis, C.; Gromaire, M.; Chebbo, G. Potential of turbidity monitoring for real time control of pollutant discharge in sewers during rainfall events. *Water Sci. Technol.* **2009**, *59*, 1471–1478.
17. Bai, S.; Li, J. Sediment Wash-Off from an Impervious Urban Land Surface. *J. Hydrol. Eng.* **2013**, *18*, 488–498.
18. Zhao, H.; Chen, X.; Hao, S.; Jiang, Y.; Zhao, J.; Zou, C.; Xie, W. Is the wash-off process of road-deposited sediment source limited or transport limited? *Sci. Total Environ.* **2016**, *563–564*, 62–70.
19. Dierschke, M.; Welker, A. Feine Feststoffe (PM63) in Dachabflüssen. *GWF Wasser Abwasser* **2013**, *154*, 1242–1249.
20. Förster, J. Variability of roof runoff quality. *Water Sci. Technol.* **1999**, *39*, 137–144.
21. Kobencic, R. *Verunreinigung des Regenwasserabflusses von Dachflächen*; Diplomarbeit, TU Graz: Graz, Austria, 2002.
22. Brombach, H.; Weiss, G.; Fuchs, S. A new database on urban runoff pollution: Comparison of separate and combined sewer systems. *Water Sci. Technol.* **2005**, *51*, 119–128.
23. Helmreich, B.; Hilliges, R.; Schriewer, A.; Horn, H. Runoff pollutants of a highly trafficked urban road—Correlation analysis and seasonal influences. *Chemosphere* **2010**, *80*, 991–997.

CHAPTER 10

Parameterization of a Hydrological Model for a Large, Ungauged Urban Catchment

Gerald Krebs [1,*], Teemu Kokkonen [1], Heikki Setälä [2] and Harri Koivusalo [1]

[1]*Department of Built Environment, School of Engineering, Aalto University, Tietotie 1E, 02150 Espoo, Finland*

[2]*Department of Environmental Sciences, University of Helsinki, Niemenkatu 73, 15140 Lahti, Finland*

ABSTRACT

Urbanization leads to the replacement of natural areas by impervious surfaces and affects the catchment hydrological cycle with adverse environmental impacts. Low impact development tools (LID) that mimic hydrological processes of natural areas have been developed and applied to mitigate these impacts. Hydrological simulations are one possibility to evaluate the LID performance but the associated small-scale processes require a highly spatially distributed and explicit modeling approach. However, detailed data for model development are often not available for large urban areas, hampering the model parameterization. In this paper we propose a methodology to parameterize a hydrological model to a large, ungauged urban area by maintaining at the same time a detailed surface discretization for direct parameter manipulation for LID simulation and a firm reliance on available data for model conceptualization. Catchment delineation was based on a high-resolution digital elevation model (DEM) and model parameterization relied on a novel model regionalization approach. The impact of automated delineation and model regionalization on simulation results was evaluated for three monitored study catchments (5.87–12.59 ha). The simulated runoff peak was most sensitive to accurate catchment discretization and calibration, while both the runoff volume and the fit of the hydrograph were less affected.

Keywords: SWMM; model regionalization; DEM delineation; ungauged; large urban catchment; LID

1. INTRODUCTION

Urbanization is associated with land-cover modification and increased imperviousness through the replacement of natural areas with features typical to urban landscapes such as rooftops, streets, and parking lots. These changes alter the hydrological catchment processes [1,2] and have harmful impacts on the receiving environment [3,4,5,6,7,8,9,10]. As a consequence, stormwater management practices have been developed with the aim of mimicking hydrological processes of natural catchments (such as infiltration, storage, and evapotranspiration) and to treat and reduce runoff at the source of its generation [11,12]. These concepts are referred to as Low Impact Development (LID, used in this paper) in the USA [12], Water Sensitive Urban Design (WSUD) in Australia [13], and Sustainable Urban Drainage Systems (SUDS) in the United Kingdom [14]. LID covers a range of tools and planning strategies, including green roofs, pervious pavement, or bio-retention, but also the minimization of required impervious surfaces through planning and layout of newly constructed areas [12,15].

Model simulations are one option to assess the effectiveness of LID tools before their implementation; the Stormwater Management Model (SWMM) [16,17] is a commonly used one, providing the required capabilities [18]. LID tools represent small-scale hydrological processes and thus require an explicit and spatially distributed modeling strategy [19]. This strategy is supported by the recent development of urban data banks [20] containing rather detailed information on the urban land-cover and drainage network properties, but in addition to such spatial characteristics the modeling also requires data on runoff. However, such data are still unavailable for many urban areas [21,22,23], implying that the hydrological assessment of particularly larger urban areas has to be conducted using less detailed information [24]. Consequently, LID tools are commonly simulated by the alteration of lumped parameters [25] rather than explicitly describing their properties. Such approaches include the curve number (CN) method used by Palla et al. [26] and Carter and Jackson [27]. Mentens et al. [28] assessed the runoff reduction potential of green roofs for the city of Brussels (Belgium) using green roof runoff retention values reported in the literature. Montalto et al. [29] estimated the effectiveness of various LID tools in reducing combined sewer overflows (CSO) for a catchment in New York (USA) using composite runoff coefficients. A more explicit approach to simulate green roof runoff at the catchment scale was undertaken by Versini et al. [30,31]. They extracted roof areas from two catchments for separate parallel simulation using their own computation module for green roof description, while other catchment hydrological processes were simulated using SWMM. While this approach could allow the detailed parameterization of larger (order of >10 km^2) urban areas, the application was so far reported only for smaller (2.4–5.5 km^2) watersheds. Amaguchi et al. [19] used vector-based catchment delineation to assess urban runoff for a 1.1 km^2 watershed in Japan. They used a very detailed surface description representing small and homogeneous elements of the urban landscape, which would allow the individual alteration of parameter values for LID simulations.

Krebs et al. [23] reported that a hydrological model that discarded inter-subcatchment surface flow and sewer pipe flow routing for conduits with a diameter smaller than 300 mm was well able to replicate the temporal dynamics of urban runoff when compared to high-resolution (HR) model setups. The low-resolution (LR) models proposed by Krebs et al. [23] omitted details on surface flow patterns and information for small conduits, which often are not readily available. Still, their LR models were based on a detailed vectorized surface discretization that allows direct parameter manipulation for LID simulations. Catchment delineations were based on the HR models, for which surface flow patterns were investigated during on-site visits and missing conduit data were manually complemented. Missing conduit data manifested as both missing conduit information (such as invert elevations) that had to be manually interpolated, and disconnected network sections, where entire conduits were missing in the data set [23]. This approach, however, is not feasible for larger urban areas and thus automated methods of catchment delineation utilizing high resolution DEMs are needed. The flow of stormwater in urban areas does not necessarily follow the topography due to obstacles such as buildings or street curbs [32]. Jankowfsky et al. [33] showed for a peri-urban catchment in France that catchment properties (such as the drainage area) differed significantly when either only a DEM was used for delineation or the topographical information was complemented with field survey data. They stated, however, that for large parts of the catchment only a DEM with a low spatial resolution (25 m) was available. Currently, with the expansion of airborne LIDAR observations, high-resolution DEMs have become available for many urban regions and have been used for several hydrological assessments in urban areas [34,35,36,37,38]. However, small-scale features, such as street curbs that influence urban surface flow, remain unsatisfactorily represented in high-resolution terrain models [22,39,40].

While topographic information is to some extent available, many urban areas are ungauged and thus rainfall–runoff data for model calibration are not always available [21,41,42,43,44,45] and consequently alternative approaches for model parameterization are needed for ungauged catchments [46,47]. Parameterization of ungauged catchments is often referred to as model regionalization [48]. Numerous studies have addressed model regionalization and the methods investigated include regression analysis [44,45,49,50], site-similarity [41,50], and the spatial proximity approach [51]. Kokkonen et al. [50] suggested that a direct parameter transfer is worthwhile when it can be assumed that two catchments behave hydrologically in a similar way. While the model regionalization concepts are well introduced and widely tested, the regionalization of model parameters is rarely accomplished and demonstrated in urban catchments, which motivates our model application in a large urban area.

In this paper we propose a methodology to simulate the effectiveness of LID tools at the city scale by complementing the results of Krebs et al. [23] for small urban study catchments. While Krebs et al. [23] presented a surface discretization approach that supports the explicit simulation of LID strategies along with the identification of a proper spatial resolution for urban hydrological modeling, the extension of the methodology presented here is based on readily available spatial data and omits many details that require on-site visits and manual data complementation. This extension, together with earlier results [23], allows the parameterization of a large urban catchment while allowing the direct alteration of model parameters for LID simulations. The impact of DEM-based catchment delineation on simulation results was evaluated against HR model results and observed runoff available from three study catchments. Subsequently, the Vesijärvi catchment (30 km^2) of the city of Lahti (Finland) was divided into subcatchments using digital elevation data. The Vesijärvi catchment is ungauged and the parameter regionalization here is based on setting the same parameter values for all areas sharing the same surface type, independent of their location. The feasibility of the regionalization approach was evaluated by inter-changing parameter sets calibrated for the three monitored study catchments and investigating the ability of each parameter set to replicate the monitored runoff in the other study catchments.

2. MATERIALS AND METHODS

2.1. Study Site and Data

The three study catchments are located within the city of Lahti, southern Finland (Figure 1). The catchments are characterized by varying degrees of impervious cover and land-use intensity, ranging from dense urban land-use in Catchment 1 to suburban residential housing in Catchment 3. The Stormwater Management Model (SWMM) [16,17] was selected as the modeling platform for this study and rainfall–runoff data collected by the University of Helsinki [52,53] between 2008 and 2010 were used for the parameterization of each catchment. Detailed results and further description of the catchments and the hydro-meteorological data are presented in Krebs et al. [23,54] and Valtanen et al. [52,53].

The hydrological impact of perturbations in the resolution of spatial discretization of the urban landscape and infrastructure were explored in Krebs et al. [23]. In the current paper we use the same catchments and data to further investigate the applicability of parameter values calibrated to HR catchments to larger ungauged urban areas and the impact of automated methods for catchment delineation on the simulated runoff.

The Vesijärvi catchment is in this study defined as the area of the city of Lahti that drains into the Enonselkä basin of Lake Vesijärvi (Figure 1). The catchment was selected as the region of interest for upscaling the model results for several reasons: (i) all three study catchments are located within the Vesijärvi catchment; (ii) the catchment covers a large part of the city of Lahti, and is therefore of interest for urban planning; (iii) the water quality of lake Vesijärvi is of substantial interest to the city of Lahti; and (iv) the catchment is ungauged and sufficiently large (30 km^2) to pose those challenges that need to be overcome in hydrological assessments of large urban areas. The impervious cover of the Vesijärvi catchment is ca. 27%, of which 25% (6.8% of the whole catchment) consists of roof tops and the remaining 75% (20.5%) is associated with traffic infrastructure, including streets, parking facilities, and sidewalks. The majority of the pervious land-cover consists of natural forested areas (37% of the whole catchment) and cultivated green areas such as yards and parks (12.5% of the catchment). Details on the land-cover of the Vesijärvi catchment and the three study catchments are given in Table 1.

The Vesijärvi catchment is drained through a separate stormwater sewer system and two open streams flowing through the city and draining into Lake Vesijärvi.

The stormwater sewer network used for model development considered conduits with a minimum diameter of 300 mm. Krebs et al. [23] proposed this diameter threshold for assessing small-scale processes in large urban areas as it allows: (i) a sufficiently accurate replication of the dynamics of urban runoff and (ii) a feasible model development considering the quality of available spatial data for large areas, where manual correction and complementation of the data is too laborious. The stormwater sewer network in the Vesijärvi catchment has a length of 115.8 km and the open stream length is 28.6 km (Table 2).

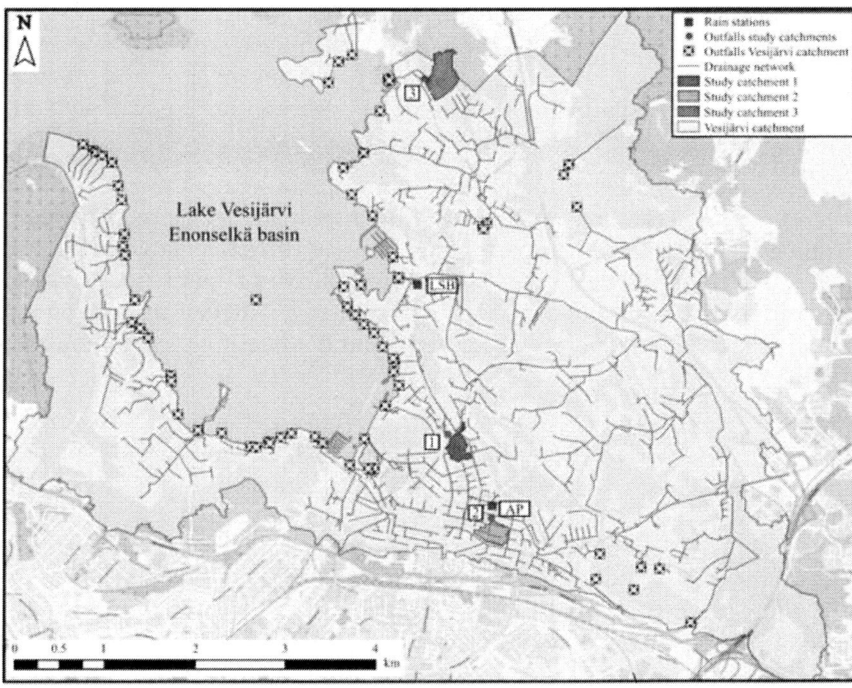

Figure 1. The Vesijärvi catchment of the city of Lahti, the drainage network (conduits with a diameter ≥300 mm and open channels), and the outfalls of the system to the receiving water bodies. The monitored study Catchments 1, 2, and 3 along with their outfalls and the two rainfall stations (AP and LSB) are shown.

Table 1. Surface and land-use types and their fractions of the catchment for HR and LR-DEM models of the study catchments 1, 2, and 3, and for the Vesijärvi catchment.

Surface Type	Surface Type CODE	Land-Use Type	1				2				3				Vesijärvi	
			Area (ha)		Fraction (%)		Area (ha)		Fraction (%)		Area (ha)		Fraction (%)		Area (ha)	Fraction (%)
			HR	LR DEM	HR	LR DEM	HR	LR DEM	HR	LR DEM	HR	LR DEM	HR	LR DEM		
Asphalt	A	Other	-	-	-	-	-	-	-	-	-	0.002	-	0.0	36.2	1.2
		Parking lot	1.505	1.016	25.6	18.7	0.783	0.888	11.8	13.1	0.140	0.163	1.1	1.4	165.3	5.5
		Pavement	0.846	1.048	14.4	19.3	0.412	0.493	6.2	7.3	0.053	0.064	0.4	0.6	83.9	2.8
		Street	1.126	1.075	19.2	19.8	0.492	0.516	7.4	7.6	0.679	0.673	5.4	6.0	157.4	5.3
Concrete	C	Other	-	0.107	-	2.0	-	0.007	-	0.1	-	-	-	-	7.7	0.3
		Parking lot	-	0.044	-	0.8	-	0.051	-	0.8	-	-	-	-	0.7	0.0
		Pavement	-	-	-	-	-	0.114	-	1.7	-	-	-	-	3.9	0.1
		Street	-	-	-	-	-	0.030	-	0.4	-	-	-	-	0.3	0.0
Gravel	G	Other	0.130	0.143	2.2	2.6	0.184	0.023	2.8	0.3	-	-	-	-	72.2	2.4
		Parking lot	0.130	0.002	2.2	0.0	0.892	0.640	13.4	9.5	0.033	0.053	0.3	0.5	23.4	0.8
		Pavement	-	0.004	-	0.1	0.002	0.049	0.0	0.7	0.010	-	0.1	-	37.8	1.3
		Railway	-	-	-	-	-	-	-	-	-	-	-	-	7.4	0.2
		Street	-	-	-	-	0.011	0.011	0.2	0.2	-	-	-	-	7.1	0.2
Natural stone paver	NS	Other	-	-	-	-	-	0.001	-	0.0	-	-	-	-	0.0	0.0
		Pavement	-	-	-	-	-	-	-	-	-	-	-	-	0.0	0.0
Cobble stone	CS	Other	0.247	0.007	4.2	0.1	0.109	0.128	1.6	1.9	0.015	-	0.1	-	3.6	0.1
		Parking lot	-	-	-	-	0.232	0.124	3.5	1.8	0.095	0.006	0.8	0.1	1.4	0.0
		Pavement	-	0.021	-	0.4	0.131	0.017	2.0	0.2	-	-	-	-	0.3	0.0
		Street	-	-	-	-	0.024	-	0.4	-	-	-	-	-	2.3	0.1
Metal sheeting	MS	Roof	1.149		19.6		1.275		19.2		-		-			
Sheeting			-	1.474	-	27.2	-	1.406	-	20.8	0.020	0.915	0.2	8.1	203.0	6.8
Tiles/Sheeting	TS		-		-		-		-		1.342		10.7			
Crop	CR	Green	-	-	-	-	-	-	-	-	-	-	-	-	33.3	1.1
Lawn	L		0.383	0.099	6.5	1.8	0.325	0.005	4.9	0.1	0.500	0.015	4.0	0.1	75.2	2.5
Lawn/Vegetation	LV		-	0.105	-	1.9	0.493	0.285	7.4	4.2	0.923	2.619	7.3	23.2	262.9	8.8
Peat	P		-	-	-	-	-	-	-	-	-	-	-	-	3.0	0.1
Vegetation/Lawn	VL		-	0.040	-	0.7	0.570	1.503	8.6	22.2	1.198	0.352	9.5	3.1	372.9	12.5
Vegetation	V		0.358	0.241	6.1	4.4	0.697	0.467	10.5	6.9	0.200	0.028	1.6	0.3	234.1	7.9
Swamp	SW		-	-	-	-	-	-	-	-	-	-	-	-	13.2	0.4
Rock	R	Rock	-	-	-	-	-	-	-	-	-	0.266	-	2.4	64.4	2.2
Forest	F	Forest	-	-	-	-	-	-	-	-	7.381	6.137	58.6	54.3	1107.6	37.2
Σ			5.874	5.427	100	100	6.632	6.756	100	100	12.588	11.292	100	100	2980.6	100

Table 2. Drainage network subdivision for the HR and LR-DEM models of the study catchments and the Vesijärvi catchment.

Drainage Type	Material	1 Length (m)		2 Length (m)		3 Length (m)		Vesijärvi Length (m)
		HR	LR-DEM	HR	LR-DEM	HR	LR-DEM	
Conduit	Concrete	762	611	835	624	1649	1041	101,805
	PVC	1851	26	2397	-	-	-	13,957
Open channel	Natural	-	-	-	-	74	-	28,620
Σ		2613	637	3232	624	1723	1041	144,382

Rainfall input data for simulations of the Vesijärvi catchment were derived from a tipping bucket gauge at Catchment 2 (AP in Figure 1) with a recording interval of 1 minute. Data from this rainfall gauge were already used for calibration and validation of the study catchments [23]. Additionally, rainfall data recorded at 10-min intervals on the roof of the Lahti Science and Business Park (LSB in Figure 1) were available and used for additional validation of the HR study catchment models (VAL-LSB) [23]. Runoff data were recorded at the outfall of each study catchment through 2008–2010 with a 1-min recording interval. The rainfall–runoff data were scanned to identify events where both the rainfall and runoff measurements were functioning [23]. The sequence of events used for calibration (CAL) consists of six, six, and three events for Catchments 1, 2, and 3, respectively. The sequence of validation events using AP rainfall data (VAL-AP) consists of six, five, and one event(s) for Catchments 1, 2, and 3, respectively. The sequence using LSB rainfall data (VAL-LSB) consists of five, three, and eight events for Catchments 1, 2, and 3, respectively. A comprehensive list of rainfall events is given in Krebs et al. [23]. These data are used in this paper to evaluate the impact of DEM-based delineation and model regionalization on simulation results. Spatial data for the discretization of the urban landscape and the drainage network included a DEM (2-m resolution, provided by the Geological Survey of Finland), data on the building roofs and the road network, and digitized maps of the drainage network.

2.2. DEM-Based Delineation (LR-DEM)

Krebs et al. [23] studied the impact of spatial resolution for urban hydrological modeling and concluded that the temporal dynamics can be sufficiently well replicated with a model that (i) discards stormwater sewer conduits with a diameter smaller than 300 mm and (ii) discards surface flow routing between subcatchments before surface runoff enters the drainage network. Reduction of network conduits represented in the model (d_{min} ≥ 300 mm) results in a replacement of conduit flow by overland flow in the discretized catchment. Additionally, the neglecting of surface flow routing affects the effective impervious area (*EIA*) [23]. Thus, the approach presented in Krebs et al. [23] implies that *EIA* equals the total impervious area (*TIA*) of the catchment. The increase of *EIA* depends on the catchment structure; the changes were reported to be minor for highly urbanized catchments (+0.3% in Catchment 1) and larger (up to 155% for Catchment 3) for suburban catchments, where impervious surfaces are sometimes not connected to the drainage network [23]. Despite the network conduit and surface flow assumptions, the concept presented by Krebs et al. [23] allows the direct application of LID tools in the model as subcatchment delineation is based on identifying homogeneous surface types (such as rooftops or street sections) at a high spatial resolution.

The LR-DEM model developed in the current study follows the principles described above to maintain the ability to replicate the temporal dynamics of urban runoff. However, the contributing drainage area for each stormwater network inlet is delineated in an automated manner using a DEM (2 m resolution) rather than manually determining the surface routing pattern as in Krebs et al. [23]. While this automated method for catchment delineation is suitable for large urban areas where on-site observation of surface flow patterns is not feasible, small-scale features of the urban landscape (such as street curbs) that are not well presented in DEMs [39,40] will be neglected in the model surface discretization. As a result, the HR and LR-DEM models differ not only in *EIA*, length of the drainage network, and representation of inter-subcatchment surface flow, but also in terms of catchment boundaries (i.e., area) and consequently the catchment *TIA*. While this approach obviously discards important information on the catchment, the suggested simplifications for model development allow the assessment of larger urban areas in a feasible manner. Subsurface drainage conduits are naturally not represented in a DEM and, thus, the DEM is commonly pre-processed by burning the conduit network into the DEM surface [22]. In this study a constant value was reduced from DEM grid cells where conduits were located and thereafter depressions filled to allow for hydraulic connectivity. Gironás et al. [22] suggested that all conduits, even though not explicitly modeled, should be included when pre-processing a DEM to allow for realistic drainage basin delineation in urban areas. Following this suggestion, all available conduit information was used to derive the pre-processed DEM, even though only conduits with a minimum diameter of 300 mm were included in the LR-DEM model.

The LR-DEM model was developed using the following steps: (i) the surface was digitized based on both surface (such as asphalt, gravel, or different vegetation types) and land-use types (such as streets, parking lots, or rooftops) derived from spatial data and aerial images (see Table 1); (ii) the stormwater drainage network was digitized using conduits with a minimum diameter of 300 mm; (iii) the contributing drainage area of each inlet present in the minimum diameter was delineated based on the pre-processed DEM; and (iv) the digitized surface map was discretized using the delineated drainage areas resulting in a number of subcatchments per drainage inlet. The described LR-DEM model was developed for the entire Vesijärvi catchment, including the three study catchments.

Krebs et al. [23] concluded that the neglecting of inter-subcatchment surface flow routing for models with lower resolution resulted in decreased model performance for the least urbanized Catchment 3, due to the fact that roof runoff in this catchment is, unlike described in guidelines by the city [55], not drained into the drainage network but onto adjacent green areas. The associated large change in hydraulically connected impervious area when neglecting inter-subcatchment flow resulted in over-prediction of runoff volumes and peaks. Thus, also for the LR-DEM model in this paper, roof runoff was diverted to adjacent green areas in this study catchment. However, this exemption applies only to roof runoff and also in Catchment 3 runoff from all other subcatchments was directly routed to the corresponding drainage network inlet. Surface and conduit related parameters were adopted from the calibrated study catchment model. Model parameters related to the subcatchment and conduit geometry were derived from spatial data. While the subcatchment flow width FW_{HR} for the HR models was determined using Equation (1), the corresponding flow width FW_{LR} for the LR-DEM model was determined using Equation (2):

$$FW_{HR} = \frac{A}{L} \tag{1}$$

$$FW_{LR} = k\sqrt{A}, \tag{2}$$

where A (m^2) is the subcatchment area, L (m) is the length of the overland flow path, and k (-) is a dimensionless coefficient. Krebs et al. [23] determined 0.7 to be a suitable value for k. This value was based on the distribution of k for 2652 subcatchments, for which FW_{HR} was calculated using Equation (1) and used herein to define FW_{LR}.

The impact of this automated model parameterization procedure that neglects details of the stormwater drainage description with respect to both the network and surface flow was first evaluated against monitored runoff in the three study catchment. The model performance was thereafter evaluated against the HR model results reported in Krebs et al. [23,54]. The HR study catchment models were used as reference, as they include all catchment details concerning both the surface and network drainage. The model evaluation was conducted both by visual inspection of simulated and observed hydrographs and using model performance statistics. The performance statistics were the Nash-Sutcliffe efficiency *E* [56], the peak flow error *PFE*, and the volume error *VE*.

2.3. Parameter Regionalization

Krebs et al. [23,54] initially developed the HR models for the three study catchments with the aim to assess the much larger ungauged Vesijärvi catchment. The use of separate surface types for catchment subdivision produces a mosaic of subcatchments with homogenous hydrological properties. To allow for the transfer of parameters from sample catchments to an ungauged area, subcatchments with similar homogenous properties (such as a stretch of asphalt street) have to be identified both in each study catchment and in the entire Vesijärvi catchment. In other words, a similar detailed subdivision of the catchment surface for all study catchments and the targeted up-scale area are a prerequisite for the regionalization approach conducted in this study. While this condition mainly concerns the technical applicability of parameter values in the ungauged area, the question remains whether the application is hydrologically justified. The calibration of the HR study catchment models was conducted independently for each catchment [23,54]. Krebs et al. [23] showed that the parameter values calibrated for the same surface types (e.g., an asphalt surface) were similar for each of the three study catchments.

To further investigate the hydrological applicability, in this paper, the parameter sets of each study catchment were applied to the remaining two study catchments to evaluate the regionalization impact on simulated flow. The HR study catchment models were used for this purpose and three parameter sets were defined based on the parameter values calibrated to the study catchments (V_1 for Catchment 1, V_2 for Catchment 2, and V_3 for Catchment 3). Additionally, a reference parameter set (V_{Ref}) was compiled based on parameter value suggestions given in SWMM user manuals [16,17]. Parameter values were either adopted as the mean value in the case of a suggested parameter value range or as a single suggested value. The application of V_{Ref} represents a common strategy for the parameterization of ungauged catchments [57] and is thus used herein to evaluate the performance of calibrated parameter values transferred to an ungauged catchment.

Table 3. The parameter sets V_1, V_2, V_3, and V_{Ref} for the depression storage *D*, the Manning's roughness for overland flow n_o, and the imperviousness *I*. Parameters that were calibrated for the catchment have a gray background, uncalibrated values have a white background, and the subscripts 1–3 indicate from which study catchment values were adopted if corresponding surfaces were missing. Parameter values of $V_{Ref}$with subscript a were derived from Rossman [17], values with subscript b were derived from Huber and Dickinson [16], and values with subscript ab were interpolated between suggestions in Rossman [17] and Huber and Dickinson [16]. A description of the "surface type codes" is given in Table 1.

Surface Type	*D* (mm)				n_o (−)				*I* (%)			
	V_1	V_2	V_3	V_{Ref}	V_1	V_2	V_3	V_{Ref}	V_1	V_2	V_3	V_{Ref}
A	0.39	0.42	0.62	1.91_a	0.011	0.011	0.013	0.007_b	100.0	88.5	100.0	100.0
C	0.39	0.42	0.62	1.91_a	0.011	0.011	0.013	0.007_b	100.0	88.5	100.0	100.0
G	2.54	2.49	2.54	1.91_a	0.020	0.030	0.024	0.021_b	0–50	33.4	69.5	0
NS	1.01	0.39	1.09	1.91_a	0.012	0.020	0.020	0.024_a	91	86.6	100.0	100.0
CS	1.01	0.39	1.09	1.91_a	0.012	0.020	0.020	0.024_a	91	86.6	100.0	100.0
MS												
S				1.91_a	0.012			0.007_{ab}	100.0	100.0	100.0	100.0
TS												
CR	4.18	4.13	7.53	7.62_a	0.300	0.667	0.790	0.600_a	0.0	0.0	0.0	0.0
L	4.98	4.82	5.07	3.81_a	0.150	0.168	0.200	0.150_b	0.0	0.0	0.0	0.0
LV	4.22_2	4.22	5.07	5.08_a	0.238_2	0.238	0.300	0.375_{ab}	0.0	0.0	0.0	0.0
P	7.39_3	7.39_3	7.39	7.62_a	0.668_3	0.668_3	0.668	0.600_a	0.0	0.0	0.0	0.0
VL	3.59_2	3.59	2.54	5.08_a	0.326_2	0.326	0.399	0.375_a	0.0	0.0	0.0	0.0
V	4.18	4.13	7.53	7.62_a	0.300	0.667	0.790	0.600_a	0.0	0.0	0.0	0.0
SW	50.00	50.00	50.00	50.00	0.500	0.500	0.500	0.500	0.0	0.0	0.0	0.0
R	7.39_3	7.39_3	7.39	7.62_a	0.668_3	0.668_3	0.668	0.600_a	100.0	100.0	100.0	100.0
F	7.39_3	7.39_3	7.39	7.62_a	0.668_3	0.668_3	0.668	0.600_a	0.0	0.0	0.0	0.0

For surface types that were not present in the study catchment used for defining the parameter values, the corresponding values were selected from one of the remaining study catchments to attain a complete set of parameters for simulation. Parameters that are specific to a subcatchment rather than a surface type were derived directly from spatial data. Surface-type specific parameters were the depression storage *D*, Manning's roughness n_o for overland flow, and the imperviousness *I*, while the subcatchment slope, the flow width FW_{HR} or FW_{LR}, and the subcatchment area depend on the subcatchment geometry. The required model parameter values for the drainage network were derived in a similar manner. However, only the Manning's roughness n_c for conduit flow is specific for a certain conduit material, while parameters such as the conduit length and slope are specific for each conduit. A full list of values for the three derived parameter sets (V_1, V_2, and V_3) and the reference parameter set V_{Ref} are given in Table 3 (surface parameters) and Table 4 (drainage network parameters). The effect of parameter transfer on simulated flow for the HR models was evaluated for the calibration and validation sequences used in Krebs et al. [23] as well as for individual events. Additionally, the hydrological applicability of calibrated parameters to other urban catchments was investigated by applying the three calibrated parameter sets (V_1, V_2, and V_3) and the reference parameter set (V_{Ref}) to the LR-DEM model to evaluate the parameter set related variation of simulated flow for both the study catchments and the entire Vesijärvi catchment. Details on the hydro-meteorological properties of selected rainfall–runoff events and sequences can be found in Krebs et al. [23].

Table 4. The parameter sets V_1, V_2, V_3, and V_{Ref} for the Manning's roughness for conduit flow n_c. Calibrated parameter values have a grey background, un-calibrated values have a white background and the subscripts 1–3 indicate from which study catchment values were adopted if corresponding surfaces were missing. Parameter values of V_{Ref} were derived from Rossman [17].

Drainage Type	Material	n_c (–)			
		V_1	V_2	V_3	V_{Ref}
Conduit	Concrete	0.011	0.015	0.015	0.013
	PVC	0.011	0.011	0.011_2	0.013
Open channel	Natural	0.049_3	0.049_3	0.049	0.05

3. RESULTS

3.1. DEM Delineation

As a result of the DEM delineation, the areas of Catchments 1 and 3 decreased, while the area of Catchment 2 increased compared to the areas determined for the HR models (Table 5, Figure 2). While the change of area was similar for the Catchments 1 and 3 (8%) it was smaller for Catchment 2 (2%). The ability of the LR-DEM approach to correctly represent the HR catchments was further evaluated by determination of the overlapping and excessive surface. The first is defined as the fraction of the HR catchment that is included in the LR-DEM catchment and the second defines the fraction of the LR-DEM catchment that is not included in the HR catchment. The best overlap was found for Catchment 2, where 97% of the HR catchment was included in the LR-DEM catchment (i.e., 3% of the HR Catchment 2 were omitted due to the simplification implied by the automated delineation process). For the same catchment, 5% of the LR-DEM catchment surface was excessive;

this area was included in the LR-DEM Catchment 2 due to the simplified delineation process. For Catchments 3 and 1 the overlap was lower at 89% and 80%, respectively, and while the excessive surface for Catchment 3 was 3% it was clearly higher (13%) for Catchment 1 (Table 5). As runoff generation in urban areas is dominated by processes on impervious surfaces, the simulated runoff volume and dynamics are not only affected by the change in catchment size, but also by the associated alteration in both *TIA* and *EIA*. The HR model of Catchment 1 had a *TIA* of 5.04 ha, of which most of the area was hydraulically effective (*EIA* 5.03 ha), while both the *EIA* and *TIA* of the LR-DEM model were lower at 4.79 ha. The fraction of hydraulically connected impervious surfaces was smaller for the HR model of Catchment 2, for which *EIA* and *TIA* were 3.23 ha and 3.56 ha, respectively. Besides the catchment area, also *EIA* increases for the LR-DEM model of Catchment 2 to 3.74 ha as most of the excessive catchment area comprised rooftops and street sections. For Catchment 3, the reduction in *TIA* is proportional to the reduction of catchment area and *TIA* reduces from 2.37 ha (HR model) to 2.13 ha (LR-DEM model).

However, as stated above, *EIA* does not equal *TIA* for this catchment due to roof runoff routing onto adjacent green areas. The neglecting of inter-subcatchment surface flow routing in the LR-DEM model, while maintaining the original surface flow pattern for roofs (onto adjacent green areas) results in *EIA* of 1.23 ha consisting of streets, pavements and driveways. This corresponds to an increase of 0.3 ha (32%) in *EIA* compared to the HR model of Catchment 3. Roofs account for 0.9 ha in Catchment 3 and also remain hydraulically disconnected from the network in the LR-DEM model setup. The boundaries, contributing drainage areas for each inlet, and subcatchments of the three study catchments are illustrated in Figure 2.

Differences in catchment area, *TIA*, and *EIA* naturally affect the simulated catchment runoff (Figure 3, Table 6). The effects were evaluated by comparison of LR-DEM simulated runoff for Catchments 1, 2, and 3 with HR simulated and observed runoff through visual hydrograph inspection and evaluation of model performance statistics. The smallest impact of DEM delineation on simulated runoff was found for Catchment 1, for which the LR-DEM model yielded a similar efficiency *E* (0.87) for the calibration sequence (CAL) than achieved with the HR model (0.88).

A similar minor degradation in efficiency was found for the LSB validation sequence where the LR-DEM model achieved an efficiency of 0.83 compared to 0.84 achieved by the HR model. The degradation in model efficiency for VAL-AP was slightly larger with an efficiency of 0.81 for the LR-DEM model compared to 0.85 achieved by the HR model.

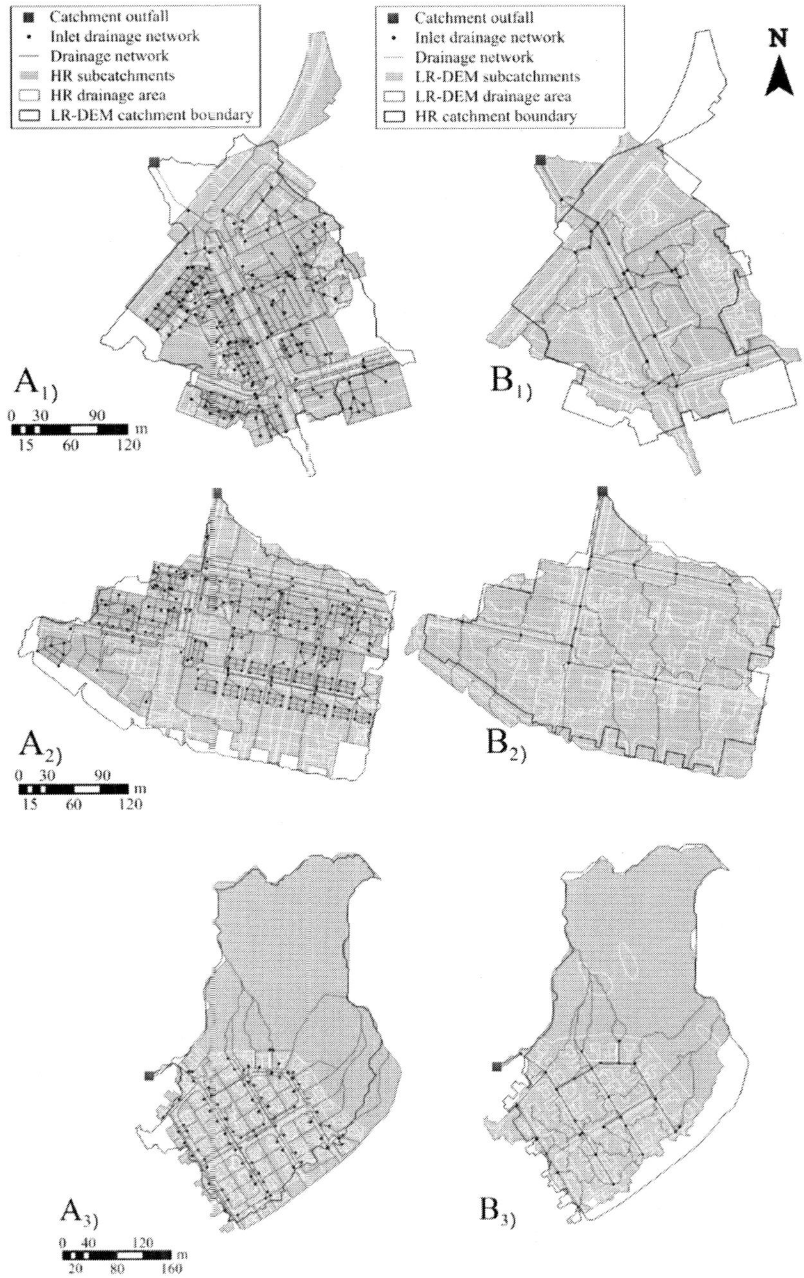

Figure 2. (**A$_1$**,**A$_2$**,**A$_3$**) Surface discretization for the HR model and (**B$_1$**,**B$_2$**,**B$_3$**) the boundaries based on DEM delineation and the surface discretization for the LR-DEM model for Catchments 1, 2, and 3.

In terms of simulated runoff volumes only small impacts were observed with the LR-DEM and HR model producing similar volume errors *VE* (3.7%–7.0% for the HR model compared to 1.4%–6.5% for the LR DEM model) for the three simulation sequences of Catchment 1.

The largest impact of the lower resolution and the DEM delineation method on model performance for Catchment 1 can be seen in the peak flow errors (*PFE*). While the HR model absolute mean *PFE* for simulated events was 12.4%, 8.1%, and 13.7%, for CAL, VAL-AP, and VAL-LSB, respectively, it increased to 23.9%, 26.0%, and 25.3%, for the same sequences of Catchment 1 using the LR-DEM model.

Table 5. Catchment properties of the HR and LR-DEM models of the three study catchments and the overlapping and excessive surface areas.

	Unit	1		2		3	
		HR	LR-DEM	HR	LR-DEM	HR	LR-DEM
Area	(ha)	5.87	5.43	6.63	6.76	12.59	11.58
TIA	(ha)	5.04	4.79	3.56	3.74	2.37	2.13
EIA	(ha)	5.03	4.79	3.23	3.74	0.93	1.23
Overlapping surface	(%)	100	80	100	97	100	89
Excessive surface	(%)	0	13	0	5	0	3
Subcatchments	(−)	690	317	784	426	821	413
Drainage basins	(−)	160	22	188	15	90	28
Drainage network	(m)	2613	637	3232	628	1722	1041

The LR-DEM model efficiency for Catchment 2 showed a larger decrease from the HR simulations than was found for Catchment 1, especially for CAL. However, the performance remained acceptable for both CAL (E = 0.82) and VAL-AP (E = 0.90) (Table 6). The performance for VAL-LSB was lower (E = 0.53), but it should be noted that for this sequence also the HR model produced lower efficiencies (0.61), which may be explained by the relatively large distance between the LSB rainfall station and Catchment 2. With an increasing distance, the temporal dynamics of recorded rainfall and runoff show lower correspondence, which negatively affects simulation results [23]. The capability of the LR-DEM model to replicate observed peak flows in Catchment 2 was better than for Catchment 1, but it can be noted also here that *PFE* is the most sensitive performance criterion in terms of accurate surface flow routing. The impact of the lower resolution on the simulated runoff volume is larger for Catchment 2 compared to Catchment 1, but *VE* remains within acceptable limits (ranging from −1 to 19.5%) for Catchment 2.

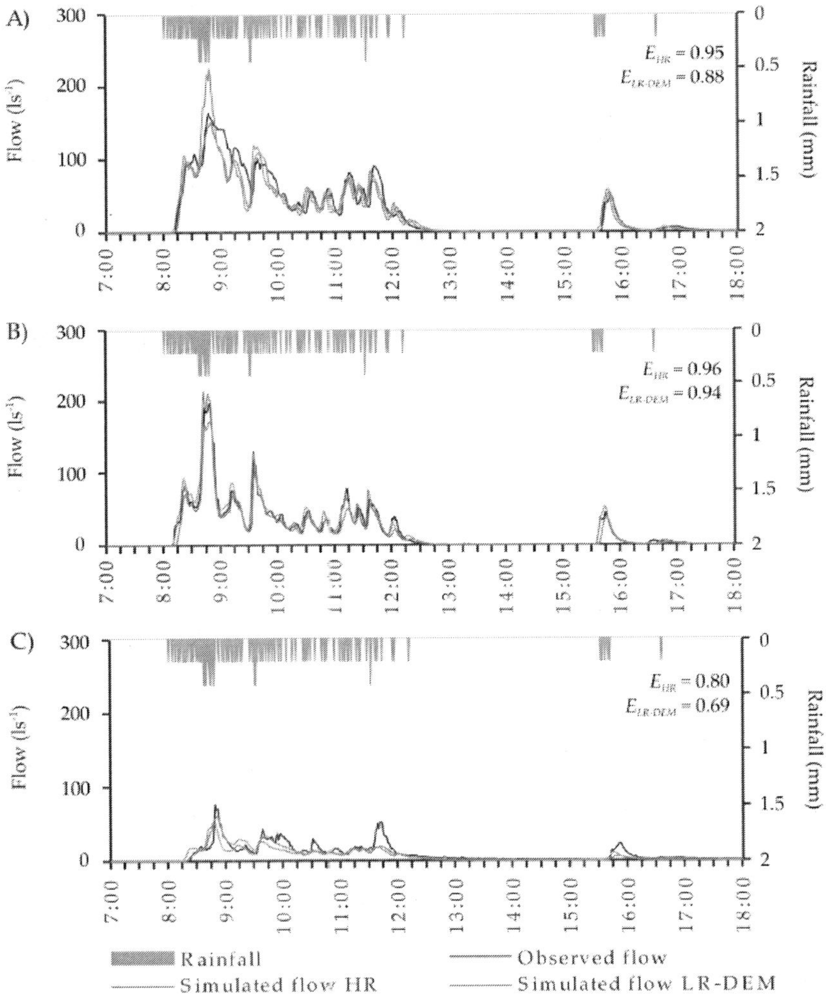

Figure 3. Observed and simulated (HR and LR-DEM) flow for (**A**) Catchment 1; (**B**) Catchment 2; and (**C**) Catchment 3, and the model efficiency E (event rainfall depth 22.0 mm, peak intensity 1.73 mm/5 min).

The impact of the LR-DEM model development procedure on simulation results for Catchment 3 was comparable to that seen for Catchment 2. The LR-DEM model achieved efficiencies of 0.70 (HR model 0.80), 0.56 (0.66), and 0.75 (0.81), for CAL, VAL-AP, and VAL-LSB, respectively. It is to be noted, however, that for Catchment 3 only a single event was available for VAL-AP, due to the remote location of the catchment related to the rainfall measurement station. The change in *PFE* was the smallest for all three catchments except for the single AP validation event. In terms of *VE*, the LR-DEM and HR model performed equally well, with identical or very similar errors for CAL and VAL-LSB. Again, a larger change was found for the AP validation event.

Table 6. Model performance statistics (efficiency *E*, absolute peak flow error *ABS PFE*, and volume error *VE*) for the HR and LR-DEM models of the study Catchments 1, 2, and 3 for the alternative parameter sets V_1, V_2, V_3, and V_{Ref} and the three studied sequences of events. Performance statistics with a grey background were produced by the parameter set specifically calibrated to the study catchment. *ABS PFE* in the first row of each parameter set represents the mean value for the sequence events while the values given in parentheses represent the range of *ABS PFE*.

Sequence	Resolution	Parameter Set	E (–)			ABS PFE (%)			VE (%)		
			1	2	3	1	2	3	1	2	3
CAL	HR	V_1	0.88	0.93	0.71	12.4 (4.1–20.4)	17.1 (4.0–37.3)	29.2 (6.4–51.5)	7.0	−1.8	26.8
		V_2	0.86	0.97	0.73	21.3 (10.3–33.8)	13.1 (3.3–24.9)	35.4 (4.4–56.5)	18.7	9.3	36.1
		V_3	0.81	0.91	0.80	24.6 (2.5–53.9)	25.8 (0.7–53.1)	35.7 (20.9–53.3)	14.6	4.0	32.2
		V_{Ref}	0.73	0.89	0.75	30.7 (2.1–99.8)	34.3 (3.4–100.0)	42.6 (17.2–79.2)	25.6	22.5	36.8
	LR-DEM	V_1	0.87	0.58	0.66	23.9 (5.9–86.7)	53.9 (19.2–77.8)	25.7 (7.9–36.3)	6.5	−29.2	31.4
		V_2	0.91	0.82	0.66	14.4 (0.3–22.6)	31.2 (5.2–51.3)	33.2 (18.7–41.5)	15.1	−19.5	38.3
		V_3	0.87	0.74	0.70	23.5 (2.2–54.6)	37.9 (24.0–49.4)	26.5 (9.0–36.3)	18.1	−18.4	32.2
		V_{Ref}	0.74	0.65	0.66	38.1 (6.5–100.0)	53.1 (26.2–100.0)	45.5 (35.3–65.9)	26.9	1.8	38.6
VAL-AP	HR	V_1	0.85	0.93	0.64	8.1 (0.0–27.6)	10.5 (1.1–30.1)	3.8 -	6.5	8.9	5.0
		V_2	0.83	0.94	0.67	17.8 (12.3–27.5)	16.1 (7.0–38.4)	16.7	18.2	17.8	13.7
		V_3	0.81	0.91	0.66	26.6 (11.8–42.0)	22.6 (1.7–43.0)	29.0 -	13.3	12.7	16.1
		V_{Ref}	0.81	0.84	0.64	21.9 (4.7–46.5)	24.9 (9.6–39.2)	27.1 -	21.0	29.5	23.3
	LR-DEM	V_1	0.81	0.79	0.40	26.0 (7.2–65.8)	37.3 (11.2–64.4)	12.2 -	4.5	−12.3	26.1
		V_2	0.86	0.90	0.54	10.6 (4.2–22.7)	18.5 (2.8–40.1)	8.6 -	14.5	−1.0	34.2
		V_3	0.85	0.92	0.56	19.3 (0.4–44.7)	12.9 (2.6–21.1)	2.1 -	14.2	−1.6	27.9
		V_{Ref}	0.76	0.85	0.52	26.6 (4.6–49.4)	12.6 (1.7–22.2)	0.3 -	20.2	12.5	37.5
VAL-LSB	HR	V_1	0.84	0.59	0.78	13.7 (0.3–26.5)	16.2 (4.7–23.3)	26.3 (1.2–69.4)	3.7	−1.0	11.4
		V_2	0.82	0.61	0.80	30.1 (18.7–38.0)	21.9 (2.9–34.6)	37.0 (2.7–75.2)	16.3	9.7	24.0
		V_3	0.80	0.54	0.81	26.8 (18.0–36.1)	15.8 (5.7–23.2)	33.8 (2.6–71.4)	11.4	0.5	19.9
		V_{Ref}	0.78	0.60	0.78	20.6 (7.8–34.0)	21.4 (6.7–28.9)	33.0 (0.1–70.9)	18.3	17.1	31.5
	LR-DEM	V_1	0.83	0.45	0.65	25.3 (0.6–46.9)	8.3 (0.2–20.6)	33.0 (11.1–96.4)	1.4	−24.6	16.4
		V_2	0.87	0.53	0.72	17.1 (2.5–32.2)	12.8 (11.5–13.8)	35.4 (10.5–67.3)	10.8	−14.9	25.4
		V_3	0.86	0.43	0.75	15.9 (2.9–35.8)	7.9 (1.0–21.2)	27.6 (1.8–64.4)	10.8	−19.9	19.0
		V_{Ref}	0.79	0.49	0.73	22.6 (4.6–32.4)	10.9 (6.1–15.0)	28.3 (0.9–64.2)	16.7	−3.4	33.8

3.2. Parameter Regionalization

Application of model parameters to larger, ungauged areas necessitates that a model parameter set calibrated to one catchment also produces acceptable simulation results for other catchments within the same region. Thus, parameter sets calibrated for the HR models of Catchment 1 (V_1), 2 (V_2), and 3 (V_3) (Table 3 and Table 4) were applied to the other two catchments to evaluate the performance difference to the parameter set specifically calibrated to the catchment. Additionally, the reference parameter set V_{Ref} was applied to each of the study catchments. As the large-scale Vesijärvi catchment model was developed using the LR-DEM methodology the performance of regionalized parameters (V_1, V_2, V_3, and V_{Ref}) was furthermore investigated for the LR-DEM models of the three study catchments. The performance of parameter transfer was evaluated using events of the calibration and validation periods for each catchment investigating both the entire sequences and individual events; the impact of parameter transfer was thus evaluated for a total of 18 simulation sequences (HR and LR-DEM models for each catchment using three sequences each, Table 6) and a total of 86 simulation events.

With respect to model efficiency *E*, the specifically calibrated parameter set achieved the best performance for 13 of the 18 sequences; while for five sequences a transferred parameter set outperformed the specifically calibrated set, V_{Ref} did not achieve a higher efficiency for any of the evaluated sequences (Table 6). For the simulation events, the specifically calibrated parameter set achieved a mean efficiency *E* of 0.76. While the transferred parameter sets achieved a mean efficiency *E* of 0.70, V_{Ref} yielded a lower mean efficiency of 0.52. The specifically calibrated parameter set achieved the best efficiency for 62% of the events. For 35% of the simulation events a transferred parameter set achieved the highest efficiency, while V_{Ref} yielded the highest efficiency for 3% of the events. While the efficiency of V_{Ref} was inferior to both transferred sets for 70% of the simulation events, it outperformed the transferred sets for 9% of the events.

With respect to the simulated runoff volume for entire simulation sequences, a better replication of sequences was achieved by the transferred parameter sets compared to specifically calibrated parameters; while the specific parameter sets produced the most accurate volume replication for seven of the 18 sequences, nine of the sequences were best simulated by a transferred parameter set, and the volume of two sequences was best replicated using V_{Ref} (Table 6). The best performing parameter set concerning runoff volume was V_1, which performed best for all sequences of Catchment 1 (for which it was calibrated) as well as for all sequences of Catchment 3. It is to be noted, however, that a low *VE* for a sequence of events may also be achieved even though the *VE* for individual events is high. This would be the case if there is a balance between events with a volume underestimation and a volume overestimation. When looking at individual simulation events, the specifically calibrated parameter sets produced a lower mean absolute volume error (17.8%) compared to the performance of the transferred sets (21.9%), while V_{Ref} yielded a clearly larger mean volume error (33.9%). The specifically calibrated parameter set produced the smallest volume errors (*VE*) for 40% of the simulation events, and while for 49% of the events a transferred parameter set produced the smallest *VE*, V_{Ref} performed best for 11% of the simulation events. While V_{Ref}produced a larger volume error than both transferred sets for 72% of the simulation events, it outperformed the transferred sets for 13% of the events.

The performance of parameter sets concerning the simulated peak flow was similar to the performance found for the runoff volume. For six sequences the best mean *PFE* was achieved using the specifically calibrated parameter set, for 10 sequences the lowest mean *PFE* was achieved by a transferred set, and for two sequences V_{Ref} showed the best performance. For 11 of the 18 sequences, V_{Ref} produced a higher mean *PFE* than both transferred parameter sets, while both transferred sets were outperformed by V_{Ref} for four sequences. Also, the results for simulation events were similar to the ones observed for the runoff volume; for 34% of the simulation events the specifically calibrated parameter set produced the smallest *PFE* and while the smallest *PFE* was achieved for the majority of events (56%) using a transferred set, V_{Ref} achieved the best results for only 10% of the events. However, while concerning *E* and *VE* V_{Ref} was clearly inferior compared to the transferred parameter sets, the results concerning *PFE* were more vague; while V_{Ref} was inferior to the transferred sets for 38% of the simulation events, the reference parameter set performed better than both transferred parameter sets for 29% of the simulation events. Also, the difference in mean absolute *PFE* was smaller than for *VE* with 21.7% for the specifically calibrated sets compared to 24.6% for the transferred sets and 29.2% achieved by V_{Ref}. Figure 4 illustrates the observed and HR simulated flow with V_1, V_2, V_3, and V_{Ref} for an event used for all three catchments and Figure 5 illustrates the observed and LR-DEM simulated flow for all parameter sets for the same event.

3.3. Vesijärvi Catchment Model

The entire Vesijärvi catchment of the city of Lahti was parameterized using the LR-DEM methodology and the parameter sets V_1, V_2, V_3, and V_{Ref}. Forested areas account for the largest fraction in the catchment

(37.2% or 1108 ha), followed by other green areas (33.4% or 995 ha). The impervious cover of the catchment is 27.3%, consisting of roofs (6.8% of the total catchment area) and traffic-related surfaces (20.5%) such as streets, parking areas, and sidewalks (Table 1).

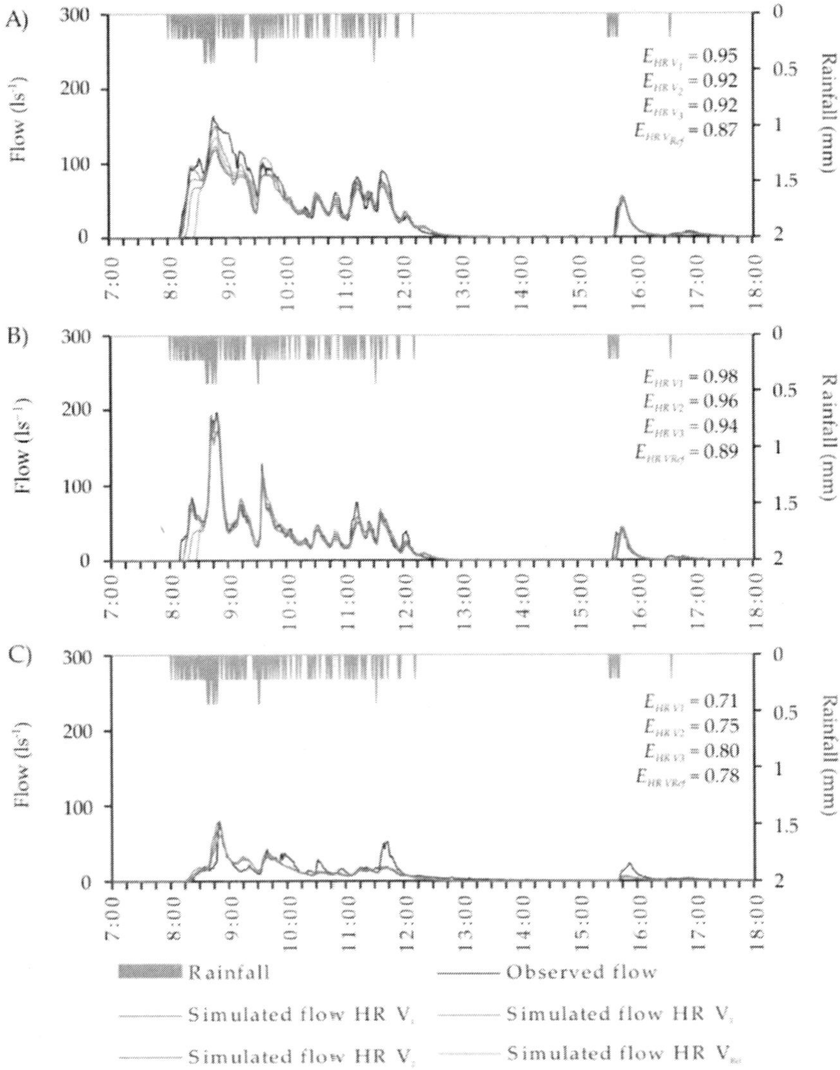

Figure 4. Observed and simulated (HR) flow for (**A**) Catchment 1; (**B**) Catchment 2; and (**C**) Catchment 3, and the model efficiency E for the parameter sets V_1, V_2, V_3, and V_{Ref} (event rainfall depth 22.0 mm, peak intensity 1.73 mm/5 min).

The catchment discretization for the model resulted in a total of 56,037 subcatchments that are drained through a drainage network of 144.4 km. The network comprises of 115.8 km of stormwater sewers, 28.6 km of streams (Table 2), and 5574 inlet nodes. The network drains into Lake Vesijärvi via 71 outfalls that include both sewer conduit and stream outfalls into the lake. A section of the model covering a fraction of the city center is presented in Figure 6 and the total surface runoff received by Lake Vesijärvi for one event is shown in Figure 7.

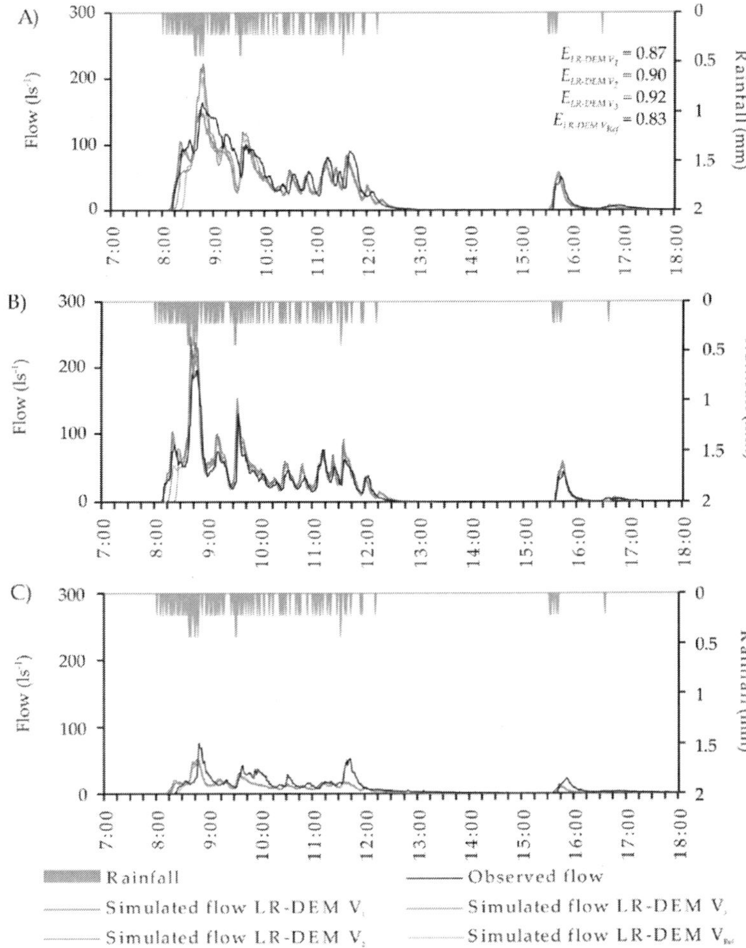

Figure 5. Observed and simulated (LR-DEM) flow for (**A**) Catchment 1; (**B**) Catchment 2; and (**C**) Catchment 3, and the model efficiency E for the parameter sets V_1, V_2, V_3,and V_{Ref} (event rainfall depth 22.0 mm, peak intensity 1.73 mm/5 min).

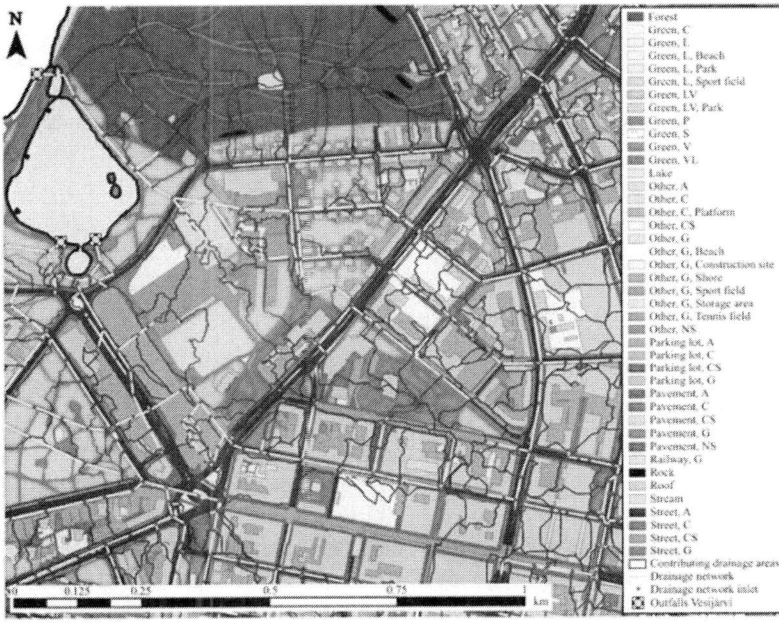

Figure 6. A section of the Vesijärvi catchment showing the surface discretization, the drainage network, and inlets along with their corresponding drainage areas (black boundary lines).

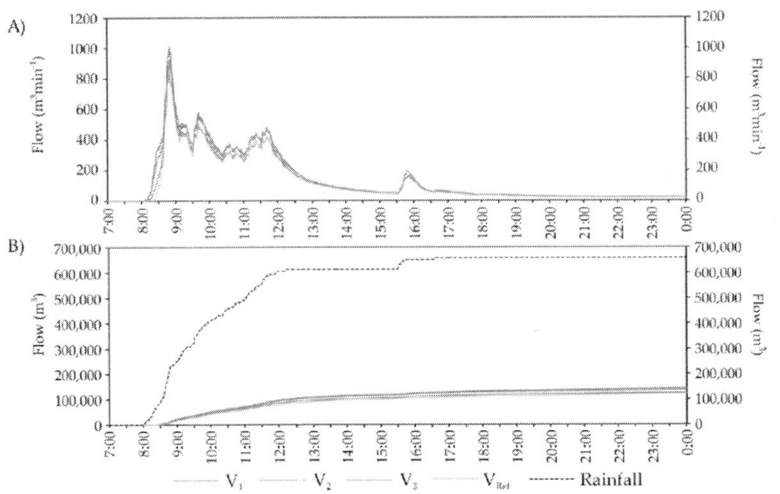

Figure 7. Simulated (**A**) flow time series and (**B**) flow accumulation for the Vesijärvi catchment for the parameter sets V_1, V_2, V_3, and V_{Ref} (Event rainfall depth 22.0 mm, peak intensity 1.73 mm/5 min).

4. DISCUSSION

In this paper we complemented the methodology presented by Krebs et al. [23] by using a DEM to define contributing drainage areas for drainage network inlets rather than information acquired during on-site visits. This addition extends the applicability of the methodology to considerably larger urban areas than the small monitored catchments. The impact of DEM delineation on model performance was assessed against results of Krebs et al. [23], where catchments were delineated using information gathered during site visits.

No correlation between the degree of urbanization and the error in catchment size was observed. While for the most urbanized Catchment 1 and the least urbanized Catchment 3 the catchment area decreased by 8% and 10%, respectively, the area of the moderately urbanized Catchment 2 increased by 2%. However, the change in *EIA* that is induced by the negligence of inter-subcatchment surface flow routing correlated with the degree of urbanization, with the smallest impact found for Catchment 1 (−5%), followed by Catchment 2 (15%) and Catchment 3 (30%). It is noteworthy, however, that inter-subcatchment surface flow was maintained for roofs in the LR-DEM model of Catchment 3, in which, contradicting city guidelines [55], roof runoff is not routed into the drainage network but rather to adjacent pervious areas. If roofs had been hydraulically connected in Catchment 3, *EIA* would have increased by 129%, producing inacceptable simulation results. While not investigated in this study, the large effect of roofs on *EIA* in Catchment 3 indicates the potential of roof runoff disconnection as a measure for stormwater management.

Jankowfsky et al. [33] studied different automated catchment delineation methods for a 3.5 km^2 peri-urban catchment in France and compared them with a method where topographical data were complemented with field observations. Their study produced subcatchment area overlaps of 85%–98% depending on the automated method used and the corresponding excessive surface fractions were 1%–15%. It has to be noted, however, that the DEM used for a part of their catchment had a much lower resolution (25 m) compared to the DEM used in our study (2 m). The results produced for the study catchments in our study are in accordance with those presented by Jankowfsky et al. [33], with the lowest errors found for Catchment 2 (97% overlap and 5% excessive surface) and the highest errors found for Catchment 1 (80% overlap and 13% excessive surface). The rather large errors in both overlap and excessive surface for Catchment 1 are due to a street area in the northeast of the catchment, which, considering sewer inlets, is outside the catchment area (see Figure 2A_1,B_1). This stretch of street consists of a rather steep curve with sewer inlets located at the outer edges of the curve. As a result, street runoff actually bypasses the designated inlets connected to a sewer line outside Catchment 1 and drains into Catchment 1 instead. This reveals how the impact of small details on catchment delineation are specific to urban areas but neglected in most automated approaches, as suggested by earlier studies [22,40].

The alteration in catchment properties that is induced by the automated DEM delineation naturally affects the simulated runoff. However, both the Nash–Sutcliffe efficiency *E* and the volume error *VE* remained within acceptable limits for all catchments in the current study. The results indicate that the simulated runoff is more affected by the alteration in *EIA*, induced by the negligence of inter-subcatchment surface flow, rather than the DEM delineation itself. The largest change in *VE* was found for simulations of Catchment 2, for which also the change in *EIA* was the largest. For Catchments 1 and 3, the LR-DEM delineation had less effect on the runoff volume, which can be explained by smaller alterations in *EIA* due to the high degree of urbanization in Catchment 1 (the difference between *TIA* and *EIA* is negligible) and the routing of roof runoff in Catchment 3. For all three catchments the peak flow error *PFE* was most sensitive to the inaccuracies arising from the DEM-based catchment delineation. These results indicate that *PFE* is more sensitive to accurate catchment delineation and overland flow routing than the volume error and the Nash–Sutcliffe efficiency *E*. Consequently, the selection of spatial and temporal model resolution and the used accuracy for catchment delineation depend on the objective of the hydrological assessment as accurate simulations of the peak flow require a more precise model description than the simulation of runoff volume or the shape of the runoff hydrograph.

Several studies have addressed the model parameterization of ungauged areas where a direct calibration of model parameters is not possible [41,44,47,50,51]. Seibert [45] used parameter–catchment relationships for model regionalization and reported a large decrease in model efficiency. Merz and Blöschl [47] used various regionalization approaches and came to the conclusion that hydrological catchment differences need to be accounted for and a mean of calibrated parameter values as well as a single "expert judgement" produced poor simulation results. Parajka et al. [51] studied a variety of regionalization methods (such as parameter mean, spatial proximity, regression, and site-similarity) and achieved the best results with a kriging approach and a site-similarity using a donor catchment with similar hydrological properties. Kokkonen et al. [50] suggested that the mean of available calibrated parameter sets is not sufficient to properly parameterize an ungauged catchment; parameters transferred from hydrologically similar catchments produced results that were not much inferior to results achieved by locally calibrated parameter sets. In this paper, we propose a methodology to parameterize large urban catchments by a regionalization of parameters calibrated to data from high-resolution study catchments. The key basis of this approach is a surface discretization that produces subcatchments with homogenous surface properties. The surface discretization was mostly based on spatial data readily available for both the small-scale study catchments and the larger, ungauged Vesijärvi catchment. Compared to earlier work, the proposed methodology can be classified as a site-similarity approach [41,50,51]. However, the site-similarity used in our approach is independent from the actual surface structure (i.e., degree of urbanization) of both the calibrated and ungauged areas and rather created by identifying homogenous urban surface types at a high spatial resolution. This approach allows the definition of parameters that can be assigned to the corresponding surfaces (e.g., a stretch of street or a roof) in the ungauged catchment. Krebs et al. [23] showed a high correlation between parameters independently calibrated to our three study catchments, which supports the applicability to an ungauged area. In this study, three parameter sets (V_1, V_2, and V_3based on the calibration to the three study catchments) were defined and inter-changed between the study catchments for both the HR and the proposed LR-DEM model setups.

The presented results indicate that while an overall better simulation performance was achieved by the specifically calibrated parameter sets, the performance of transferred parameter sets was not much inferior. Concerning both the shape of the hydrograph and the simulated runoff volume, transferred parameter sets performed better than the compiled reference parameter set V_{Ref}; transferred sets achieved a mean efficiency *E* of 0.70 and a mean absolute *VE* of 21.9% compared to *E* of 0.52 and *VE* of 33.9% for V_{Ref}. While a better performance of transferred sets (*PFE* of 24.6%) compared to V_{Ref} (*PFE* of 29.2%) was also observed for the simulated peak flow, the advantage of calibrated parameters over a parameter set based on literature suggestions was less clear. It is to be noted that while the model efficiency *E* was used as an objective function during calibration, neither the runoff volume nor the peak flow rate were optimized during calibration. Therefore, while the site-specific parameter sets clearly outperformed the transferred sets with respect to model efficiency, the results concerning the runoff volume and peak flow rate are less clear. Of the evaluated objectives, the peak flow error was found to be the most sensitive evaluation criterion to the transfer of model parameter values. While earlier regionalization studies reported varying success, the surface type-based model regionalization approach produced overall good results. However, none of the earlier studies were explicitly focusing on urban areas. Runoff processes in urban areas are dominated by impervious surfaces whose high fraction together with an efficient drainage system make the rainfall–runoff response more straightforward than in natural catchments. Regionalization of hydrological model parametrizations is likely to be easier in the urban setting, where the runoff response is easier to describe as long as the input rainfall data have sufficiently high temporal resolution and accuracy.

Several studies have shown that a high spatial resolution can improve simulation results and reduce the number of calibration parameters [9,22,54,58]. However, a high spatial resolution can also produce a large number of calibration parameters implying a model over-parameterization. In other words, the impact of an inappropriate parameter value is compensated by an inappropriate value of another parameter, resulting in poor predictive model capability [59] and affecting the model reliability and robustness [60]. While the methodology presented in this paper uses a high spatial resolution, the number of calibration parameters, due to the use of surface type subcatchments with homogeneous hydrological properties, is relatively low (8–25 calibration parameters depending on the catchment) considering the detailed description of the surface and drainage network. Furthermore, the identifiability of parameter values is indicated by (i) similar calibrated parameter values for same surfaces in the three study catchments despite the independent calibration; (ii) good performance of transferred parameter sets in all three catchments despite the differences in catchment characteristics (e.g., imperviousness as well as the composition of the impervious cover); and (iii) good model performance for a large number of rainfall events with varying rainfall depth, intensity, and duration. The same observations also indicate that the model calibrations produced reliable parameter sets. As stated above, two rainfall stations were available for the hydrological assessments in this study. In particular, the distances between Catchment 3 and the AP station (4.9 km) and Catchment 2 and the LSB station (2.7 km) are relatively large. With increasing distance between the measurement stations the recorded rainfall and runoff become less consistent, i.e., the measured rainfall differs from the rainfall that generated the measured runoff. This might lead to a model calibration that compensates for an inadequate rainfall input rather than producing a robust parameter set. However, the stated observations show that the measured rainfall was, despite the partly relatively large distance, sufficiently similar to the rainfall that generated the monitored runoff.

The methodology proposed in this paper to parameterize a large, ungauged, urban area, was applied to the Vesijärvi catchment of the city of Lahti to test the feasibility of the approach. Despite the reduction in the data requirement manual work remained necessary to conceptualize the catchment. However, the work load required remained acceptable considering the size of the catchment (30 km^2) and the spatial detail used for the surface description. This detail allows for an explicit spatial description of LID tools also in the city-scale modeling studies. While it is acknowledged that the hydrological evaluation of various LID scenarios requires an explicit and spatially highly distributed modeling approach [19], LID assessments are still commonly simulated by alteration of lumped parameters [25]. Besides for assessments of LID strategies, the proposed methodology is also suited for assessing the hydrological effects of urban land-cover changes, such as urban densification or the construction of new sites. The large Vesijärvi catchment model parameterized in this study allows urban planners and engineers to evaluate potential LID benefits on the city scale, concerning both the optimal placement and the types of LID tools, especially in the light of a given financial budget.

5. CONCLUDING REMARKS

Based on the presented results, we conclude that:

1) while the simplifications induced by the automated, DEM-based delineation process affect the simulated urban runoff, the simulation results remain acceptable;

2) a transfer of parameter values that were calibrated to high-resolution study catchments to ungauged areas is preferable to a parameterization using the available literature;

3) simulation results for three different urban catchments and a large number of rainfall events with varying characteristics show that the proposed catchment discretization produces identifiable model parameters; and

4) the detailed surface discretization allows for the explicit spatial description of various LID scenarios (e.g., green roofs, pervious pavers, or disconnection of impervious surfaces), supporting the evaluation of alternative urban water management strategies.

ACKNOWLEDGMENTS

Funding for this research was provided by the Doctoral Program in the Built Environment (RYM-TO), Maa-ja vesitekniikan tuki ry (MVTT), and Aalto University School of Engineering. Spatial data were provided by the city of Lahti, Lahti Aqua, and Teemu Heusala (TEKLA). Furthermore, data were provided by the STORMWATER research program lead by the University of Helsinki and funded by the European Regional Development Fund.

REFERENCES

1. Schueler, T.R. The importance of imperviousness. *Watershed Prot. Tech.* **1994**, *1*, 100–111.
2. Arnold, C.L., Jr.; Gibbons, C.J. Impervious surface coverage: The emergence of a key environmental indicator. *J. Am. Plan. Assoc.* **1996**, *62*, 243–258.
3. Bannerman, R.T.; Owens, D.W.; Dodds, R.B.; Hornewer, N.J. Sources of pollutants in Wisconsin stormwater. *Water Sci. Technol.* **1993**, *28*, 241–259.
4. Beighley, E.; Kargar, M.; He, Y. Effects of impervious area estimation methods on simulated peak discharges. *J. Hydrol. Eng.* **2009**, *14*, 388–398.
5. Booth, D.B.; Jackson, C.R. Urbanization of aquatic systems: Degradation thresholds, stormwater detection, and the limits of mitigation. *J. Am. Water Resour. Assoc.* **1997**, *33*, 1077–1090.
6. Brabec, E.A. Imperviousness and land-use policy: Toward an effective approach to watershed planning. *J. Hydrol. Eng.* **2009**, *14*, 425–433.
7. Guan, M.; Sillanpää, N.; Koivusalo, H. Modelling and assessment of hydrological changes in a developing urban catchment. *Hydrol. Process.* **2015**, *29*, 2880–2894.
8. Haase, D. Effects of urbanisation on the water balance-A long-term trajectory. *Environ. Impact Assess. Rev.* **2009**, *29*, 211–219.
9. Lee, J.G.; Heaney, J.P. Estimation of urban imperviousness and its impacts on storm water systems. *J. Water Resour. Plan. Manag.* **2003**, *129*, 419–426.
10. Ruth, O. The effects of de-icing in Helsinki urban streams, Southern Finland. *Water Sci. Technol.* **2003**, *48*, 33–43.
11. Maksimovic, C. General overview of urban drainage principles and practice. In *Urban Drainage in Specific Climates, Volume 2: Urban Drainage in Cold Climates*; Saegrov, S., Milina, J., Thorolfsson, S., Eds.; UNESCO: Paris, France, 2000; pp. 1–21.
12. The United States Environmental Protection Agency (US EPA). *Low Impact Development (LID)—A Literature Review*; US EPA Office of Water: Washington, DC, USA, 2000.
13. Roy, A.H.; Wenger, S.J.; Fletcher, T.D.; Walsh, C.J.; Ladson, A.R.; Shuster, W.D.; Thurston, H.W.; Brown, R.R. Impediments and solutions to sustainable, watershed-scale urban stormwater management: Lessons from Australia and the United States. *Environ. Manag.* **2008**, *42*, 344–359.
14. Butler, D.; Davies, J.W. *Urban Drainage*, 3rd ed.; Spon Press: Oxon, UK, 2011.
15. Cheng, M.-S.; Coffman, L.S.; Clar, M.L. Low-impact development hydrologic analysis. In Proceedings of the Specialty Symposium Held in Conjunction with the World Water and Environmental Resources Congress, Orlando, FL, USA, 20–24 May 2001; Brashear, R.W., Maksimovic, C., Eds.; ASCE: Orlando, FL, USA, 2001; pp. 659–681.
16. Huber, W.C.; Dickinson, R.E. *Storm Water Management Model Manual Version 4: User's Manual*; US EPA: Athens, GA, USA, 1988.
17. Rossman, L.A. *Storm Water Management Model User's Manual Version 5*; US EPA National Risk Management Research Laboratory: Cincinnati, OH, USA, 2010.
18. Elliott, A.H.; Trowsdale, S.A. A review of models for low impact urban stormwater drainage. *Environ. Model. Softw.* **2007**, *22*, 394–405.
19. Amaguchi, H.; Kawamura, A.; Olsson, J.; Takasaki, T. Development and testing of a distributed urban storm runoff event model with a vector-based catchment delineation. *J. Hydrol.* **2012**, *420–421*, 205–215.

20. Rodriguez, F.; Andrieu, H.; Creutin, J.-D. Surface runoff in urban catchments: morphological identification of unit hydrographs from urban databanks. *J. Hydrol.* **2003**, *283*, 146–168.
21. Cantone, J.P.; Schmidt, A.R. An innovative approach for modeling large urban hydrologic systems. In Proceedings of the World Environmental and Water Resources Congress, Kansas City, MO, USA, 17–21 May 2009; ASCE: Reston, VA, USA, 2009; Volume 342, pp. 824–839.
22. Gironás, J.; Niemann, J.D.; Roesner, L.A.; Rodriguez, F.; Andrieu, H. Evaluation of methods for representing Urban Terrain in storm-water modeling. *J. Hydrol. Eng.* **2010**, *15*, 1–14.
23. Krebs, G.; Kokkonen, T.; Valtanen, M.; Setälä, H.; Koivusalo, H. Spatial resolution considerations for urban hydrological modelling. *J. Hydrol.* **2014**, *512*, 482–497.
24. Jacqueminet, C.; Kermadi, S.; Michel, K.; Béal, D.; Gagnage, M.; Branger, F.; Jankowfsky, S.; Braud, I. Land cover mapping using aerial and VHR satellite images for distributed hydrological modelling of periurban catchments: Application to the Yzeron catchment (Lyon, France). *J. Hydrol.* **2013**, *485*, 68–83.
25. Eric, M.; Fan, C.; Joksimovic, D.; Li, J.Y.; Walters, M. Urban hydrological response units for modelling low impact developments. In Proceedings of the Annual Conference of Canadian Society for Civil Engineering 2012: Leadership in Sustainable Infrastructure (1), Edmonton, AB, Canada, 6–9 June 2012; Curran Associates: Edmonton, AB, Canada, 2012; Volume 1, pp. 665–674.
26. Palla, A.; Berretta, C.; Lanza, L.G.; La Barbera, P. Modelling storm water control operated by green roofs at the urban catchment scale. In Proceedings of the 11th International Conference on Urban Drainage (ICUD), Edinburgh, Scotland, UK, 31 August–5 September 2008; Oldenbourg Industrieverlag: Munich, Germany, 2008; pp. 1–10.
27. Carter, T.; Jackson, C.R. Vegetated roofs for stormwater management at multiple spatial scales. *Landsc. Urban Plan.* **2007**, *80*, 84–94.
28. Mentens, J.; Raes, D.; Hermy, M. Green roofs as a tool for solving the rainwater runoff problem in the urbanized 21st century? *Landsc. Urban Plan.* **2006**, *77*, 217–226.
29. Montalto, F.; Behr, C.; Alfredo, K.; Wolf, M.; Arye, M.; Walsh, M. Rapid assessment of the cost-effectiveness of low impact development for CSO control. *Landsc. Urban Plan.* **2007**, *82*, 117–131.
30. Versini, P.-A.; Jouve, P.; Ramier, D.; Berthier, E.; de Gouvello, B. Use of green roofs to solve storm water issues at the basin scale-Study in the Hauts-de-Seine County (France). *Urban Water J.* **2016**, *13*, 372–381.
31. Versini, P.-A.; Ramier, D.; Berthier, E.; de Gouvello, B. Assessment of the hydrological impacts of green roof: From building scale to basin scale. *J. Hydrol.* **2015**, *524*, 562–575.
32. Smith, M.B.; Vidmar, A. Data set derivation for GIS-based urban hydrological modeling. *Photogramm. Eng. Remote Sens.* **1994**, *60*, 67–76.
33. Jankowfsky, S.; Branger, F.; Braud, I.; Gironás, J.; Rodriguez, F. Comparison of catchment and network delineation approaches in complex suburban environments: Application to the Chaudanne catchment, France. *Hydrol. Process.* **2013**, *27*, 3747–3761.
34. Brown, J.D.; Spencer, T.; Moeller, I. Modeling storm surge flooding of an urban area with particular reference to modeling uncertainties: A case study of Canvey Island, United Kingdom. *Water Resour. Res.* **2007**, *43*, W06402.
35. Daniel, E.B.; Camp, J.V.; le Boeuf, E.J.; Penrod, J.R.; Abkowitz, M.D.; Dobbins, J.P. Watershed modeling using GIS technology: A critical review. *J. Spat. Hydrol.* **2010**, *10*, 13–28.
36. Fewtrell, T.J.; Bates, P.D.; Horritt, M.; Hunter, N.M. Evaluating the effect of scale in flood inundation modelling in urban environments. *Hydrol. Process.* **2008**, *22*, 5107–5118.
37. Mason, D.C.; Horritt, M.S.; Hunter, N.M.; Bates, P.D. Use of fused airborne scanning laser altimetry and digital map data for urban flood modelling. *Hydrol. Process.* **2007**, *21*, 1436–1447.
38. Neal, J.C.; Bates, P.D.; Fewtrell, T.J.; Hunter, N.M.; Wilson, M.D.; Horritt, M.S. Distributed whole city water level measurements from the Carlisle 2005 urban flood event and comparison with hydraulic model simulations. *J. Hydrol.* **2009**, *368*, 42–55.

39. Fewtrell, T.J.; Duncan, A.; Sampson, C.C.; Neal, J.C.; Bates, P.D. Benchmarking urban flood models of varying complexity and scale using high resolution terrestrial LiDAR data. *Phys. Chem. Earth* **2011**, *36*, 281–291.
40. Sampson, C.C.; Fewtrell, T.J.; Duncan, A.; Shaad, K.; Horritt, M.S.; Bates, P.D. Use of terrestrial laser scanning data to drive decimetric resolution urban inundation models. *Adv. Water Resour.* **2012**, *41*, 1–17.
41. Kay, A.L.; Jones, D.A.; Crooks, S.M.; Kjeldsen, T.R.; Fung, C.F. An investigation of site-similarity approaches to generalisation of a rainfall-runoff model. *Hydrol. Earth Syst. Sci.* **2007**, *11*, 500–515.
42. Rodriguez, F.; Cudennec, C.; Andrieu, H. Application of morphological approaches to determine unit hydrographs of urban catchments. *Hydrol. Process.* **2005**, *19*, 1021–1035.
43. Rodriguez, F.; Bocher, E.; Chancibault, K. Terrain representation impact on periurban catchment morphological properties. *J. Hydrol.* **2013**, *485*, 54–67.
44. Sefton, C.E.M.; Howarth, S.M. Relationships between dynamic response characteristics and physical descriptors of catchments in England and Wales. *J. Hydrol.* **1998**, *211*, 1–16.
45. Seibert, J. Regionalisation of parameters for a conceptual rainfall-runoff model. *Agric. For. Meteorol.* **1999**, *98–99*, 279–293.
46. Andréassian, V.; Hall, A.; Chahinian, N.; Schaake, J. Introduction and Synthesis: Why should hydrologists work on a large number of basin data sets? In *Large Sample Basin Experiments for Hydrological Model Parameterization: Results of the Model Parameter Experiment-MOPEX*; Andréassian, V., Hall, A., Chahinian, N., Schaake, J., Eds.; IAHS Red Books: Wallingford, UK, 2006; pp. 1–5.
47. Merz, R.; Blöschl, G. Regionalisation of catchment model parameters. *J. Hydrol.* **2004**, *287*, 95–123.
48. Blöschl, G.; Sivapalan, M. Scale issues in hydrological modelling: A review. *Hydrol. Process.* **1995**, *9*, 251–290.
49. Götzinger, J.; Bárdossy, A. Comparison of four regionalisation methods for a distributed hydrological model. *J. Hydrol.* **2007**, *333*, 374–384.
50. Kokkonen, T.S.; Jakeman, A.J.; Young, P.C.; Koivusalo, H.J. Predicting daily flows in ungauged catchments: Model regionalization from catchment descriptors at the Coweeta Hydrologic Laboratory, North Carolina. *Hydrol. Process.***2003**, *17*, 2219–2238.
51. Parajka, J.; Merz, R.; Blöschl, G. A comparison of regionalisation methods for catchment model parameters. *Hydrol. Earth Syst. Sci.* **2005**, *9*, 157–171.
52. Valtanen, M.; Sillanpää, N.; Setälä, H. Effects of land use intensity on stormwater runoff and its temporal occurrence in cold climates. *Hydrol. Process.* **2014**, *28*, 2639–2650.
53. Valtanen, M.; Sillanpää, N.; Setälä, H. The effects of urbanization on runoff pollutant concentrations, loadings and their seasonal patterns under cold climate. *Water. Air. Soil Pollut.* **2014**, *225*, 1–16.
54. Krebs, G.; Kokkonen, T.; Valtanen, M.; Koivusalo, H.; Setälä, H. A high resolution application of a stormwater management model (SWMM) using genetic parameter optimization. *Urban Water J.* **2013**, *10*, 394–410.
55. Lahti Aqua OY Vesihuollon Yleiset Toimitusehdot. Available online: http://www.lahtiaqua.fi/Asiakaspalvelu/Lahti/Asiakkaalle/Hinnastot%20sek%C3%A4%20sopimus-%20ja%20toimitusehdot (accessed on 18 January 2016).
56. Nash, J.E.; Sutcliffe, J.V. River flow forecasting through conceptual models part I-A discussion of principles. *J. Hydrol.***1970**, *10*, 282–290.
57. Huong, H.T.L.; Pathirana, A. Urbanization and climate change impacts on future urban flooding in Can Tho city, Vietnam. *Hydrol. Earth Syst. Sci.* **2013**, *17*, 379–394.
58. Zhou, Y.; Wang, Y.; Gold, A.J.; August, P.V. Modeling watershed rainfall-runoff relations using impervious surface-area data with high spatial resolution. *Hydrogeol. J.* **2010**, *18*, 1413–1423.
59. Petrucci, G.; Bonhomme, C. The dilemma of spatial representation for urban hydrology semi-distributed modelling: Trade-offs among complexity, calibration and geographical data. *J. Hydrol.* **2014**, *517*, 997–1007.

60. Perrin, C.; Michel, C.; Andréassian, V. Does a large number of parameters enhance model performance? Comparative assessment of common catchment model structures on 429 catchments. *J. Hydrol.* **2001**, *242*, 275–301.

CHAPTER 11

Applications of SuDS Techniques in Harvesting Stormwater for Landscape Irrigation Purposes: Issues and Considerations

Andrew B. Shuttleworth, Ernest O. Nnadi, Fredrick U. Mbanaso, Stephen J. Coupe, Joris G.W.F. Voeten and Alan P. Newman

1 SEL Environmental Ltd, Blackburn, Lancashire, United Kingdom

2 School of Energy, Construction and Environment, Coventry University, Coventry, United Kingdom

3 Centre for Agroecology, Water and Resilience, Coventry University, Coventry, United Kingdom

4 Permavoid Ltd, Amsterdam, Netherlands

ABSTRACT

While urbanization and increasing population has put much pressure on natural drainage channels and resulted in increase in flooding, there is increased pressure on available water resources due to climate change, reduction in frequency of rainfall events and drought. The emergence of a sustainable drainage system (SuDS), also known as best management practice (BMP) and low impact development (LID), has changed the management strategy of drainage from conventional to sustainable. SuDS techniques seek to deliver the three cardinal paradigms of sustainable drainage: quantity, quality and amenity and as such, they can offer an additional benefit for applications such as landscape irrigation. Most SuDS techniques have the potential for water storage with minimal or no modifications required. This chapter, while covering the capabilities of SuDS systems, explores SuDS devices such as pervious pavements equipped with excess storage capacity, cisterns and tanks harvesting roofwater, infiltration systems aimed at supporting the growth of urban plants and green roofs with the potential to store water in order to maintain water demanding planting scheme even during dry periods. It also covers systems where SuDS is the main driver to device installation and address issues and considerations surrounding applications of such systems in water harvesting for irrigation.

Keywords: Irrigation, Drainage, SuDS, Stormwater, rain harvesting

1. INTRODUCTION

Urbanization can initiate undesirable local modifications to the water cycle. In particular, the effects of the spread of impermeable area both within and at the margins of towns and cities have highlighted the need to modify the way that society deals with stormwater.

The problems of surface flooding in cities and overloading of foul sewers (when combined systems are in place) are the most headline-grabbing aspects of increasing impermeability but reduction in aquifer recharge can also be important when water demand rises due to increase in local population.

In many western societies, particularly in the United Kingdom, there is a preference for housing schemes with gardens (including, often, very water demanding planting regimes). This increases the demands for water well beyond traditional domestic use. Alongside these domestic developments, hotels, hospitals, office blocks, light industrial units and retail developments often try to show a supposedly green face to the world in the form of extensive landscaped areas around their equally extensive impermeable parking surfaces.

We are thus faced with the dual problem of a demand for water for landscaping accompanied by rapid and wasteful run off from impervious parking areas. This can cause flooding problems in response to short heavy rain events. The problem is particularly important in summer months when the demand for watering domestic, commercial and municipal planted areas is at its highest, at the same time as the supply from surface sources is at a minimum.

The term SuDS was originally used in the United Kingdom as an acronym for 'sustainable urban drainage systems' but this term has more recently lost favour and is commonly replaced by SuDS (sustainable drainage systems). In the USA, the terms low impact development and best management practices (LID and BMP) cover the same approach to drainage and in Australia the term 'water-sensitive urban design' is favoured.

While all of the above terms are used, sometimes interchangeably, to indicate a holistic approach to stormwater management, they are all dominated in their philosophy by the concept of stormwater source control.

This entails controlling both the quality and quantity of stormwater as close to its site of deposition as possible. Commonly, parts of the United Kingdom can be under drought orders (which impose summer hosepipe bans which limit landscape irrigation) and still suffer from localized flooding during short-term storms. These occurrences are becoming an increasing feature in some parts of the United Kingdom and this adds extra incentive to retrofitting of SuDS systems that can also make a contribution to the reduction of water use.

The principles of SuDS are often presented as follows:

1) Storing runoff so that it can be released slowly or used beneficially (attenuation).

2) Allowing water to soak into the ground so as to mimic the processes on an undeveloped surface (infiltration).

3) Where necessary to allow the water to move doing so at the surface and at a controlled velocity.

4) Using the processes of collection, storage and transportation to facilitate the removal of pollutants by sedimentation, precipitation, adsorption (in its widest sense) and degradation.

Most SuDS devices have the potential to be used for water storage with minimal or no modifications required. Table 1 summarises the irrigation water resource potential of a range of hard SuDS devices that have been used for water harvesting, whereas Table 2 outlines a number of green SuDS devices that themselves place a potential demand on water resources but which can be modified to harvest water for self-irrigation and sometimes for other uses.

Table 1. Irrigation water resource potential of a range of hard SuDS devices.

Device	Quantity issues	Quality issues
Underground and above ground barrels, tanks and cisterns harvesting roof water only	The sizing of the system needs to be done carefully and the usage of the water be well established if both stormwater attenuation and water harvesting are to be achieved in the same system. With modern computer control the sizing can be reduced by actively draining tanks in response to predicted rain events.	The quality of water will be affected mainly by atmospheric fallout and the nature of materials used in roof construction. While this is not really considered an issue new UK guidance requires that even roof water should have some treatment before discharge to a watercourse
Underground and above ground storage harvesting both surface runoff and roof water	The additional resource will require a greater storage volume for it to be utilisable. If there is attenuation available upstream of the storage tanks this can be minimized as a problem	Surface water will generally be of much poorer quality than roof water alone (see above) but provided the system is correctly designed to retain day to day pollutant releases it will generally only be an issue if major pollutants releases overcome the pollution attenuation mechanisms. Salt applications in temperate zones can be a problem.
Pervious pavements and similar without off line additional storage	Unless taking runoff from impervious surfaces too the total inputs are limited to the water falling on the surface. Unless designed with extra storage capacity even an attenuation based system will have limited capacity to store water for more than short periods of time	Input of potentially harmful organisms from faecal contamination and chemical pollutants from atmospheric fallout. If exposed to traffic there will also be day to day input of automotive based pollutants and if not provided with upstream protection the water can be subject to contamination from the very rare losses of engine oil and fuel. In temperate areas de-icing salts can be an issue.
Pervious pavements with off- line additional storage	By directing water into off line underground or above ground tanks the storage required for attenuation or to overcome limited rates of infiltration will not be compromised by the need to recover water	

Table 2. Green SuDS devices that can be modified to harvest water for self-irrigation and other uses.

Device	Quantity issues	Quality issues
Green roofs with additional off line storage Green roofs with integrated storage.	Both need careful sizing taking into account predicted available rainfall and demand placed by both growing plants and any additional off-roof demands. Storage needs to take into account sufficient temporary storage to provide stormwater source control when substrate is close to saturated. Provision may need to be made for rapid dumping of stored water when large storms are predicted in a 'reservoir full' situation.	If used for self-irrigation of the roof or for additional off-roof irrigation there will be little problem but if, as in some installations, green roof drainage water is pooled with other water and used for such as toilet flushing the colour can become a source of user complaint. If fertiliser is used on the roof, then recycling the water via the roof will retain the nutrients where they are needed.
Infiltrating urban tree planters with subsidiary methods of storage.	Maintaining sufficient moisture for the trees can be a problem. Proprietary systems which incorporate water storing foams (as used by florists) can offer a solution. Species selection for the variability of available water needs to be done carefully.	There will be day-to-day input of automotive based pollutants and if not provided with upstream protection the water can be subject to contamination from the very rare losses of engine oil and fuel. In temperate areas de-icing salts can be an issue.

This chapter, while covering the capabilities of all SuDS systems, gives particular emphasis to those that are particularly suitable for retrofitting. In this chapter, the greatest attention is given to devices such as pervious pavements equipped with excess storage capacity, cisterns and tanks harvesting roofwater, infiltration systems deliberately aimed at improving the growth of urban plants (including trees) and green roofs, which can offer the potential to store water in excess of the capacity of the substrate so as to maintain a water demanding planting scheme during dry periods. Inevitably, there will be some overlap with harvesting for purposes other than irrigation. Competing use for harvested water includes toilet flushing and washing machine use and with appropriate treatment, many of the types of use that normally demand potable water. We must also recognize that rainwater harvesting systems are commonly provided without involvement of the SuDS philosophy, being designed without any attempt to enhance the control of stormwater (but sometimes doing so by accident). This chapter only covers systems where SuDS is the main driver to device installation.

Harvested rainwater from urban environments (even water directly from roofs) will inevitably contain pollutants that would not be expected if the water was collected from rural upland catchments or extracted from protected aquifers. If water is to be used for irrigation after being in contact with pollutant materials inevitably generated by both the daily activities of urban living and by fallout from industry, a first question must be whether the quality of the water is good enough for purpose. Irrigation water limits vary from place to place and not all organisations include all the possible pollutants (see Ref. [1]). A useful comparative table is provided by Nnadi et al. [2] which presents the irrigation water limits provided by a range of authorities.

2. ROOFWATER HARVESTING: RAIN BARRELS AND CISTERNS

While many SUDS systems have been used to harvest rainwater for a combination of both irrigation and indoor uses (such as toilet flushing and washing machine pre-wash) it is the direct harvesting of roof water (and sometimes water from non-trafficked paved areas) into barrels and cisterns that has seen the greatest uptake as a component of the LID programmes in the USA. The relatively high quality of this source of harvest (although well short of potable water standards in most jurisdictions (see Ref. [3]) that makes this source attractive for internal use as well as for irrigation. The primary differences between a cistern or rain barrel which forms part of a SuDS system and one which is simply a water supply device are the way the container is sized and the way the water is managed. Whether managed manually by an individual householder (possibly backed up by an advice and education program) or through an automated system, unless efforts are made to maintain enough available storage to contain at least a significant proportion of a design storm and unless the stored water is managed either by using the water or by releasing a proportion before the next storm then this is not SuDS at all but purely a water harvesting exercise. Jones and Hunt [4] made the point that the greatest problem with roofwater harvesting as a stormwater control mechanism is that many of the cisterns and barrels remain full all of the time because water is not utilized. Managing the cistern by using the water regularly or releasing during a dry period is necessary if roofwater storage is to be seen as SuDS.

North Carolina State University (NCSU) offers online a very comprehensive review of rainwater harvesting [5] and also one of the best sources of information provided for home owners on water cisterns and their use [4]. Their document usefully attempts to guide the owners in relation to the need for additional ground support for cisterns in gardens, an often overlooked factor. NCSU also provides an informative guide to roof water quality [6]. Many other American universities provide a similar extension literature and the various States of the USA represent such a broad diversity of climates that the guidance from either state regulators or that provided through local university extension programs should be selectable if someone is looking for information that suits their climate elsewhere in the world. That is not to say that authors from outside the USA have not been active in reporting their work in a variety of jurisdictions (although in many cases, the stormwater source control element is missing).

Cisterns can be either above ground or below ground, but the land take and unattractive appearance of reasonably sized water cisterns are often a barrier to adoption. Underground tanks can be expensive and offer significant structural problems if they are to be covered with a trafficked surface or even one with a reasonable dead load. Unlike above ground storage, where irrigation use can often be achieved under gravity flow, they will normally need a pump.

Underground tanks are not normally suitable for occupier retrofit even by the most capable householders. An option that addresses this is the Skeletank®, where the thin polyethylene tanks get their structural strength from internal interlocking polypropylene skeleton. This provides structural strength while maintaining a system that can be carried, if necessary, on the roof of a car and can be lifted easily by two people. They are easily linked together to achieve the size required. The system was originally designed to provide stormwater attenuation and soakaway tanks (used upside down) within individual domestic curtilages. The system has since been adopted for rainwater harvesting and lightweight pump and filter chambers are available as accessories. Apart from the wiring for the pumps, the system is totally suitable for installation by an amateur. Figure 1 shows an installation in Preston (Lancashire, UK) which provides an interesting example of a householder built retrofit water harvesting SuDS project particularly since the primary aims of harvesting the water was to provide irrigation water to landscaped areas and to support a small pond on site while allowing total disconnection of drainage from the combined sewer. The owners of the property in question were undergoing conversion works which included the installation of a new patio area. The tanks were installed underneath the proposed patio area. The concept of the system begins with the collection of rainwater in a downpipe chamber. On this site, three downpipe chambers were installed, two chambers for a 94 m^2 roof area at the main house and the final chamber positioned to collect water from a 55 m^2area (serving an adjacent outbuilding). These chambers are intended to pre-clean the rainwater and allow silt to deposit within the sump of the chamber. Water passes through a fabric filter on the outlet spigot of the chamber before running into the tanks which were connected together using standard underground drainage pipe to provide a combined reservoir. For this site, sixteen tank units, providing a total storage capacity of 4800 litres, were used. The required number of units was determined by a modelling to fit available storage to the various demands including sufficient temporary storage, with controlled discharge to accommodate stormwater during rain events when the temporary storage is full. The rainwater pump chamber is the final component in the chain and is primarily used to pump rainwater on demand for irrigating flower beds and grassed areas.

Figure 1. Left: Skeletanks® in place and connected together. Right: lightweight pump chamber.

When available in excess stored rainwater is also pumped into a header tank for internal household use, such as, flushing toilets and water feed for washing machine pre-wash. The primary re-use of the harvested rainwater is to irrigate the flower beds and grassed areas surrounding the property and also for supporting the flora and fauna of the pond. Since the installation of the system, both use of mains water for irrigation and for topping up the pond has been minimal. Checks have shown the public health microbiological qualities (including legionella) are not a problem.

3. PERVIOUS AND MACRO-PERVIOUS PAVEMENTS AS A WATER HARVESTING SYSTEM

Pervious pavement systems (PPS) have been around as SuDS elements since the 1970s and are extremely well documented (see Refs. [7–12]). In the United Kingdom, their origin is largely associated with Chris Pratt's design which utilized concrete block pavers provided with infiltration channels as the wearing course [13].

The pervious wearing course surface can also be porous asphalt or porous poured concrete. The wearing course is usually laid on a bedding layer of, typically, 10 mm material and there is often a geotextile as a separation and filtration layer between the bedding layers and either the subbase of aggregate (typically a uniform 50 mm, with a void ratio of about 30%) or on a subbase replacement of plastic load-bearing boxes (a typical void ratio of 90%). If the pavement is to be used for water harvesting, it requires a reliable under-sealing layer which can be a folded or welded polyolefin membrane or, particularly within individual curtilages, can be conveniently provided by one or more of the sub-surface tanks discussed in the example above. These can also be used below pervious pavements to capture a proportion of the stormwater with the number of tanks selected in relation to demand with both unharvested and overflow water disposed of by infiltration or at a controlled rate to a surface water body or a combined sewer.

The most common application of PPS has been parking areas, whether they are extensive parking lots or individual house driveways and courtyards. The efficacy of pervious pavements in intercepting and immobilising or degrading traffic-derived pollutants has been the subject of a large body of research in many countries including England [14, 15], Scotland [16], Spain[17], Australia [18, 19] and the USA [20].

One of the most extensive experimental studies of PPS-derived water quality, with respect to irrigation, from pervious parking surfaces was instigated by Nnadi [21]. This was an indoor simulation experiment on a relatively large scale. The experimental outline was to apply oil drops (simulating oil dripping onto pavements from parked vehicles) and street dust to pervious pavement system models built into 1200 mm × 1000 mm × 600 mm heavy-duty polypropylene boxes (used for heavy industrial storage). Models were created with a selection of wearing courses including several cross-sections of new porous asphalt, a 10-year aged asphalt from a quarry parking surface, porous poured concrete and concrete paving blocks with infiltration channels. Thirteen millimetre simulated rainfall events were applied to the models at 15 mm/hour (52 min rain events) twice monthly. These were applied a day before and a day after the monthly application of used motor oil and dust. The effluents were captured and analysed for total suspended solids (TSS), metals and total petroleum hydrocarbons (TPH). The results were considered in relation to water quality as applied to irrigation and other forms of re-use as detailed in Nnadi et al. [2]. The results indicated that the porous asphalt and porous concrete are as good as the block paved permeable pavements in the treatment of stormwater pollution with all surfaces delivering water quality that is suitable for irrigation and clearly demonstrated the capability of the porous asphalt and concrete pavement system to trap hydrocarbons within the system.

While the use of pervious pavements for control of day-to-day automotive-derived pollutants is well established, it must be remembered that the traditional PPS has shown a poor pollution retention performance when challenged by major hydrocarbon releases [22]. Modification of the permeable pavement to incorporate a gravity separator just below the laying course has been reported for both aggregate-based [15] and plastic box-based pavements [23] showing a capability of retaining sufficient oil should every car in a parking lot simultaneously lose the content of their sump.

The first pot trials to investigate the acceptability of water quality from pervious pavements for irrigation originated at Coventry University as a PhD programme by one of the authors [21]. The source of water was a series of laboratory models. The models used were based on the modified designs by Newman [24] which incorporates a gravity separation system to ensure that free oil can never escape from the system and were artificially fertilized by adding Osmocote® slow release fertilizer and brushing it into the infiltration channels. This was to enhance the biodegradation rate of the hydrocarbons. The pot trial investigation showed that where the effluent water could be used for irrigation the excess nutrients provided distinct advantages to the growing plants [25]. In all cases, the plants irrigated with effluent grew much better than those irrigated with tap water. In effect, the system was being used for fertigation.

There has also been more recent work using a live car park, this case based on a highly modified design called a macro-pervious pavement system (MPPS) (Figure 2). In 2011, an investigation was initiated [26] on an MPPS installed as a prison car park in Scotland.

The parking lot consists of two major sub-catchments of around 1350 m^2 and a minor one of 300 m^2(which was not included in the irrigation water study). The majority of the surface of the parking lot consists of traditional asphalt with stormwater entering the subbase/storage and attenuation layer of crushed limestone through miniaturized, linear, gravity separator units (which look like normal channel drains from the surface) through a chamber containing a floating mat of oil-sorbing textile. It then flows into the subbase, which drains towards separate flow control chambers (with orifice plate flow control) in each of the sub-catchments. These provided convenient sampling points and in-line storage chambers which would be available for landscape irrigation. Samples were analysed for a wide range of determinants.

For the heavy metals, all results were below the most stringent of irrigation water limits. For suspended solids, all measurements were below 20 mg/l, a concentration below which blockage of drip irrigation systems is not a problem. For TPH, the irrigation limits that were adopted are dependent on the solubility of the various carbon chain fractions (except for C_9–C_{14} where a limit of 1.8 mg/l applies). Since no measurements of TPH ever exceeded 1.8 mg/l and in the absence of any observed free product (indicating that solubility was never exceeded for any fraction), it can be concluded that hydrocarbons should have no detrimental effect if this water was harvested and used for irrigation.

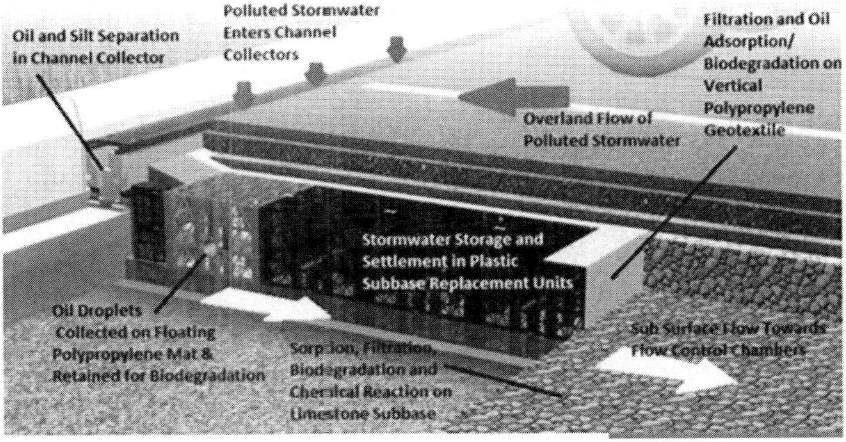

Figure 2. The macro-pervious pavement system as installed at Perth Prison, Scotland. The line of 'Channel Drain' is actually a series of oil and silt separating infiltration points, which direct stormwater into the sub surface storage and attenuation layer.

BTEX concentrations were all below the 1 μg/l limits of detection for these compounds. The irrigation water limits for benzene and substituted benzenes proposed by the New Zealand Government [27] range from 800 μg/l for benzene to as high as 39 mg/l for toluene. MTBE was also invariably below its detection limit (10 μg/l). Irrigation water standards for MTBE could be found in the literature but since the State of Florida mandates a limit of 50 μg/l in drinking water, this value could be seen as a very conservative irrigation water limit. Boron was always below the 40 μg/l limit of detection and thus well below the 500 μg/l soil pore water limit for the most sensitive plants [28]. Examining the major cation concentrations, it was shown that calcium, magnesium and potassium were low compared to any irrigation water limits and most authors would expect that at least magnesium and potassium would need to be supplemented for optimum plant growth. For most of the year, neither the sodium (Figure 6) or sodium absorption ratio (SAR) values were a concern at this site, but in the winter months, following the application of de-icing salt (which will vary from year to year) the values of both these parameters show the water to be unsuitable for irrigation with sodium values in January up to 1200 mg/l. By March, the sodium concentrations were below 100 mg/l but above the 50 mg/l irrigation water limit until May. It was concluded that if water is to be stored for irrigation purposes it would be necessary to divert the meltwater (and the rainwater falling onto a salt contaminated pavement) away from the storage tanks during the winter months after filling them as much as possible during October and November and then topping them up in April and May before the need for irrigation is established in the summer. If the worst of the salt contaminated water can be diverted, the direct effect on plants will be minimized. Another issue from this particular site was the pH that was higher than optimal, but this was due to the limestone used in the subbase and would have been much lower if a granite subbase had been chosen. Adjustment of pH with addition of calcium sulphate was proposed as a potential solution, but species selection for high pH tolerance would be an alternative.

Pot trial experiments were carried out using both ryegrass and tomato plants (using the experimental protocol previously proposed by Nnadi [21] irrigated with effluent collected from this pavement in September with initial results being reported to the SuDSnet conference in 2015 [29] and continued experiments (and tissue analysis) have shown that at this site, the water is perfectly suitable for irrigation, if harvested before the salt application. However, the data represents only relatively short-term use of the water. Hence, further work in this area is required and this provides a good research opportunity for anyone interested in this area of study.

One problem with block paving surfaced PPS is that they can become a habitat for weeds and chemical methods of weed control still dominate as the most preferred in the United Kingdom [30] and most European cities [31, 32]. Although several herbicides are in use, glyphosate-containing herbicides (GCH) are by far the most widely used herbicide for weed control on hard surfaces [32–35]. Recent work on the impact of GCH on pollution attenuation and biodegradation in a PPS indicated that GCH is not retained in the PPS structure and is subject to rapid wash through in response to water movement thus increasing the potential risk of reaching receiving environments [36]. Hence, not only does one need to consider the herbicide itself, but the breakdown of other pollutants such as hydrocarbons within the PPS can be potentially affected [35] with potential impact on the quality of irrigation water. In a recent study to determine the suitability of stormwater harvested from PPS for reuse purposes in conditions where GCH was applied as part of PPS maintenance procedure, Mbanaso et al. [37] observed that effluent from the test models including those dosed with high GCH concentration of 7200 mg/l do not pose infiltration or salinity problems when used for irrigation. They, however, indicated that high dosage of the herbicide could lead to an elevated electrical conductivity of the recycled water. Hence, if PPS-derived water is used for irrigation; chemical methods of weed control should be avoided until the dynamics of breakdown of the particular herbicides are better understood. Further work needs to be done on establishing the lifetime of such compounds in both the PPS and the stored irrigation water.

3.1. Green roofs

As well as the need to control rapid water runoff in urban areas, there are several other problems to which SuDS can make a contribution. The urban heat island effect [38, 39] and reduction in green space with associated loss of both amenity and biodiversity [40] are important issues. Much of this is being exacerbated by climate change. A contribution to the mitigation of such problems is to apply green SuDS techniques and foremost amongst these is the attempt to utilise the roof spaces of buildings to create green roofs. Green roofs can be either extensive, usually planted with a mat of relatively drought-resistant species such as sedum or intensive [41], or being more 'garden like' where the aim goes beyond the immediate environmental benefits with aesthetic and social aims being contributed to by a more varied and attractive planting scheme with the potential to contribute more to biodiversity.

The potential of green roofs to contribute to the mitigation of rainwater runoff is well established, see Refs. [42–45]. What is less often stressed, however, are the problems in certain climates associated with the fact that after a heavy rainstorm, the water in the saturated substrate can take a considerable time to drain or evaporate and a subsequent storm will be offered significantly reduced attenuation.

The capability of green roofs to contribute to the cooling of buildings and combating the heat island effect has been widely reported [45–50]. Green roofs can also play a role in improving biodiversity within urban areas [40, 51] and, the aesthetic value of a green roof can often be a dominant factor in its adoption [52] even though aesthetic planting schemes can come into conflict with biodiversity aims [53].

Particularly where the green roof is intended to be accessible and provide, in part, the function, for example, of an urban park, it is important that the provision of water, to maintain adequate growing conditions (for a wide range of plants), during dry periods is recognised. This helps to enhance the amenity value. One of the factors is the maintenance of adequate soil moisture content. The limited load-bearing capacity (LBC) of the roofs of (existing) buildings often dictates the amount of water that can actually be safely retained on the roof itself. Designing traditional intensive green roofs to satisfy the load-bearing capacity of a roof thus involves finding a balance between storing as much rainwater as possible, maintaining conditions for plant growth and respecting the LBC of the construction [54, 55]. Adding more substrate typically achieves a water-stored-to-weight ratio (WSWR) of just 0.2 l of water per added kg of soil for loamy sand, or 0.4 l/kg for a typical, specialised, extensive green roof substrate [56].

An alternative to storing the water solely within the substrate is to drain the water into a cistern to allow the water to be used to irrigate the roof during dry periods and, if correctly sized, can offer temporary storage during storms occurring when the substrate is close to saturation. A green roof with a cistern for reuse was the subject of a modelling study by Hardin et al. [57]. Their system included a green roof with its drainage system connected to a cistern which in turn supplied irrigation water to the roof via a pump. A supplemental water source is also connected to the cistern to provide water should there not be sufficient water to perform the irrigation event. It was proposed that the irrigation should be managed via a controller, similar to what is widely used for home lawn irrigation, which only irrigates on the prescribed times unless sufficient rain has fallen within 24 h of the intended irrigation event.

Building rainwater cisterns inside or adjacent to the building and using pumps and irrigation systems is thus an option which will provide both runoff attenuation and on-site rainwater reuse. However, the running costs, capital costs and building space/land take combined with the need for maintenance and the propensity for active systems to break down are factors which would militate against their choice. There is also a need to consider the trade-off between energy used to pump the water from ground or basement level and the loss of space if the cistern is maintained in one of the higher floors of the building.

An alternative to using a separate cistern is to maintain 'ponding elements' under the substrate. Green roofs.com [58] presents a useful summary of some of the systems available, some of which are dependent on a slightly sloping roof, one which incorporates active pumped elements and a system that stores up to 40 mm of water in a plastic drainage layer equipped with an overflow device (although little extra capacity is available for temporary stormwater storage during rain events). However, without a separate cistern, the volume available for storage will be relatively small unless special steps are taken. One such approach has been developed in the Netherlands [56]. This modified green roof stores the water directly under the substrate within modified load bearing plastic void formers originally developed for pervious pavement applications. The water is not required to be pumped up to the substrate, this being achieved by capillary actions through 'capillary cones' inserted into the load-bearing vertical struts [59].

An excellent example of an application of this type of system is adjacent to the elevated railway station at Orlyplein in Amsterdam [56], shown in Figure 3. This is a remarkable installation which has transform the former rooftop bus station (vacant for some time after it being moved to ground level) into a popular public park area with a resulting increase in economic development into the form of many additional rooms constructed in adjacent hotels.

The system of capillary irrigation outlined above has also found application in ground level applications. A very recent example of this is the 'Green Stream' in Zuidas an Amsterdam city district in the Netherlands. Zuidas is a very densely build urban area and therefore prone to urban flooding during intense rain events. With the city district being under re-development, aiming to become the major business hub in the Amsterdam Metropolitan region, the city is developing innovative multifunctional designs to improve urban quality, reduce the urban heat-island effect by improved evaporative cooling by plants and increase water retention (and reuse) to prevent urban flooding. The 'Green Stream' is a project in which rainwater from rooftops and adjacent sidewalks is collected in a 2 m wide planting strip that runs along the houses and sidewalk (Figure 4). The strip is deeper than its surroundings, making it a natural water collection point in the street design. To get from the sidewalk to the houses, bridges are used to cross the 'green stream'. This planting strip is designed to be dry throughout the year, but is allowed to flood during rain events. Planting species are selected to be able to withstand occasional flooding. Innovative in the design is the 150 mm high water attenuation system 40 cm below the planting, which is fed with water from the roofs. A continuous chain of plastic void forming units, as used in the Orlyplein roof park, is placed in a waterproof liner to create a subsoil water-tank and features the same capillary irrigation system, capable of returning water to the soil when plants are using water without the use of pumps (and thus energy). The improved water availability for plants maintains their evapotranspiration rates at close to the potential evapotranspiration generated by the local weather, improving their urban cooling capacity. Surplus of water can drain freely to ground water level alongside the water drainage and capillary irrigation system. To prevent the Green Stream from overflowing onto the sidewalk, extra emergency overflows are created at the maximum fill level, connected to the conventional sewer.

Figure 3. The roof park built on the former Orlyplein Bus Station *deck*, before, during and after construction.

The previously lonely and threatening exit from the rail station has been transformed into a place that is popular and well-populated whenever the station is open.

Figure 4. Left to right: Schematic of the green stream system, under construction and on the day of completion.

Another application of this technology has been on sports surfaces such as football pitches both on rooftops and at ground level. The system provides water to the growing grass turf, while the void space can be made sufficiently deep to satisfy the most stringent stormwater attenuation requirements for new stadium construction. Even if supplemental water has to be used to maintain the playing surface in very hot dry countries, the application of water from below by capillary action is more efficient than spray irrigation from above. Currently, trials are underway in preparation for the soccer world cup in Qatar.

3.2. Irrigation of urban trees

Novel below ground water storage options provide a significant opportunity for the reliable delivery of acceptable water to trees in the urban environment. In a non-urban environment and under natural establishment conditions, a mature tree have grown partly due to a favourable water regime that has provided sufficient water reliably over the life of the tree. Historically, urban trees have faced significant challenges that are not faced by non-urban trees which include:

- Establishment in unfavourable environments
- A deficiency in total water volume provision
- Runoff of water from non-permeable surfaces away from tree roots, even where local rainfall totals are adequate
- Low retention rates of applied water, where water evaporates or percolates away from tree roots before uptake
- Insufficient mature and productive soil/substrate for nutrient regeneration and incomplete establishment of beneficial microbial processes (e.g. limited growth of mycorrhizae)
- Tree removal or inappropriate management resulting in damage to trees if land use changes regard existing trees as an obstacle

Thus, water supply considerations provide some of the major challenges faced by urban trees and depending on the tree species, a mature tree with a 500 mm trunk diameter could require 860l of water to be provided for survival and normal function during drought conditions. It is important to acknowledge the local hydrological benefits of urban trees in times of intense and or prolonged rainfall where the drainage system may be at capacity. The positive impacts of trees include rainfall detention, retention and uptake, which may remove significant volumes of water from runoff totals, or delay local peak discharge in comparison with a location without trees [60].

In theory, it would be acceptable to deliberately divert stormwater from a new development to supplement existing urban trees to meet the total water need. This could include disconnection of downpipes into soil and conveyance to areas where trees are located. However, this retrofit option could result in standing water in the case of intense events if infiltration and percolation are not sufficient and a bypassing of the tree if the water is not retained by the soil. Hence, solutions are required that provide the required volume of water, but allow this to be retained close to the roots and provide useful water. In practice, this is necessarily targeted at new developments and is usually provided by tree pits, where a selected soil medium is placed into a structural chamber. Landscaping solutions can incorporate this type of structure into bioretention schemes as shown by the image below of a parking area for a shopping Centre drained by bioretention cells, which often successfully incorporate trees (Figure 5).

Figure 5. A bioretention cell with trees and shrubs in Raleigh, North Carolina, USA.

The medium can be an engineered soil or a replacement medium such as absorbent foam-filled structurally resistant void forming boxes. An ultra-absorbent foam that is available commercially as a component in a tree box system was tested independently in a laboratory for its impact on hydraulic conditions and was shown to be a promising solution for runoff management and water absorption [61]. This material also has the potential for water transfer to woody plants.

The use of engineered materials, including allochthonous soils, in providing water to trees has not been proven to fully replicate the functioning of the biological component of mature natural soils, nor have an equivalent of the nutrient recycling and regeneration of soil. However, it has been established that drainage systems that include trees and soil can improve water quality in sustainable drainage applications, including the absorption and retention of heavy metals nitrogen and phosphorus [62]. The addition of urban pollutants in the form of sediment moved by stormwater is shown in Figure 6. In such circumstances, however, the blinding of the surface can become an issue and a means of pre-filtration or sedimentation to take out the worst of the sediment may be advisable and if not available close control of maintenance may be required.

Figure 6. A tree pit receiving urban sediment from an adjacent parking area, Greensborough, North Carolina, USA.

The understanding of the requirements for successful irrigation of urban trees has improved in recent years, particularly as the recognition of the full range of benefits provided by urban trees has become apparent and the use of trees in the urban landscape has been more readily encouraged. As shown in the plates in this section, care must be taken to prevent waterlogging of supporting soils, partly due to the confined nature of tree pits and also the provision of water in excess from large impermeable areas, which may exclude the air that is found in natural soils and may, if serious enough, lead to anoxic conditions and the production of greenhouse gases. Where possible, water should be directed to a sub-root zone position without passing through the root ball itself. Consideration of aeration and an under drain in design should prevent this, alongside a consideration of local rainfall totals and the nearby landscape conditions.

4. CONCLUSION

SuDS elements, by their nature, store water but it is not always simple to either make a SuDS system provide a suitable supply of irrigation water or to make a water harvesting system contribute to stormwater source control. Both quality issues and quantity issues are important and in areas where road salt is used ground level collection needs to be carefully managed to prevent stored water becoming heavily contaminated with salt that could make the water unsuitable for irrigation and the collection mechanism needs to provide attenuation of automobile-derived pollutants sufficient to allow plants to grow without inhibition. Where the prime aim is to harvest a good quality of water such as on a roof, whether this is used directly to irrigate plants on the roof or is collected for offline irrigation sizing of tanks needs to allow enough excess storage to deal with storm events and the tanks need to be managed to ensure such volume is routinely available.

REFERENCES

1. FAO, (2008), Wastewater Quality Guidelines for Agricultural Use. Food and AgriculturalOrganisation Corporate Document Repository: Natural Resources Management andEnvironmental Department [online]. http://www.fao.org/ docrep/ T0551E/t0551e04.htm<25 September 2016>
2. Nnadi, E.O., Coupe, S.J., Newman, A.P. and Mbanaso, F.U., (2015) 'Stormwater harvestingfor irrigation purposes: an investigation of chemical quality of water recycled inpervious pavement system' Journal Environmental Management, 147, 246–256.
3. Meera V. and Mansoor Ahammed M. (2006) Water quality of rooftop rainwater harvestingsystems: a review, Journal of Water Supply: Research & Technology, 55 (4)257–268.
4. Jones, M.P. and Hunt, W. F. (2008) Rainwater Harvesting (2008) 'Urban Waterways: Guidance for Homeowners', North Carolina Cooperative Extension Service AGW-588-11.https://content.ces.ncsu.edu/rainwater-harvesting-guidance-for-homeowners
5. DeBusk, K.M. and Hunt, W.F. (2014) 'Rainwater Harvesting: a Comprehensive Reviewof Literature', Report No. 425 NC Water Resources Research Institute. http://repository. lib.ncsu.edu/dr/bitstream/1840.4/8170/3/NC-WRRI-425.pdf
6. DeBusk, K.M, Hunt, W.F, Osmond, D.L and Cope, G.W. (n.d.) 'Water Quality ofRooftop Runoff: Implications for Residential Water Harvesting Systems', North Carolina Cooperative Extension, https://content.ces.ncsu.edu/water-quality-of-rooftop-runoff
7. Booth, D. B. and Leavitt, J. (1999), 'Field evaluation of permeable pavement systemsfor improved stormwater management', Journal of the American Planning Association,65(3), 314–325.

8. Gomez-Ullate, E., Bayon, J. R., Coupe, S. and Castro-Fresno, D. (2010). 'Performance ofpervious pavement parking bays storing rainwater in the north of Spain', Water Scienceand Technology. 62(3), 615–621.
9. Klein, R. D. (1979), 'Urbanization and stream quality impairment' Water Resources Bulletin, 15(4), 948–963.
10. Rushton, B. T. (2001), 'Low-impact parking lot design reduces runoff and pollutantloads', Journal of Water Resources Planning and Management-ASCE, 127(3), 172–179.
11. Sañudo-Fontaneda, L.A., Charlesworth, S., Castro-Fresno, D. andrés-Valeri, V.C.A. and
12. Rodriguez-Hernandez J. (2014) 'Water quality and quantity assessment of perviouspavements performance in experimental car park areas', Water Science and Technology,69(7), 1526–1533.
13. Scholz, M. and Grabowiecki, P. (2007), 'Review of permeable pavement systems',Building and Environment, 42(11), 3830–3836.
14. Pratt, C.J. (1996) 'Permeable Pavements for Stormwater Quality Enhancement'Proceeding Paper Urban Stormwater Quality Enhancement: Source Control, Retrofitting and Combined Sewer Technology American Society of Civil Engineers.
15. Newman, A. P., Pratt, C. J., Coupe, S. J. and Cresswell, N. (2002), 'Oil bio-degradationin permeable pavements by microbial communities', Water Science and Technology,45(7), 51–56.
16. Newman, A., Puehmeier, T., Shuttleworth, A. and Pratt, C.J. (2014) Performance of anenhanced pervious pavement system loaded with large volumes of hydrocarbons. WaterScience and Technology, 70(5), 835–842.
17. Macdonald, K. and Jefferies, C. (2001), 'Performance Comparison of Porous Pavedand Traditional Car Parks' Proceedings of First National Conference on Sustainable Drainage, Coventry University, ISBN 1 903818 06 0, pp. 170–181.
18. Bayón J.R., Castro D., Moreno-Ventas X., Coupe S.J. and Newman A.P., Pervious pavementresearch in Spain: Hydrocarbon degrading microorganisms, Proceedings of 10thInternational Conference on Urban Drainage, Copenhagen/Denmark, 21–26 August2005. CD-ROM
19. Jayasuriya, N., Kadurupokune, N., Othman, M. and Jesse, K. (2007) 'Contributing tothe sustainable use of stormwater: the role of pervious pavements' Water Science &Technology, 56 (12), 69–75.
20. Fletcher, T.D., Deletic, A., Mitchell, V.G. and Hatt, B.E (2008) 'Reuse of urban runoffin Australia: a review of recent advances and remaining challenges', Journal ofEnvironmental Quality, 37(5), 116–127.
21. Brattebo, B. O. and Booth, D. B. (2003), 'Long-term stormwater quantity and quality performanceof permeable pavement systems', Water Research, 37(18), 4369–4379.
22. Nnadi, E.O. (2009) 'An evaluation of modified pervious pavements for water harvestingfor irrigation purposes' Unpublished PhD Thesis. Coventry: Coventry University.
23. Newman, A. P., Puehmeier, T., Kwok, V., Lam, M., Coupe, S. J., Shuttleworth, A. andPratt, C. J. (2004), Protecting groundwater with oil-retaining pervious pavements:Historical perspectives, limitations and recent developments, Quarterly Journal ofEngineering Geology and Hydrogeology, 37(4), 283–291.
24. Wilson, S., Newman, A.P., Puehmeier, T. and Shuttleworth, A. (2003). Performanceof an oil interceptor incorporated into a pervious pavement, ICE Proc: EngineeringSustainability, 156(1), 51–58.
25. Newman, A. P. (2003), Pollutant containment system, WO03040483 [patent]
26. Nnadi, E.O., Newman, A.P. and Coupe, S.J., (2014), 'Geotextile incorporated permeable pavement system as potential source of irrigation water: effects of re-used water on thesoil, plant growth and development', Clean-Soil Air Water, 42(2), 125–132.
27. Newman, A.P., Aitken, D. and Antizar-Ladislao, B. (2013), Stormwater quality performanceof a macro-pervious pavement car park installation equipped with channel drainbased oil and silt retention devices, Water Research, 47(2), 7327–7336.

28. New Zealand Govt. (2011), Guidelines for Assessing and Managing PetroleumHydrocarbon Contaminated Sites [online]. http://mfe.govt.nz/ publications/ hazardous/Oil-guide-jun99/module-5.pdf
29. Bauder, T.A., Waskom, P.L., Sutherland, P.L. and Davis, J.G. (2014) 'Irrigation WaterQuality Criteria' Factsheet 0.506 Colorado State University: http://www.ext. colostate.edu/ pubs/crops/00506.html
30. Newman A.P., Nnadi, E.O., Mbanaso, F.U., Shuttleworth, A., Aitken D., Antizar-LadislaoB. and Theophilus S.C. (2015) 'Potential of Pervious and MacroPervious Pavement as Harvesting Systems for Irrigation of Adjacent Lawns and Flower Beds' SUD SnetInternational Conference 2015 3rd & 4th September 2015 Techno-Centre, CoventryUniversity, UK. http://sudsnet.abertay.ac.uk/documents/SUDSnet 2015_ Newmanetal_ PotentialofPerviousandMacroPerviousPavementsasHarvesting Systemsfo_ 000. pdf
31. Tuffnell, N. and Britt, C. (2008) Determining the Usage and Usage Patterns of AmenityPesticides across the UK (PS2230) [online]. Available from http://www. pesticides.gov.uk/uploadedfiles/Web_Assets/PSD/RPA_Amenity_Report_2008.pdf [08/07/2016]
32. Rask, A.M. (2012) Non-chemical weed control on hard surfaces: An investigation oflong-term effects of thermal weed control methods. Forest & Landscape Research No.
33. 52-2012. Forest & Landscape Denmark, Frederiksberg. 156.
34. Kempenaar, C. (2010) Best practice for chemical weed control on hard surfaces. EuropeanGlyphosate Environmental Information Source, Wageningen: Wageningen Universityand Research Centre - Plant Research International.
35. Kolpin, D.W., Thurman, E.M., Lee, E.A., Meyer, M.T., Furlong, E.T. and Glassmeyer, S.T.(2006), 'Urban contributions of glyphosate and its degradate AMPA to streams in theUnited States', Science of the Total Environment, 354, 191–197.
36. Kristoffersen, P., Rask, A.M., Grundy, A.C., Franzen, I., Kempenaar, C., Raisio, J.,Schroeder, H., Spijker, J., Verschwele, A. and Zarina, L. (2008), 'A review of pesticidepolicies and regulations for urban Amenity areas in seven European countries', WeedResearch 48(3), 201–214.
37. Woodburn, A.T., (2000), 'Glyphosate: production, pricing and use worldwide', PestManagement Science, 56, 309–312.
38. Mbanaso, F.U., Coupe, S.J., Charlesworth, S.M. and Nnadi, E.O. (2012), Laboratory-basedexperiments to investigate the impact of glyphosate-containing herbicide on pollution attenuation and biodegradation in a model pervious paving system, Chemosphere, 90 (2013), 737–746.
39. Mbanaso, F.U., Nnadi, E.O., Coupe, S.J. and Charlesworth, S.M. (2016), Stormwater harvesting from landscaped areas: effect of herbicide application on water quality andusage, Environmental Science and Pollution Research, 23, 15970.
40. Baik, J., Kwak, K., Park, S. and Ryu, Y. (2012), 'Effects of building roof greening on airquality in street canyons', Atmospheric Environment, 61, 48–55.
41. Bowler, D.E., Buyung-Ali, L., Knight, T.M. and Pullin, A.S. (2010). Urban greening tocool towns and cities: a systematic review of the empirical evidence. Landscape andUrban Planning, 97, 147–155.
42. Brenneisen S., (2006), Space for urban wildlife: designing green roofs as habitats inSwitzerland, Urban Habitats, 4(1), 27–36.
43. Hunt, W.F. and Szpir, L.L. (n.d.,) Permeable Pavements, Green Roofs and CisternsStormwater Treatment Practices for Low-Impact Development, Pub. North CarolinaCooperative Extension Service https://www.bae.ncsu.edu/extension/ext-publications/ water/protecting/ag-588-06-permeable-pavements-green-roofs-and-cisterns.pdf
44. Stovin V. (2010), 'The potential of green roofs to manage urban stormwater', Water andEnvironment Journal, 24(3), 192–199.
45. Carson, T.B., Marasco, D.E., Culligan, P.J. and McGillis, W.R.(2013), 'Hydrological performance of extensive green roofs in New York City: observations and multi-year

modelingof three full-scale systems', Environmental Research Letters, 8, online at; http://iopscience.iop.org/article/10.1088/1748-9326/8/2/024036/pdf

46. Fioretti, R., Palla, A., Lanza, L.G. and Principi, P. (2010) Green roof energy and water relatedperformance in the Mediterranean climate, Building and Environment, 45, 1890–1904.
47. Wong N.H. , Chen Y., Ong C.L. and Sia A. (2003), Investigation of thermal benefits ofrooftop garden in the tropical environment, Building and Environment, 38(2), 261–270.
48. Palomo Del Barrio, E.(1998) ,Analysis of the green roofs cooling potential in buildings,Energy and Buildings, 27(2), 179–193.
49. Castleton, H.F., Stovin, V., Beck, S.B.M. and Davison, J.B. (2010), 'Green roofs: buildingenergy savings and the potential for retrofit', Energy and Buildings, 42(10), 1582–1591.
50. Alexandri, E. and Jones P.(2008), Temperature decreases in an urban canyon due to greenwalls and green roofs in diverse climates, Building and Environment, 43(4), 480–493.
51. Takebayashi, H. and Moriyama, M.(2007), Surface heat budget on green roof and high reflection
52. Susca T., Gaffin S.R. and Dell'Osso G.R.(2011), Positive effects of vegetation: urban heatisland and green roofs, Environmental Pollution, 159(8–9), 2119–2126.
53. Kadas G. (2006), 'Rare invertebrates colonizing green roofs in London', Urban Habitats,4, 66–86.
54. Matsuoka, R. H. and Kaplan, R. (2008), "People needs in the urban landscape: analysis oflandscape and urban planning contributions", Landscape Urban Planning, 84(1), 7–19.
55. Dunnett, N. (2006) 'Green Roofs for Biodiversity: Reconciling Aesthetics with Ecology',Proc. Fourth Annual Greening Rooftops for Sustainable Communities Conference, Awards and Trade Show Proceedings May 10–12, 2006, Boston, MA.
56. Molineux, C.J., Fentiman, C.H. and Gange, A.C.(2009), Characterising alternativerecycled waste materials for use as green roof growing media in the UK, EcologicalEngineering, 35, 1507–1513.
57. Snodgrass, E. and McIntyre, L.(2010), The Green Roof Manual, Portland, OR: TimberPress.
58. Voeten J.W.G.F., van de Werken L. and Newman A.P. (2016) Demonstrating the Use ofBelow-Substrate Water Storage as a Means of Maintaining Green Roofs - Performance Dataand a Novel Approach to Achieving Public Understanding, Proc. ;World Environmentaland Water Resources Congress 2016, Professional Development, Innovative Technology,International Perspectives and History and Heritage, Pathak CS and Reinhart D (Eds)American Society of Civil Engineers , Reston.
59. Hardin, M., Wanielista, M. and Chopra, M. (2012) 'A mass balance model for designing green roof systems that incorporate a cistern for re-use', Water, 4, 914–931.
60. Greenroofs.com (n.d.) Water Storage & Irrigation. http://www.greenroofs.com/Greenroofs101/water-storage.htm
61. Van Raam C.H. Shuttleworth A.B. and Culleton P.D. (2014) Plant surface structure andmodules and method for forming the same, Patent. WO2014006180 A1
62. Stovin, V.R., Jorgensen, A. and Clayden, A. (2008) 'Street trees and stormwater management'Arboricultural Journal, 30(4), 297–310.
63. Nnadi, E.O., Coupe, S.J., Sañudo-Fontaneda, L.A. and Rodriguez-Hernandez, J. (2014),'An evaluation of enhanced geotextile layer in permeable pavement to improve stormwaterinfiltration and attenuation', International Journal of Pavement Engineering,15(10), 925–932.
64. Hunt, W.F., Smith, J.T., Jadlocki, S.J., Hathaway, J.M. and Eubanks, P. R. (2008) Pollutantremoval and peak flow mitigation by a bioretention cell in urban Charlotte, N.C., Journalof Environmental Engineering, 134(5), 403–408.

CHAPTER 12

Controlling Stormwater Quality with Filter Soil—Event and Dry Weather Testing

Karin Cederkvist [1,*], Marina Bergen Jensen [2], Simon Toft Ingvertsen [3] and Peter E. Holm [1]

[1] *Department of Plant and Environmental Sciences, Faculty of Science, University of Copenhagen, Thorvaldsensvej 40, DK-1871 Frederiksberg, Denmark*

[2] *Department of Geosciences and Natural Resource Management, Faculty of Science, University of Copenhagen, Rolighedsvej 23, DK-1958 Frederiksberg, Denmark*

[3] *EnviDan A/S, Fuglebækvej 1A, 2770 Kastrup, Denmark*

ABSTRACT

The use of filter soil is increasing for control of quality of stormwater runoff prior to infiltration or discharge. This study aimed to gain knowledge about treatment efficacy of filter soils at field scale. Percolate samples from swale-trench systems with filter soil based on agricultural till with/without limestone were monitored for 15 and 9 rain events respectively. Further, two curb extensions with filter soil based on landfill soil were monitored for 10 and 8 events. Pollutant concentrations in percolate were compared to influent samples from the catchment area. Additionally one of the curb extensions was tested twice by adding high-dose synthetic influent containing runoff pollutants of concern. Despite generally low influent pollutant levels, phosphorus, copper, zinc, lead and some polyaromatic hydrocarbons exceeded guiding criteria for protection of groundwater and freshwater. Concentrations in the percolate were in most cases reduced, but phosphorus increased and despite reduced concentrations copper, lead and benzo(a)pyrene still exceeded guiding criteria. Pollutants from the synthetic influent were efficiently retained, except the pesticide MCPA. Filter soil based on landfill soil tended to perform better than agricultural till. No impact of limestone was observed. Overall the filter soils performed well in retaining pollutants, despite simultaneous processes of mobilization and immobilization.

Keywords: bioretention; bioswales; heavy metals; infiltration; organic carbon; phosphorous; stormwater runoff; water quality

1. INTRODUCTION

For more than 25 years on-site management of stormwater runoff, often referred to as SUDS (sustainable urban drainage systems), LID (low impact development) or LSM (landscape-based stormwater management), has been practiced and developed as a supplement to conventional pipe-based drainage of urban areas [1]. Within this concept infiltration elements such as swales, rain gardens, and curb extensions are some of the most widespread and well-known SUDS [2]. Infiltration elements are designed to contribute to urban drainage by retaining stormwater runoff locally and thereby decreasing downstream flow rates and volumes. Additionally, the infiltration elements may function as stormwater treatment facilities because pollutants are retained by processes such as sedimentation, filtration, adsorption and in some cases degradation of pollutants present in the stormwater runoff that have accumulated during dry weather on the urban surfaces [3]. The infiltration capacity as well as the treatment capacity depends on the media used in the facility, and often an engineered soil mixture is used. The engineered soil layer can be based on the local soil in case of new developments, i.e., on bare land. In case of retrofitting an existing area the soil will have to be transported to the site. To ensure proper treatment performance, filter soil guidelines have been developed in several countries, with recommendations on e.g., soil texture, soil layer thickness, pH and organic matter content. These guidelines differ among countries, and typically reflect the parameters that have driven the process for the on-site management such as hydraulic efficiency, runoff treatment or plant growth [4,5,6,7,8].

Germany has a long record of roadside infiltration facilities that also intentionally address water quality by utilizing an engineered soil media as described in the German DWA-guidelines [5]. The guidelines include, among other technical specifications, recommendations of a 30 cm top soil layer where pH should be 6–8, the saturated hydraulic conductivity no less than 10^{-5} m/s, the clay content not exceeding 10% and the organic content between 1% and 3% [9]. Since 2008 the practice of local infiltration of stormwater runoff by means of an engineered soil, based on the German guidelines, has gained ground in Denmark and is termed filter soil.

When increasing the drainage capacity of a city by means of on-site management of stormwater runoff, the risk of recipient contamination increases, both for surface waters like streams and rivers, and groundwater. Therefore the Danish environmental authorities increasingly require that the effluent leaving the filter soil does not deteriorate the recipient. According to Danish legislation, which is aligned with the EU legislation, stormwater runoff is considered wastewater, and it is compulsory to obtain a permit from the relevant environmental authorities to discharge (to sewer, marine or freshwater recipients) or infiltrate (to the soil) the runoff. No specific quality criteria exist that stormwater runoff has to comply with, and with the use of certain standard measures a permit will normally be granted as long as the runoff originates from normal urban surfaces such as roofs, minor roads and parking sites, squares, and parks. Authorities can only require a specific quality of the discharged or infiltrated water if the runoff is considered a risk for meeting the targeted environmental goals for a specific water body, i.e., good ecological and chemical quality as defined under the European Water Frame Directive (WFD) [10]. In such cases the authorities can define emission limit values that the runoff has to comply with before discharge or infiltration. Such limit values are local and individually decided, but the authorities may take into consideration the quality criteria for surface water bodies and the groundwater quality criteria summarized in Table 1. The groundwater quality criteria has been developed by the national environmental authorities in Denmark [11], to aid local authorities in defining requirements for when to release contaminated soil from further remedial measures like pumping of groundwater. Likewise, the quality criteria for surface freshwater bodies as listed in the departmental order 1070 under WFD [12] are not concerning discharge of stormwater runoff, but may be used as guidance in case there is a need to control a certain discharge of stormwater runoff.

Table 1. Danish and European Union quality criteria for groundwater and surface water, which may provide guidance when assessing filter soil treatment performance.

	Groundwater Quality Criteria [1] [μg/L]	Short-Term Conc. Criteria for Surface Freshwater Bodies [2] (in Parenthesis the Long-Term Conc. Criteria) [μg/L]
Reference	[11]	[12]
Cadmium (Cd) [3]	0.5	<0.45–1.5 (<0.08–0.25)
Chromium (Cr)		
Cr(III) + Cr(VI)	25	124 (17)
Cr(VI)	1	4.9 (3.4)
Copper (Cu)	100	2 (1)
Nickel (Ni)	10	34 (4)
Lead (Pb)	1	2.8 (1.2)
Zinc (Zn)	100	8.4 (7.8)
Acenaphthene	No criterion	3.8 (3.8)
Benzo(a)pyrene	0.01	0.027 (0.00017)
Fluoranthene	0.1	0.12 (0.0063)
Naphthalene	1	130 (2)
Phenanthrene	No criterion	4.1 (1.3)
Pyrene	No criterion	0.023 (0.0046)
Sum of polyaromatic hydrocarbons (PAHs)	0.1 [4]	No criterion
Single pesticide	0.1	No criterion
Sum of pesticides	0.5	No criterion
Linear alkylbenzene sulfonate (LAS)	No criterion	160 (54)

Notes: [1] Danish threshold values (total concentrations) for groundwater under polluted soils; [2] Dissolved concentrations. Short-term is defined as discharge in maximum 24 h once per month with minimum 6 days in between; [3] Depending on water hardness; [4] Sum of benzo(b)fluoranthene, benzo(k)fluoranthene, benzo(g,h,i)perylene, indeno(1,2,3-cd)pyrene.

Since authorities tend to assess stormwater discharges by concentrations rather than mass fluxes, it makes sense to evaluate stormwater treatment facilities based on what percolate concentrations they can achieve. As an example, the environmental authority from the city of Copenhagen has announced that stormwater runoff discharged to lakes in the municipality may contain no more than 0.08 mg phosphorus (P) per liter [13].

The retention of stormwater pollution in filter soils is controlled by precipitation, complexation and sorption processes. In most cases sorption processes control the distribution of pollutants between the solid and the solution phases. Sorption is used here as a collective term for various soil surface-related processes including surface complexation, ion exchange, and other processes related to binding by chemical or physical properties. The soil system is too complex to unravel the actual reactions controlling the specific pollutants.

Studies both under laboratory [14,15,16,17] and field conditions [18,19,20,21,22] have investigated the ability of engineered soils to retain pollutants from percolating water, focusing on heavy metals, polyaromatic hydrocarbons (PAHs), nutrients and suspended solids. Overall these studies show that the soil media perform well in retaining pollutants; however, there can be leaching of dissolved organic carbon (DOC) from the systems, as well as pollutants such as phosphorus (P) and copper (Cu). Further, the leaching of P and Cu is likely to be linked to DOC mobility [16,23]. No data on pesticides have been found.

Recognizing that there may be significant variations between laboratory and field scale as well as between local and regional field studies, the Danish environmental authorities have sought more documentation on performance of filter soil in field scale facilities. Thus, the aim of this study was to add to our knowledge about the treatment efficacy of filter soils at field scale in a Danish context. Improved understanding of treatment capacity and behavior of infiltration facilities will lead to better design guidance and hopefully also contribute to more sustainable urban water management internationally.

The specific objectives were:

a. To monitor the treatment performance of Danish road runoff infiltration systems with filter soil during storm events, and to look for differences between filter soil based on modified local soil with/without limestone and unpolluted soil from a landfill, and

b. To test the treatment performance of one of the systems under more extreme conditions using a synthetic road runoff solution with high yet realistic concentrations of a range of relevant pollutants.

2. MATERIALS AND METHODS

2.1. Locations

Two types of field-scale road runoff infiltration systems with filter soil were studied, including swale-trenches in the city of Odense and curb extensions in the Copenhagen area (Table 2). At the Odense site the filter soil was placed as the top layer in two infiltration swale-trenches receiving stormwater runoff from a newly established parking lot. For both trenches the filter soil was made from the local till soil by adding sand, to meet the DWA standard. The filter soil in one of the trenches was further modified by adding limestone, though not exceeding the prescribed pH. At the Copenhagen sites the filter soil was included as the top layer in curb extensions used to disconnect two streets located 10 km apart, Moellebakken and Lindevang, from the sewer system. At Moellebakken, there were four curb extensions, and at Lindevang, six curb extensions. Here the filter soil was made from unpolluted soil transported to the site from landfill storage (RGS 90 A/S).

Table 2. Description of test sites and sampling methods.

Swale-Trenches—Odense

Photo of One of the Swale-Trenches

Curb Extensions—Copenhagen

Photo of One of the Curb Extensions

	East (+limestone)	West	Moellebakken	Lindevang
Year of construction	Spring 2011		Fall 2012	
Catchment typology UTM	Parking lot for 200 cars 55°22′4.22″ N 10°25′26.74″ E		Residential road <5000 cars per day 55°42′16.83″ N 12°27′44.72″ E	Residential road <5000 cars per day 55°37′8.84″ N 12°24′59.75″ E
Design	0.3 m deep filter soil layer on top of a thin gravel layer and 0.5 m of infiltration trench made from stormwater boxes		Curb extension with a 0.4 m deep filter soil layer on top of 1.6 m soakaway made from stormwater boxes	
Infiltration area	1000 m^2	1180 m^2	20 m^2	16 m^2
Catchment area: infiltration area	5:1	5:1	33:1	21:1
Filter soil composition	Local topsoil mixed 1:1 with sand in grain size 0.02–2 mm and 3 kg of bryozo calcite [1] per m^3 of soil-sand mixture	Local topsoil mixed 1:1 with sand in grain size 0.02–2 mm	Made from RGS90 growth-topsoil (mixture 4) [2] from a landfill storage	
No. of influent samples	12		10	
Influent sampling method	Runoff samples collected via line drainage placed along the edge of parking lot connected to a sampling well		Runoff samples were collected from nearby road with similar traffic load (55°41′49.71″ N 12°27′50.74″ E) by means of a fraction collector placed in a road well equipped with 24 PE-containers of 0.5 L (ISCO 3700)	
No. of percolate samples	15	9	10	8
Percolate sampling method	From sampling wells (Quadro Control Fränkische) connected with the stormwater boxes via a longitudinal perforated pipe mounted along the top of the stormwater boxes, just below the filter soil, capturing fractions of the percolating water		From bottom of the stormwater boxes by vacuum pumping in a tube (Teflon™) lowered into the stormwater boxes through a vertical pipe bypassing the filter soil. As percolate ponded in the bottom of the trench for some time before ex-filtrating into surrounding soil it was considered mixed and representative of the filter soil percolate	
Monitoring period	June 2014–October 2014		July 2013–December 2014	October 2013–December 2014

Notes: [1] Mined bryozoan calcite from Faxe Limestone Quarry, Denmark, diameter range: 1–3 mm, bulk weight: 1250 kg/m^3, bulk porosity: 54%; [2] Homogeneous mixture of topsoil, sand and compost, according to the supplier's specification: Humus: 3%, clay: 6%, silt: 8%, fine sand: 38%, course sand: 44%, pH: 8.

At Moellebakken the percolate from one out of a total of four curb extensions was monitored and at Lindevang the effluent from one out of a total of six- curb extensions was monitored.

2.2. Soil Samples

To document that the filter soils complied with the German DWA guidelines and were originally unpolluted, representative samples were obtained for analysis. From the swale-trench systems 20 samples of topsoil mixed with sand (prior to mixing with limestone) were obtained before the soil was placed in the trenches. These 20 soil samples were analysed for pollutant levels by an accredited laboratory. The 20 samples were then bulked to one sample for texture analysis. After mixing half of the topsoil with limestone, further samples were obtained for pH and plant-available P-measurements of the two filter soil types: with and without limestone Regarding the curb extensions, filter soil was sampled from all four in Moellebakken and all six in Lindevang shortly after construction, by collection of 10 samples from 0 to 40 cm depth distributed over the surface area. The 10 subsamples from each curb extension were bulked into one sample used for texture analyses. Hereafter the four bulk samples from Moellebakken and the six bulk samples from Lindevang were bulked into one sample for each street, from which two subsamples were analysed for pollutant levels.

2.3. Soil Analyses

Texture was analysed by the hydrometer method [24] and the pH of the soil samples was determined using a combined glass electrode (Metrohm, 6.0228.000) in a solution of soil and a 0.01 M $CaCl_2$ solution with a ratio of 1:2.5. An accredited laboratory performed analysis of plant-available P in the swale-trench soil by the Olsen-method [25], and total P on curb extension soil, by the method SM 3120 on ICP-OES. Heavy metals (Cd, Cr, Cu, Ni, Pb, Zn) were analysed with ICP-OES after standard method 3120, total petroleum hydrocarbons with GC-FID after REFLAB method 1:2010 and PAHs (fluoranthene, benzo(b+j+k) fluoranthene, benzo(a)pyrene, indeno(1,2,3-cd)pyrene, dibenzo(a,h) anthracene) with GC-MS after REFLAB method 4.

The hydraulic conductivity was measured by means of in situ single ring infiltration test with constant head following the method by [26]. In the swale-trenches infiltration measurements were done four times throughout 2013–2014 (two to three years after construction) at three different places in the trenches each time. The curb extensions were subjected to the infiltration tests at two places in each curb extension six months after construction.

2.4. Water Sampling Procedure

Precipitation data for the swale-trenches were recorded by a rain gauge nearby (UTM 55°20′36.5″ N 10°25′59.3″ E). For the curb extensions precipitation data were obtained from a rain gauge placed near the influent fraction collector (UTM 55°42′08.2″ N 12°27′55.2″ E), both operated by the Danish Water Pollution Committee and the Danish Meteorological Institute [27].

Pressure transducers were used to monitor the water level in the stormwater boxes underneath the filter soil and thus to control that stormwater generated percolate was sampled (swale-trenches: MJK Expert™ 7060 hydrostatic digital/analog logger; curb extensions: HOBO® Water Level Loggers).

In the swale-trenches influent samples were obtained from a line drainage placed in terrain along the parking lot edge. Due to practical reasons it was not possible to collect inlet samples at the curb extensions in Copenhagen. However, flow weighted samples of stormwater runoff were collected from a nearby road with a similar traffic load (Table 2). This was used to represent influent for both curb extensions.

Sampling procedures for percolate samples are described in Table 2. The samples were all collected in glass and plastic containers (dependent on compounds to be analysed), stored in the cold and dark, and sent to an accredited laboratory within 24 h.

2.5. Water Analyses

Table 3 presents the parameters included in the monitoring program at each location. Samples collected in the swale-trenches were all analysed after the same program. For the curb extensions 4–5 samples were analysed after a smaller program referred to as a basic program (B) and the remaining samples after a more extended program (E). A synthetic influent applied to the curb extension in Moellebakken was analysed after a third program (S).

Due to practicalities only N, P and metals were measured as both total and dissolved; however, it should be noted that [28] has shown that this speciation can also be significant for the organic compounds such as PAHs.

Table 3. Schematic presentation of the compounds included in the monitoring program for each location. B = basic program (*n* = 5 for Moellebakken, *n* = 4 for Lindevang), E = extended program (*n* = 5 for Moellebakken, *n* = 4 for Lindevang), S = synthetic stormwater program. Compounds analysed are marked with X.

	Odense	Moellebakken			Lindevang		Method	Detection Limit
Monitoring program		B	E	S	B	E		
pH	X	X	X	X	X	X		
Electric conductivity	X	X	X	X	X	X	EN 27888	0.5 mS/m
Suspended solids	X	X	X		X	X	EN 872	0.5 mg/L
TOC/DOC	X		X		X	X	EN 1484	1 mg/L
P-total and dissolved	X	X	X	X	X	X	SM 4500-P (F)	0.005 mg/L
N-total	X		X		X	X	EN ISO 11905 auto mod Skalar	0.05 µg/L
NH_3+NH_4-N			X		X	X	SM 4500-NH3 (H)	0.005 mg/L
NO_3-N			X		X	X	SM 4500-NO3 (H)	0.1 mg/L
Cl		X	X		X	X	SM 4500-Cl (E)	1 mg/L
Al—total and dissolved		X	X	X	X	X	EN ISO 17294m: 2005 ICP/MS	30 µg/L
Cd—total and dissolved	X	X	X	X	X	X	EN ISO 17294m: 2005 ICP/MS	0.05 µg/L
Cr—total and dissolved	X	X	X	X	X	X	EN ISO 17294m: 2005 ICP/MS	0.5 µg/L
Cu—total and dissolved	X	X	X	X	X	X	EN ISO 17294m: 2005 ICP/MS	1.0 µg/L
Ni—total and dissolved[1]		X	X	X	X	X	EN ISO 17294m: 2005 ICP/MS	1.0 µg/L
Pb—total and dissolved	X	X	X	X	X	X	EN ISO 17294m: 2005 ICP/MS	0.5 µg/L
Zn—total and dissolved	X	X	X	X	X	X	EN ISO 17294m: 2005 ICP/MS	5.0 µg/L
Total petroleum hydrocarbons (TPH)		X	X				ISO 9377-2 mod.GC/FID	C6H6-C10: 2 µg/L C10-C25: 8 µg/L C25-C35: 10 µg/L
PAHs		X	X	X		X	GC/MS	0.01 µg/L
Pesticides: 2-(4-chloro-2-methylphenoxy) acetic acid (MCPA), Glyphosate				X			GC/MS, LC/MS/MS	0.01 µg/L
Anionic detergents				X			DS 237	0.15 mg/L

2.6. Application of Synthetic Influent to One Curb Extension

To test the filter soil under conditions where a number of relevant contaminants (S-program in Table 3) are present at high but still realistic concentrations, a synthetic contaminant dose was added to the monitored curb extension at Moellebakken in dry weather. In all, 10–11 m^3 of contaminated water was added to the curb extensions during approximately six hours (detailed procedure and considerations are described in [29]). Bromide was included in the dose as a conservative tracer. The test with synthetic effluent was conducted twice two months apart.

3. RESULTS

3.1. Filter Soil Characteristics

Texture, hydraulic conductivity and pollution levels in the filter soils of swales and curb extensions are presented in Table 4 and Table 5 respectively. Since the same type of local soil was used for the two swale-trenches and the same type of landfill soil for the curb extensions, the results are bulked respectively. The only exception is pH and P for the swale-trenches, where separate results are given for the west and east ends.

Table 4. Soil characteristics (with standard deviation) of filter soils in the soils in Odense (texture n = 1, infiltration n = 15) and Copenhagen (texture n = 4, infiltration n= 20), compared with requirements in the DWA guidelines.

	pH	Texture (%)					Hydraulic Conductivity (m/s)
		Organic C	Clay	Silt	Fine Sand	Sand	
DWA [5]	6–8	1–3	Clay + Silt < 10				10^{-5}–10^{-3}
Odense Swale-Trenches	East 7.9 West 7.5	1.7	2.8	1.5	25.2	68.7	2×10^{-5} (1.6×10^{-5})
Copenhagen Curb Extensions	7.8 (0.04)	2.0 (0.3)	4.9 (1.1)	3.9 (1.1)	30.4 (1.8)	61.1 (2.2)	2.7×10^{-5} (3.0×10^{-6})

In the swale-trenches, plant-available P was 0.8 mg/kg for the limestone-modified eastern trench (East) and 1.3 mg/kg for the western trench (West). In the curb extensions, the average of total P was 438 mg/kg, with a standard deviation of 34 mg/kg dry matter.

Table 5. Average soil pollutant concentrations (with standard deviations in parenthesis) for Copenhagen (*n* = 4) and Odense (*n* = 20), compared with Danish values for clean soil. b.d. = below detection limit. n.i. = not included.

Parameter	Odense Swale-Trenches	Copenhagen Curb Extensions	Soil Quality Criteria [1]
		[mg/kg DM]	
Cd	0.12 (0.02)	0.20 (0.03)	0.5
Cr	5.0 (0.6)	4.1 (0.4)	500 (20)
Cu	2.0 (0.4)	9.5 (0.4)	500
Ni	3.8 (0.4)	4.4 (0.4)	30
Pb	4.4 (0.7)	11 (0)	40
Zn	18 (2)	45 (3)	500
Sum of hydrocarbons	b.d.	25 (2.6)	100
Fluoranthene	b.d.	0.09 (0.03)	n.i.
Benzo(b+j+k)fluoranthene	b.d.	0.14 (0.04)	n.i.
Benzo(a)pyrene	b.d.	0.05 (0.02)	0.3
Indeno(1,2,3-cd)pyrene	b.d.	0.05 (0.02)	n.i.
Dibenzo(a,h)anthracene	b.d.	0.01 (0.004)	0.3
Sum of 7 PAHs [2]	b.d.	0.34 (0.1)	4

Notes: [1] If the soil quality criteria are met the soil is considered safe for use without fear of exposure in e.g., private gardens and on playgrounds [11]; [2] Sum of 7 PAHs fluoranthene, benzo(b)fluoranthene, benzo(j)fluoranthene benzo(k)fluoranthene, benzo(a)pyrene, dibenzo(a,h)anthracene and indeno(1,2,3-cd)pyrene.

When compared all tested filter soils initially complied with the Danish soil quality criteria [11] (Table 5). These data can be used in the future to evaluate the accumulation of pollution in filter soil. It is noticed that the Copenhagen filter soil based on landfill soil contained slightly higher levels of Pb, Cu, Zn, hydrocarbons and PAHs compared to the farmland-based filter soil in Odense. The plant-available P-level in the landfill soil used in Copenhagen is in the range of 3–6 (information from the landfill soil provider, RGS90) and thus higher compared with the measured value of plant-available P of 0.8 in swale-trench East and 1.3 in swale-trench West in Odense.

3.2. Characterization of Rain Events

The rain events sampled represented significant but not extreme events. In Table 6 the rain depths for the sampled events are seen. As the arrival of percolate in the stormwater boxes is delayed compared to the stormwater runoff, and the percolate may originate from series of rains, it was decided to calculate rain depth in the 24 h preceding the percolate sampling time for the Odense sites. For the Copenhagen sites the rain depth corresponding to the influent water-sampling period was recorded. To look for correlations between the preceding dry weather period, the number of dry weather days before the event as well as the accumulated rain depth in the 7 days antecedent to the period were calculated.

Table 6. Rain characteristics for sampled events at the swale-trench systems in Odense (above) and the curb extensions in Copenhagen (below).

Odense												
Date (all 2014)	28/6	14/7	10/8	14/8	18/8	31/8	15/9	22/9	30/9	13/10	17/10	20/10
Rain depth antecedent 24 h [mm]	1.2	21.8	0.8	0.6	1.4	16.4	2.6	11.4	15.0	7.8	2.6	27.6
Antecedent dry days [d]	0	3	0	1	0	0	0	0	4	0	0	1
Antecedent 7 days rain depth [mm]	28.8	1.6	11.4	15.8	32.8	12.8	16.6	2.3	1.6	15.4	15.4	40.2

Copenhagen										
Date (all 2014)	7/5	9/5	8/8	29/8	30/9	7/10	13/10	24/11	5/12	12/12
Rain depth inlet sample [mm]	15.1	11.9	19.8	8.8	2.5	5.0	3.3	6.2	4.0	11.9
Antecedent dry weather [d]	13	0	4	7	6	6	1	1	11	2
Antecedent 7 days rain depth [mm]	0	17.9	30.4	5.2	7.0	9.2	11.2	8.2	0	14.0

3.3. Influent and Percolate Concentrations

The average values for influent compared to the concentrations in the percolate during rain events are listed in Table 7. Compounds that were below detection limit in all events in both influent and percolate (volatile organic carbon, total petroleum hydrocarbons, acenaphthene, naphthalene and fluorine) are left out.

Suspended solids and metals were plotted as a function of rain depths as well as antecedent dry weather periods. Only weak linear correlations could be observed (R^2-values between 0.1 and 0.4) and it is concluded that in this case there was little or no coherence between influent pollution levels and rain patterns.

The EC measured in the influent water is comparable to other studies [30,31,32] and was seen to increase from influent to percolate, reflecting that the water mobilized various ions as it passed through the soil.

The observed concentration of less than 50 mg/L of SS in the influent from both Odense and Copenhagen (Table 7) is relatively low compared to other studies [30,31,32], but was most likely due to the low traffic level in the areas. In the swale-trench systems in Odense the SS measured in the percolate is half of what is observed in the influent. Comparing the influent concentrations of SS in Copenhagen with concentrations in the percolate from the Moellebakken curb extension, a five-fold reduction in SS is seen.

Phosphorus in influent exceeded the Copenhagen criterion of 0.08 mg P/L in 4 out of 12 events in Odense and in 7 out of 10 events in Copenhagen. In percolate, P concentrations were in general higher than in influent at all sites, and in all cases exceeding 0.08 mg P/L. Furthermore a larger fraction of P was found in the dissolved form compared to influent.

Table 7. Characterization of inlet and percolate samples from filter soil systems. Average value is shown for each parameter, standard deviation given in brackets (detection limit used for samples with concentrations below detection limit; if all samples were below detection limit, there was no standard deviation). *n* = number of samples, n.a. = not analysed.

	Swale-Trench Systems			Curb Extensions		
	Odense Influent n = 12	Odense East Percolate n = 15	Odense West Percolate n = 9	Copenhagen Influent n = 10 or n = 5 [1]	Moellebakken Percolate n = 10 or n = 5 [1]	Lindevang Percolate n = 8 or n = 4 [1]
pH	7.0 (0.3)	8.2 (0.2)	8.1 (0.2)	7.5 (0.3)	8.5 (0.6)	8.0 (0.2)
EC [mS/m]	4.1 (1.6)	32.0 (9.4)	28.4 (14.4)	11.4 (3.8)	24.5 (8.0)	20.3 (3.7)
Susp. solids [mg/L]	16 (11)	8.6 (6.2)	8.9 (5.5)	26.7 (23.5)	5.5 (4.2)	n.a.
OC [2]	4.3 (2.7)	8.5 (3.4)	8.2 (3.6)	12.3 (14.5)	5.3 (1.1)	6.2 (2.1)
N tot. [mg/L]	1.2 (0.7)	1.2 (0.6)	1.2 (0.8)	1.6 (0.6)	1.3 (0.4)	1.1 (0.3)
NH_3+NH_4-N diss. [mg/L]	n.a.	n.a.	n.a.	0.3 (0.2)	1.5 (3.3)	0.02 (0.02)
NO_3-N diss.[mg/L]	n.a.	n.a.	n.a.	0.7 (0.4)	0.9 (0.4)	0.8 (0.3)
P tot. [mg/L]	0.1 (0.03)	0.2 (0.1)	0.2 (0.1)	0.1 (0.1)	0.2 (0.1)	0.2 (0.1)
P diss. [mg/L]	0.03 (0.02)	0.1 (0.1)	0.1 (0.1)	0.03 (0.02)	0.2 (0.1)	0.2 (0.1)
Cl diss. [mg/L]	n.a.	n.a.	n.a.	7.9 (9.1)	9.6 (5.7)	3.9 (1.6)
Al tot. [µg/L]	n.a.	n.a.	n.a.	793 (643)	348 (166)	271 (146)
Al diss. [µg/L]	n.a.	n.a.	n.a.	69.2 (113.0)	113 (68.2)	54 (37)
Pb tot. [µg/L]	1.8 (1.7)	1.8 (1.5)	1.2 (0.5)	2.4 (1.8)	0.8 (0.4)	0.8 (0.3)
Pb diss. [µg/L]	0.6 (0.5)	0.6 (0.1)	0.6 (0.2)	0.5	0.5 (0.03)	0.5
Cd tot. [µg/L]	0.1 (0.02)	0.1 (0.1)	0.2 (0.3)	0.1 (0.01)	0.1 (0.01)	0.1 (0.1)
Cr tot. [µg/L] [3]	5.0 (13. 6)	4.8 (3.1)	4.9 (5.4)	2.1 (1.5)	8.1 (10.3)	5.3 (10.4)
Cr diss. [µg/L]	2.5 (6.5)	1.8 (1.7)	1.3 (0.7)	0.7 (0.3)	7.2 (10.0)	4.9 (10.6)
Cu tot. [µg/L]	7.2 (4.6)	7.1 (3.6)	6.5 (3.1)	11.7 (5.1)	9.6 (6.9)	6.0 (2.5)
Cu diss. [µg/L]	5.1 (5.7)	5.7 (3.5)	6.4 (4.1)	6.2 (3.3)	6.6 (5.5)	4.3 (3.0)
Ni tot. [µg/L]	n.a.	n.a.	n.a.	7.2 (14.4)	2.2 (1.7)	1.4 (0.7)
Ni diss. [µg/L]	n.a.	n.a.	n.a.	1.4 (0.4)	2.0 (1.9)	1.2 (0.4)
Zn tot. [µg/L]	25.9 (15.3)	11.5 (9.9)	11.1 (7.7)	27.9 (12.1)	5.7 (1.0)	8.1 (5.4)
Zn diss. [µg/L]	18.4 (20.7)	6.1 (2.1)	5.8 (2.0)	12.1 (10.3)	5.0	5.8 (2.1)
Acenaphthene [µg/L]	n.a.	n.a.	n.a.	0.01	0.01	0.01 (0.01)
Naphthalene [µg/L]	n.a.	n.a.	n.a.	0.01	0.01	0.01 (0.005)
Phenanthrene [µg/L]	n.a.	n.a.	n.a.	0.02 (0.01)	0.01	0.01
Fluoranthene [µg/L]	n.a.	n.a.	n.a.	0.03 (0.03)	0.01 (0.004)	0.01
Pyrene [µg/L]	n.a.	n.a.	n.a.	0.03 (0.03)	0.01 (0.005)	0.01
Benzo (b+j+k)fluoranthene [µg/L]	n.a.	n.a.	n.a.	0.05 (0.06)	0.02 (0.01)	0.01
Benzo(a)pyrene [µg/L]	n.a.	n.a.	n.a.	0.02 (0.02)	0.01 (0.003)	0.01
Indeno(1,2,3-cd)pyrene [µg/L]	n.a.	n.a.	n.a.	0.02 (0.02)	0.01 (0.004)	0.01

Notes: [1] *n* depended on after which program samples were analysed by, see Table 3. Further, n deviated for SS Odense West: n = 10, Odense East: n = 4, naphthalene Møllebakken: n = 8 samples, and all PAH's Lindevang: n = 3, except naphthalene: n = 2; [2] DOC for Odense, TOC for Copenhagen; [3] Chromium showed significantly higher levels for the first samplings from both curb extensions, probably originating from concrete reinforcing walls surrounding the stormwater boxes. Furthermore a total Cr concentration of 48 µg/L was observed in one event in Odense, but considered a contamination of the sample since the dissolved Cr was below detection limit in the same sample. The value of 48 µg/L is not included in the calculations.

The observed levels of heavy metals and organic contaminants in influent and percolate from rain events are compared with the guiding values in Table 1. In some cases the concentration level of a compound is not critical compared to values in Table 1, which is true for Cd, Ni, acenaphthene, naphthalene, phenanthrene and fluoranthene. In other cases either the groundwater quality criteria or the criteria for discharge to freshwater may be the more critical, which is discussed below. Other pollutants included in the monitoring program, such as chloride (Cl), aluminum (Al), benzo(b,j,k)fluoranthene, and indeno(1,2,3-cd)pyrene have no guiding criteria. For heavy metals only, in a few cases, patterns for the distribution between dissolved and particulate forms are observed.

The Cr concentrations in the influent in most cases exceeded the value of 1 μg/L for Cr(VI), but not the criterion for total Cr. The percolate from all sites had higher Cr concentrations than the influent, and the observed level was at or exceeded the quality criterion for long-term discharge. Chromium leaching from the swale-trenches was bound to particles, whereas the Cr leaching from curb extensions was mostly on the dissolved form.

The criterion for Cu in groundwater is 100 μg/L, since the potential harmful effects of Cu to humans are limited. However, the short-term criterion for dissolved Cu in freshwater is 2 μg/L and the influent concentrations of Cu at both sites in all events exceeded this level. Half of the Cu detected in the influent was in the dissolved form at both sites. In the percolate from swale-trenches a large part of the Cu in the percolate was in the dissolved form and from both trenches exceeded 2 μg/L during every rain event. In the percolate from the curb extension in Moellebakken, Cu exceeded the freshwater criterion in all samples. In the percolate from the curb extension in Lindevang, Cu was detected in 7 out of 8 events, and all exceeded the freshwater criterion. Almost all Cu found in the percolate from the curb extensions were in the dissolved form.

In general, Pb was detected in the majority of the influent samples and particle bound (Odense: 10 out of 12 events, with 7 exceeding groundwater quality criteria; Copenhagen: 10 out of 10 events, with 9 exceeding same criterion). There was only a weak Pb reduction from influent to percolate, and several samples still exceeded the 1 μg/L criterion (lime-modified swale-trench East: all samples with 12 out of 15 exceeded groundwater quality criterion; swale-trench West: 7 out of 9 samples, with 5 exceeding same criterion; Moellebakken: 7 of 10 samples, with 3 exceeding same criterion; Lindevang: 4 of 8 samples, with 3 exceeding same criterion).

At both sites total Zn in all influent surpassed the short-term criterion for emissions to freshwater, whereas the dissolved Zn exceeded the criteria in 5 events in Copenhagen and 9 out 12 events in Odense.

Polyaromatic hydrocarbons were only measured in influent in Copenhagen and in percolate from the curb extension in Moellebakken. Some of the influent samples contained PAH in problematic concentrations, when comparing with the guidelines in Table 1. Pyrene was detected in 6 out of 10 influent samples and 4 of them exceeded the short-term freshwater emission criterion, but pyrene was only observed in 2 percolate samples, both below the criterion. Benzo(b,j,k)fluoranthene was detected in 7 out of 10 events where the groundwater quality criterion was exceeded in 2, but not in any of the percolate samples. Benzo(a)pyrene (groundwater quality criterion of 0.01 µg/L) was detected in 4 out of 10 influent samples all with concentrations above the criterion, and was detected in the percolate in 2 out of 10 samples, both exceeding the criterion.

3.4. Test with Synthetic Influent

The results from the replicate experiment with addition of synthetic runoff with high yet realistic concentrations of selected pollutants are presented in Table 8. Since bromide is inert, the loss in bromide cannot be ascribed to the removal mechanisms operating in filter soil, but rather to diffusion of the solute into volumes of the filter soil or the soil matrix surrounding the facilities with non-mobile water [29,33]. Similar diffusion losses must be expected for other solutes, and without accounting for this loss, an over-estimation of pollutant removal efficiencies would occur. Adjustment to the observed effluent concentrations were made prior to calculating the contaminant removal efficiency by reducing the influent concentrations with the same reduction factor as observed for the conservative tracer bromide (named ‘expected concentration’ in Table 8).

To distinguish between percolate from rain events and percolate from this experiment with addition of synthetic runoff the term effluent is used for the latter.

As for the rain events, the SS were higher in effluent than in influent. Comparing the expected effluent concentration with the observed concentrations it is seen that P was not reduced from influent to effluent. In the second addition P in the effluent was higher than in influent. The cationic metals Cd, Cr, Cu, Ni, Pb and Zn were in all cases reduced from influent to effluent. The cationic Cr(VI) was not retained in the soil. PAHs and glyphosate concentration levels were reduced from influent to effluent as well as LAS. The pesticide MCPA was not reduced in the first addition, but retained in the second addition.

Table 8. Characterization of synthetic influent added twice to the inlet of curb extension at Moellebakken, and resulting effluent. 'Expected effluent concentration' is the calculated concentration if the compound had been retained to the same degree as bromide, i.e., if there had been no impact of the filter soil. For P and heavy metals, dissolved fraction is given in italics. *n* = number of samples, n.a. = not analysed.

Parameter	First Addition			Second Addition		
	Influent	Expected Effluent Concentration	Observed Effluent Concentration	Influent	Expected Effluent Concentration	Observed Effluent Concentration
pH	7.8		8.2	8		8.3
Suspended solids	1.1		3.2	0.7		3.7
Br [mg/L]	3000		2100	3300		1600
P [µg/L]	390 *350*	273	250 *200*	380 *390* [1]	184	300 *300*
Cd [µg/L]	3.9 *3.8*	2.7	0.06 *<0.05*	4.6 *4.5*	2.2	<0.05 *<0.05*
Cr(VI) [µg/L]	36 *33*	25.2	26 *19*	23 *19*	11.2	9.8 *9.5*
Cu [µg/L]	51 *41*	35.7	8 *<1*	54 *45*	26.2	8.1 *2.7*
Ni [µg/L]	*No data* [2]	-	*No data*	20 *17*	9.7	1.2 *1.1*
Pb [µg/L]	18 *12*	12.6	1.6 *<0.5*	19 *15*	9.2	<0.5 *<0.5*
Zn [µg/L]	*No data* [1]	-	*No data*	190 *180*	92.1	6.4 *<5*
Acenaphthene [µg/L]	1.0	0.7	0.05	0.7	0.3	0.03
Naphthalene [µg/L]	0.8	0.6	0.02	0.8	0.4	0.02
Phenanthrene [µg/L]	0.8	0.5	<0.01	0.6	0.3	0.01
Glyphosate [µg/L]	0.9	0.6	0.09	0.8	0.4	0.08
MCPA [µg/L]	0.5	0.4	0.5	1.4	0.7	0.1
Linear alkylbenzene sulfonates [µg/L]	*No data* [1]	-	*No data*	130	63.0	<0.03

Notes: [1] The insignificant difference indicates that all P is dissolved; [2] For the first addition, contaminant dose interpretations cannot be made on Zn, Ni and LAS as these compounds apparently formed a precipitate in the stock solution used for preparing the synthetic influent. For the results of the second addition this is not an issue since the stock solutions were prepared differently.

4. DISCUSSION

The tested filter soil complied with the German DWA guidelines at all locations (Table 4). As the pH of the till soil is already in the more alkaline part of the spectrum, addition of limestone to one filter soil in the swale-trench only raised pH slightly from 7.5 to 7.9. Regarding the hydraulic conductivity it should be noted that in the curb extensions the measurements were completed shortly after construction, while in the swale-trenches they were carried out repeatedly throughout the first two years after construction. With time, soil can be more compacted and the hydraulic conductivity decreased, but results from [9] showed that German swale-trench systems up to 16 years old still complied with the guidelines. Here the vegetation is assumed to play a role in maintaining a good soil structure.

The observed influent concentrations at monitored sites in Odense and Copenhagen (Table 7) were generally low when compared with other studies [30,31,34]. Worth mentioning is that many organic contaminants such as pesticides, which proved to be problematic when added synthetically, were not included in the monitoring of rain events. While pH, EC, N-total, Cd, Cr, and Zn of the influent were at the same level, the concentrations of SS, OC, Pb and Cu tended to be higher in Copenhagen than in Odense (Table 7), which may be explained by more traffic in the residential streets of the Copenhagen site compared to the parking lot in Odense, where also the atmospheric deposition may be lower due to its more rural surroundings.

The correlation between pollution level in stormwater runoff and antecedent dry weather period that is often observed [35,36] was weak in this study with R^2 values below 0.4 when using 7-day antecedent rain depths (Table 6) as a measure of dry weather conditions. The lack of significant correlation is ascribed to the character of the catchment with low pollution levels in general. Other reasons may be that too few events were sampled, not capturing the annual variation. As shown by [37,38], a minimum of between 15 and 20 runoff events are necessary to have a statistically representative sample dataset of influent samples, which was not the case for any of the locations in this study. Further, no samples were collected during the winter period due to a dry winter, which could be the reason for no observations of high Cl concentrations.

When infiltrating runoff through the filter soil of 30–40 cm, a direct comparison between influent and percolate of an individual event does not make sense, since a part of the percolate will stem from previous rain events and furthermore from equilibrium reactions with the soil during the dry weather period. This is why only the average influent and percolate concentrations presented in Table 7 should be compared. Furthermore, when the initial soil moisture level is below field capacity, the soil will retain a fraction of the water which will later be lost through evapotranspiration. This loss can vary significantly depending on the pore size distribution, the soil moisture level, the plants present, as well as spatial and temporal climatic conditions. Nonetheless, this loss of water makes direct comparison of average influent and effluent concentrations somewhat inadequate as a clear measure of pollutant retention. Considering pollutant mass fluxes instead of concentrations alone would result in more positive retention results, due to the reduced water volume. However, quantification of the water budget in field-pilot facilities is a complex matter. Also, as water quality regulation is based on threshold concentrations in the water bodies, it makes sense to limit this study to what effluent concentrations the soils in these particular facilities were able to produce.

The percolate pH was slightly more alkaline than the influent pH which is ascribed to the buffering ability of the filter soils (Table 7). Due to the original farmland soil having an alkaline pH, the addition of limestone to the soil in swale-trench East did not result in a significant difference compared to percolate from swale-trench West. When adding the synthetic stormwater runoff solution the pH in the effluent increased slightly, but not dramatically and the pH-values observed in these two experiments are in the same range as percolate pH observed during rain events (Table 8).

Removal of SSis crucial in treatment of urban stormwater runoff since the solids often carry a large part of the pollutant load [3]. Studies performed by [16] showed that even small particles are well retained in filter soil, and the main part of the particles present in the percolate at both locations are probably particles that have been mobilized in the filter soil, rather than the particles entering with the stormwater runoff. This mechanism of internal mobilization of particles is confirmed in the present study when the addition of particle-free synthetic influent resulted in effluent concentrations of SS in the same range as observed during rain events. When comparing the farmland-based swale-trench system with the landfill soil-based curb extensions there is a tendency that the concentration of SS in swale-trench percolate was higher than in curb extension percolate. The reason for this is unknown, but could either be due to the fact that the catchment area to filter soil area is much lower in the swale-trenches compared to the curb extensions (Table 2), providing more soil contact per percolate volume, or due to differences in aggregate stability between the local till soil and the oft-handled landfill soil.

The higher OC concentration in the percolate in Copenhagen compared to Odense (Table 7) may reflect slightly higher organic matter content in the filter soil of Copenhagen (Table 2), or it could be due to the fact that OC was measured as DOC in Odense and as TOC in Copenhagen. The OC found in percolates is likely to stem from mobilization of the internal organic pool in the soil [16], while incoming OC is retained. Unfortunately OC was not measured in effluent from the synthetic pulse addition, which could have given an idea of the mechanisms. When comparing the soil from swale-trenches and curb extensions it is observed that the local soil in the swale-trenches was leaching more OC than the landfill soil in curb-extensions, similar to what was observed for SS. Leaching of organic matter was also observed by [16] from German filter soil, 15 years old, but even young systems like the ones monitored here leach organic matter. Organic matter is not a pollutant as such, but can be a source of P and complex metals such as Cu and cause leaching of these e.g., [39,40]. When designing filter soil it is therefore relevant to look at the opportunities to improve the composition of the filter soil in order to retain organic matter.

It seems like the filter soil is removing particulate P while at the same time leaching dissolved P. As for SS and OC, this points to mobilization of the internal P-pool in the filter soil, as was also shown in the studies by Ingvertsen et al. [16]. In parallel with this, swale-trenches leached more particle-bound P than the low SS-leaching curb extensions, where P was mostly in dissolved form, probably reflecting the higher plant-available P in the landfill soil. The higher plant-available P found in swale-trench East did not result in more leaching of P compared to West. Besides the fact that particulate P and dissolved P are removed by different mechanisms, dissolved P will be more bioavailable for aquatic organisms and thereby pose a bigger risk for undesirable growth of algae [41]. Looking at the synthetic addition (Table 8), no significant reduction in P was observed either. Filter soil is concluded to retain P inefficiently.

Processes of immobilization of incoming P from the runoff could take place simultaneously with mobilization of the internal P-pool in the soil. Linear regression was conducted to elucidate whether the leached P was correlated with the leached organic matter. The correlation factors in Odense East of 0.6 for total P and 0.5 for dissolved P, with p-values from Pearson's test at 0.001 and 0.004 respectively indicated that there was a correlation. However, in Odense West a correlation factor of 0.5 for total P and 0.3 for dissolved P with p-values >0.05 showed no correlation. In Moellebakken the correlation factor was 0.3 for total P and 0.6 for dissolved P, but the correlation was negative and in Lindevang the correlation between the organic matter was 0.2 for total P and 0.3 for dissolved for dissolved P. Thus no direct conclusions can be made about the correlation between leaching of organic matter and P. However, Odense East where the correlation is observed was also the location with the highest amount of samples (n = 15) and thus forms the best statistical basis.

Since it is not likely that Cr is present in runoff as Cr(VI) [42], the influent concentrations of Cr are considered to meet the criterion. Concentrations of Cr in the percolate from the swale-trenches were lower than Cr from the curb extensions. This is, however, likely to be due to some significantly higher levels of Cr in the very first samples from both curb extensions, probably originating from concrete reinforcing walls surrounding the stormwater boxes, which could be Cr(VI) [43]. Potential introduction of Cr during the construction of curb extensions is therefore something to be aware of. The retention of the applied Cr was very low and the majority of the Cr in the effluent was found as dissolved Cr, which is similar to what has been observed in laboratory column experiments [44,45]. Since the Cr added with the synthetic pulse was added as Cr(VI), no direct comparison with the stormwater data can be made.

The filter soils were not able to reduce the level of dissolved Cu to below the 2 μg/L level for any of the events. Percolate and influent concentrations were similar, and there seems to be a limited effect of the filter soil on total Cu and virtually no effect on dissolved Cu. However, when adding dissolved Cu in a concentration of approximately 50 μg/L, the effluent concentrations of dissolved Cu was significantly reduced to below the detection limit and to 2.7 μg/L, but with a total Cu concentration of 8 and 8.1 μg/L, indicating that the Cu found in the effluent stems from the soil. A linear correlation of R^2 of 0.5 (Pearson's correlation test p-value of 0.04) for swale-trench West, and 0.9 (Pearson's correlation test p-value < 0.001) for East between total Cu and DOC points to complexation of Cu by the mobile DOC as the mechanism, as also observed by [16,40]. This trend was, however, not observed at Moellebakken and Lindevang where the correlation between TOC and dissolved Cu was 0.2 and 0.4 respectively (Pearson's correlation test with both p-values of 0.27 and 0.3 respectively). It should be noted that in Moellebakken the TOC was only measured at five rain events, and thus the statistical basis is weak at this location.

For Pb, similar to the suspended solids pattern, the highest inlet concentrations and the lowest percolate concentrations were observed for the curb extensions compared to the swale-trenches, despite filter soils with lower original Pb content at the swale-trenches (Table 4). Knowing that Pb is often found associated with SS and OC, the leaching from the swales and especially from the limestone-amended swale is in good accordance with the leaching of SS and OC. Almost all Pb was retained by the filter soil, when adding synthetic influent with Pb in dissolved form in concentrations much higher than those in the monitored events. In the first addition, total Pb was observed in a concentration of 1.6 μg/L with no dissolved Pb and, in the second addition, no Pb could be observed in the effluent, which is somewhat unexpected considering the fact that particles were leached. Comparing the concentrations of Pb in the percolate with the low criterion of 1 μg/L, it is concluded that the Pb concentration can be critical.

Percolate concentrations of Zn from both swale-trenches and curb extensions were below the criterion in all events. Adding almost 200 µg dissolved Zn/L to the curb extension demonstrated a significant reduction to a value below 8.4 µg/L in the effluent. Thus the filter soil can retain Zn.

Of the three PAHs surpassing the guiding criteria in influent, only benzo(a)pyrene was not in all cases sufficiently reduced in the filter soil. A good retention of PAHs in filter soil was also observed in the test based on synthetic influent. Here acenaphthene, naphthalene and phenanthrene were applied in dissolved form in high concentrations compared to reports on stormwater runoff concentration levels, but low concentrations compared to the criteria, which are relatively high for these three PAHs (Table 1). Therefore the effluent concentrations of these 3 PAHs were in no case critical in relation to the criteria.

The LAS and pesticides were not measured during normal rain events in this study, but included in the synthetic runoff, since they frequently are observed in runoff [29,46,47]. The pattern for LAS is somewhat biased by the fact that the chemical precipitated in the stock solution in the first addition. In the second addition, 130 µg/L LAS was added, but nothing was observed in the effluent, which indicates a good retention of LAS in the filter soil. For the pesticides, a good retention of glyphosate, which is known to sorb strongly to soil iron and aluminum oxides [48] was observed with concentrations in the effluent of 0.08 and 0.09 µg/L just below the groundwater quality criterion of 0.1 µg/L. MCPA was not retained at all in the first addition, having an effluent concentration of 0.5 µg/L, whereas in the second addition the effluent concentration could comply with the criterion of 0.1 µg/L. The reason for these contradictive results is not known. At the pH conditions present in the filter soil MCPA is primarily found as an anion, known to be poorly sorbed in soils, and sorption is negatively correlated with soil pH and positively correlated with organic matter content [49]. A way to improve the filter soil towards sorption of MCPA could be to add more stable organic matter; however, this has to be considered jointly with the risk of increasing DOC leaching and associated P and Cu. In general the fate of pesticides and their degradation products in filter soil is not well described. More research into retention efficacy and fate of pesticides and their degradation products in filter soil is needed.

An overall comparison of the two different systems showed that the percolate concentrations of SS, OC and Pb in Table 7 from the landfill-based filter soils in Copenhagen tended to be slightly lower than in percolate from the farmland-based filter soils in Odense possibly due to the larger infiltration area:catchment ratio at the swale-trenches or higher stability of the filter soil used in the curb extensions. It should be noted that both the influent concentration levels and the original contaminant content of the filter soil tended to be highest in the Copenhagen sites. The lime amendment of filter soil in swale-trench East in Odense did not result in improved percolate concentrations compared to swale-trench West, which could be ascribed to potential DOC mobilizing effects of the slightly increased pH of the soil.

5. CONCLUSIONS

Influent concentrations during rain events were low and several of the compounds analysed could not be detected (TPH, anionic detergents, pesticides) or were at similar low levels in influent and percolate e.g., (Cd). Despite the generally low influent levels the following contaminants were observed to exceed one or more of the guiding quality criteria for groundwater and freshwater protection: P, Cu, Pb, Zn benzo(a)pyrene. This may call for the use of filter soil or other treatment mechanisms before infiltration of discharge of stormwater runoff, even in low traffic catchments.

Filter soil reduced the inlet levels of SS, Cu, Ni, Pb, Zn and PAHs, while the concentration of P was increased in the percolate. For Zn, filter soil reduced the percolate concentration to below the guiding criteria; however, Cu was not sufficiently retained for discharge to freshwater and Pb and benzo(a)pyrene exceeded the groundwater quality criterion. Phosphorous levels in the percolate exceeded the Copenhagen limit of 0.08 mg P/L for discharge to lakes. Thus, filter soil is best suited for protecting groundwater and less suited for protecting freshwater.

The addition of synthetic influent with high concentrations of dissolved compounds confirmed the generally good retention of cationic heavy metals in filter soil, and some reduction in organic contaminants (PAH, LAS, pesticides). It also confirmed that filter soil can be a source of P. The observed patterns of SS and OC suggest internal mobilization processes rather than influent concentrations to account for occurrences of SS and OC in percolate.

No marked difference between the swale-trenches built from local farm soil and the curb extensions made from landfill soil was observed; however, there was a tendency for higher concentrations of SS and OC in percolate from swale-trenches, possibly due to the larger infiltration area:catchment ratio at the swale-trenches, although differences in aggregate stability could also play a role. The amendment of one swale-trench filter soil with limestone did not increase the retention capacity compared to the non-amended filter soil.

The replicated experiment with addition of synthetic runoff with high but still realistic concentrations of selected pollutants showed that filter soil efficiently retains the added contaminants with percolate concentrations for most compounds at the same level as observed for rain events. This indicates that the treatment capacity of the soil is sufficient, and that a filter soil could be used in highly trafficked areas as well. However, the pesticide MCPA was not well-retained and in future studies more focus should be put on retention of pesticides in filter soil. As for rain events, dissolved Cu was at a critical level compared to the criterion, and P was not sufficiently retained. DOC remains a complicating factor and no general conclusion can be drawn. Generally we conclude that filter soil used in raingardens, curb extensions and trenches is a step in the right direction when it comes to protecting water resources by local management of stormwater runoff.

ACKNOWLEDGMENTS

The authors wish to thank Per Eduard Robert Bjerager and Anita Schjødt Sandager for assisting in the field, laboratory and with data processing. Furthermore we wish to thank Mark Taylor Randall for proofreading and giving valuable inputs. The data presented in this paper was generated as part of the projects 'Cities in Water Balance' and 'Water in Cities', both financed by The Danish Council for Technology and Innovation, and performed in collaboration with Technical University of Denmark and Danish Technological Institute. Several partners have contributed with their expertise, materials and additional funding and we wish to thank HOFOR A/S, BIOFOS, Municipalities of Copenhagen, Brøndby, Odense, Albertslund, Egedal and Gladsaxe, Aarsleff A/S, Wavin, Orbicon A/S, Nordvand A/S, Middelfart Spildevand A/S, Vandcenter Syd A/S, Gottlieb Paludan Architects ApS, Nyrup Plast A/S, and MJK automation ApS.

REFERENCES

1. Vogel, R.J.; Moore, T.L.; Coffman, R.R.; Rodie, S.N.; Hutchinson, S.L.; McDonough, K.R.; McLemore, A.J.; McMaine, J.T. Critical review of technical questions facing low impact development and green Infrastructure: A perspective from the great plains. *Water Environ. Res.* **2015**, *87*, 849–862.
2. Grebel, J.E.; Mohanty, S.K.; Torkelson, A.A.; Boehm, A.B.; Higgins, C.P.; Maxwell, R.M.; Nelson, K.L.; Sedlak, D.L. Engineered infiltration systems for urban stormwater reclamation. *Environ. Eng. Sci.* **2013**, *30*, 437–454.

3. Ingvertsen, S.T.; Jensen, M.B.; Magid, J. A minimum data set of water quality parameters to assess and compare treatment efficiency of stormwater facilities. *J. Environ. Qual.* **2011**, *40*, 1488–1502.
4. Auckland Regional Council (ARC). *Stormwater Management Devices: Design Guidelines Manual*; Revision of Technical Publication: Auckland, New Zealand, 2003.
5. German Association for Water, Wastewater and Waste (DWA). *Standard DWA—A 138E, Planning, Construction and Operation of Facilities for the Percolation of Precipitation Water*; DWA: Hennef, Germany, 2005. (In English)
6. Minnesota Pollution Control Agency (MPCA). *The Minnesota Stormwater Manual*; State of Minnesota Stormwater Manual: Saint Paul, MN, USA, 2005.
7. Construction Industry Research and Information Association (CIRIA). *The SUDS Manual*; CIRIA: London, UK, 2007.
8. Davis, A.P.; Hunt, W.F.; Traver, R.G.; Clar, M. Bioretention technology: Overview of current practice and future needs. *J. Environ. Eng.* **2009**, *135*, 109–117.
9. Ingvertsen, S.T.; Cederkvist, K.; Régent, Y.; Sommer, H.; Magid, J.; Jensen, M.B. Assessment of existing roadside swales with engineered filter soil: I. Characterization and lifetime expectancy. *J. Environ. Qual.* **2012**, *41*, 1960–1969.
10. Ministry of Environment and Food of Denmark. *Departmental Order Concerning Requirements for Discharge of Contaminants to Streams, Lakes or the Sea*; Bek. Nr. 1725 af 22/12/2015; Ministry of Environment and Food of Denmark: Copenhagen, Denmark, 2015. (In Danish).
11. Danish Environmental Protection Agency. *List of Quality Criteria in Relation to Polluted Soil and Quality Criteria for Drinking Water (Liste over Kvalitetskriterier i Relation til Forurenet Jord og Kvalitetskriterier for Drikkevand*; Danish Environmental Protection Agency: Copenhagen, Denmark, 2015.
12. Ministry of Environment and Food of Denmark. *Departmental Order Concerning Establishment of Environmental Goals for Streams, Lakes, Transition Waters, Coastal Waters and Groundwater*; Bek. Nr. 1070 af 09/09/2015; Ministry of Environment and Food of Denmark: Copenhagen, Denmark, 2015. (In Danish).
13. Gerwin, L.; (Administration for Technology and Environment, Municipality of Copenhagen, Denmark). Personal communication, 2016.
14. Hatt, B.E.; Fletcher, T.D.; Deletic, A. Hydraulic and pollutant removal performance of fine media stormwater fi ltration systems. *Environ. Sci. Technol.* **2008**, *42*, 2535–2541.
15. Thompson, A.M.; Paul, A.C.; Balster, N.J. Physical and hydraulic properties of engineered soil media for bioretention basins. *Trans. ASABE* **2008**, *51*, 499–514.
16. Ingvertsen, S.T.; Cederkvist, K.; Jensen, M.B.; Magid, J. Assessment of existing roadside swales with engineered filter soil: II. Treatment efficiency and in situ mobilization in soil columns. *J. Environ. Qual.* **2012**, *41*, 1970–1981.
17. Lim, H.S.; Lim, W.; Hu, J.Y.; Ziegler, A.; Ong, S.L. Comparison of filter media materials for heavy metal removal from urban stormwater runoff using biofiltration systems. *J. Environ. Manag.* **2015**, *147*, 24–33.
18. Davis, A.P. Field performance of bioretention: Water quality. *Environ. Eng. Sci.* **2007**, *24*, 1048–1064.
19. Diblasi, C.J.; Li, H.; Davis, A.P.; Ghosh, U. Removal and fate of polycyclic hydrocarbon pollutants in an urban stormwater bioretention facility. *Environ. Sci. Technol.* **2009**, *43*, 494–502.
20. Passeport, E.; Hunt, W.F.; Line, D.E.; Smith, R.A.; Brown, R.A. Field study of the ability of two grassed bioretention cells to reduce storm-water runoff pollution: Technical note. *J. Irrig. Drain. Eng.* **2009**, *135*, 505–510.
21. David, N.; Leatherbarrow, J.E.; Yee, D.; McKee, L.J. Removal efficiencies of a bioretention system for trace metals, PCBs, PAHs, and dioxins in a semiarid environment. *J. Environ. Eng.* **2015**.
22. Lucke, T.; Nichols, P.W.B. The pollution removal and stormwater reduction performance of street-side bioretention basins after ten years in operation. *Sci. Total Environ.* **2015**, *536*, 784–792.

23. Liu, J.; Sample, D.J.; Bell, C.; Guan, Y. Review and research needs of bioretention used for the treatment of urban stormwater. *Water* **2014**, *6*, 1069–1099.
24. Gee, G.W.; Bauder, J.W. Particle-size analysis. In *Methods of Soil Analysis. Part 1. Physical and Mineralogical Methods*, 2nd ed.; Klute, A., Ed.; ASA and SSSA: Madison, WI, USA, 1986; pp. 383–411.
25. Olsen, S.R.; Cole, C.V.; Watanabe, F.S.; Dean, L.A. *Estimation of Available Phosphorus in Soils by Extraction with Sodium Bicarbonate*; Government Printing Office: Washington, DC, USA, 1954; pp. 1–19.
26. Bouwer, H. Intake rate: Cylinder infiltrometer. In *Methods of Soil Analysis. Part 1. Physical and Mineralogical Methods*, 2nd ed.; Klute, A., Ed.; ASA and SSSA: Madison, WI, USA, 1986; pp. 825–844.
27. Jørgensen, H.K.; Rosenorn, S.; Madsen, H.; Mikkelsen, P.S. Quality control of rain data used for urban runoff systems. *Water Sci. Technol.* **1998**, *37*, 113–120.
28. Zgheib, S.; Moilleron, R.; Saad, M ; Chebbo, G. Partition of pollution between dissolved and particulate phases: What about emerging substances in urban stormwater catchments? *Water Res.* **2011**, *45*, 913–925.
29. Cederkvist, K.; Jensen, M.B.; Ejerager, P.E.R.; Holm, P.E. Method for assessment of stormwater treatment facilities—Synthetic road runoff application including micro-pollutants and tracer. *Environ. Monit. Assess.* **2016**. in press.
30. Kayhanian, M.; Singh, A.; Suverkropp, C.; Borroum, S. Impact of annual average daily traffic on highway runoff pollutant concentrations. *J. Environ. Eng. ASCE* **2003**, *129*, 975–990.
31. Göbel, P.; Dierkes, C.; Coldewey, W.C. Stormwater runoff concentration matrix for urban areas. *J. Contam. Hydrol.* **2007**, *91*, 26–42.
32. Helmreich, B.; Hilliges, R.; Schriewer, A.; Horn, H. Runoff pollutants of a highly trafficked urban road—Correlation analysis and seasonal influences. *Chemosphere* **2010**, *80*, 991–997.
33. Lloyd, S.D.; Fletcher, T.D.; Wong, T.H.F.; Wooton, R.M. Assessment of pollutant removal in a newly constructed bioretention system. In Proceedings of the 2nd South Pacific Stormwater Conference, Auckland, New Zealand, 27–29 June 2001; pp. 20–30.
34. Kayhanian, M.; Fruchtman, B.D.; Gulliver, J.S.; Montanaro, C.; Ranieri, E.; Wuertz, S. Review of highway runoff characteristics: Comparative analysis and universal implications. *Water Res.* **2012**, *46*, 6609–6624.
35. Maestre, A.; Pitt, R. *The National Stormwater Quality Database, Version 1.1. A Compilation and Analysis of NPDES Stormwater Monitoring Information*; U.S. Environmental Protection Agency: Washington, DC, USA, 2005.
36. Huber, M.; Welker, A.; Helmreich, B. Critical review of heavymetal pollution of traffic area runoff: Occurrence, influencing factors, and partitioning. *Sci. Total Environ.* **2016**, *541*, 895–919.
37. Thomson, N.R.; McBean, E.A.; Snodgrass, W.; Mostrenko, I. Sample size needs for characterizing pollutant concentrations in highway runoff. *J. Environ. Eng. ASCE* **1997**, *123*, 1061–1065.
38. Drapper, D.; Tomlinson, R.; Williams, P. Pollutant concentrations in road runoff: Southeast Queensland case study. *J. Environ. Eng.* **2000**, *126*, 313–320.
39. Li, H.; Davis, A. Water quality improvement through reductions of pollutant loads using bioretention. *J. Environ. Eng. ASCE* **2009**, *135*, 567–576.
40. Mullane, J.M.; Flury, M.; Iqbal, H.; Freeze, P.M.; Hinman, C.; Cogger, C.G.; Shi, Z. Intermittent rainstorms cause pulses of nitrogen, phosphorus, and copper in leachate from compost in bioretention systems. *Sci. Total Environ.* **2015**, *537*, 294–303.
41. Correll, D.L. The role of phosphorus in the Eutrophication of receiving water: A review. *J. Environ. Qual.* **1998**, *27*, 261–266.
42. Cederkvist, K. Chromium in Stormwater Runoff—Determination and Control Options. Ph.D. Thesis, Faculty of Life Sciences, University of Copenhagen, Copenhagen, Denmark, 2012.

43. Butera, S.; Christensen, T.H.; Astrup, T.F. Composition and leaching of construction and demolition waste: Inorganic elements and organic compounds. *J. Hazard. Mater.* **2014**, *276*, 302–311.
44. Jardine, P.M.; Fendorf, S.E.; Mayes, M.A.; Larsen, I.L.; Brooks, S.C.; Bailey, W.B. Fate and transport of hexavalent chromium in undisturbed heterogeneous soil. *Environ. Sci. Technol.* **1999**, *33*, 2939–2944.
45. Cederkvist, K.; Ingvertsen, S.T.; Jensen, M.B.; Holm, P.E. Behaviour of chromium(VI) in stormwater soil infiltration systems. *Appl. Geochem.* **2013**, *35*, 44–50.
46. Birch, H.; Mikkelsen, P.S.; Jensen, J.K.; Lutzhöft, H.C.H. Micropollutants in stormwater runoff and combined sewer overflow in the Copenhagen area, Denmark. *Water. Sci. Technol.* **2011**, *64*, 485–493.
47. Gasperi, J.; Sebastian, C.; Ruban, V.; Delamain, M.; Percot, S.; Wiest, L.; Mirande, C.; Caupos, E.; Demare, D.; Kessoo, M.D.; et al. Micropollutants in urban stormwater: Occurrence, concentrations, and atmospheric contributions for a wide range of contaminants in three French catchments. *Environ. Sci. Pollut. Res.* **2014**, *21*, 5267–5281.
48. Borggaard, O.K.; Gimsing, A.L. Fate of glyphosate in soil and the possibility of leaching to ground and surface waters: A review. *Pest Manag. Sci.* **2008**, *64*, 441–456.
49. Paszko, T.; Muszynski, P.; Materska, M.; Bojanowska, M.; Kostecka, M.; Jackowska, I. Adsorption and degradation of phenoxyalkanoic acid herbicides in soils: A review. *Environ. Toxicol. Chem.* **2016**, *35*, 271–286.

CHAPTER 13

Redox Dynamics and Oxygen Reduction Rates of Infiltrating Urban Stormwater beneath Low Impact Development (LID)

Mays N. Danfoura and Jason J. Gurdak *

Department of Earth & Climate Sciences, San Francisco State University, San Francisco, CA 94132, USA

ABSTRACT

Low impact development (LID) best management practices (BMPs) collect, infiltrate, and treat stormwater runoff, and increase recharge to aquifers. Understanding the controls on reduction/oxidation (redox) conditions within LID BMPs is important for groundwater management because outflow from some LID BMPs can recharge aquifers and affect groundwater quality. Here we evaluate redox conditions of urban stormwater runoff in a LID infiltration trench in San Francisco, California, and quantify the relation between water saturation (%) and temperature (°C) and resulting dissolved oxygen (DO) concentrations, redox dynamics, and O_2 reduction rates. The DO fluctuations have an inverse response to the duration of saturation of the trench. Anoxic (<0.5 mg/L) conditions often occurred within hours of stormwater events and persisted from a few hours to two days, which indicate that microbial respiration can be a limiting factor for DO. Temperature of stormwater runoff was not a statistically significant control on DO. The estimated O_2 reduction rate is 0.003 $mg \cdot L^{-1} \cdot min^{-1}$, which is two to five orders of magnitude higher than in groundwater from previous studies. Higher rates of O_2 reduction are a function of the more oxic and organic-rich stormwater runoff that drives faster microbial O_2 reduction. Our findings have important implications for the design of infiltration trenches and other LID BMPs to achieve desired redox conditions for infiltrating stormwater toward minimizing groundwater contamination.

Keywords: stormwater; low impact development (LID); redox; oxygen reduction; recharge; urban hydrogeology

1. INTRODUCTION

Urbanization has adversely affected the quantity and quality of urban water resources [1,2] because of increased impervious surfaces, compaction of soils, and modifications to vegetation [3,4], which has reduced infiltration rates and increased stormwater runoff [5]. Impervious surfaces like pavement, rooftops, and parking lots increase stormwater runoff volume and contaminant loads in stormwater. Stormwater runoff from the impervious surfaces can be contaminated with suspended sediments, heavy metals, hydrocarbons, nutrients, and pathogens [6,7,8].

A major challenge for stormwater engineering and management in the 21st century is to develop strategies that encompass climate variability and change [9,10] and increasing impervious surfaces from the expansion of urbanized areas, while maintaining the health of the watershed and meeting regulatory requirements for stormwater quality [11]. In response to this challenge, low impact development (LID) planning and related best management practices (BMPs) have been increasingly used to reduce the adverse effects of urbanization on surface-water resources. LID is a stormwater management and planning strategy with a goal of mimicking the predevelopment hydrologic regime through the use of design techniques to create a functionally hydrologic landscape [12]. LID design techniques include micro-scale and spatially distributed BMPs such as bioretention-swales, rain-gardens, and infiltration trenches [13]. LID BMPs are constructed based on a combination of site characteristics such as land use, soil texture, drainage area, and key hydrological functions such as precipitation, infiltration, surface runoff, and groundwater recharge [12,13,14].

LID BMPs collect, infiltrate, and treat urban stormwater and improve the quality of receiving surface-water bodies [15,16]. LID has been shown to treat a range of pollutants in stormwater, including the removal of nitrate (NO_3^-) and phosphate [17,18,19,20], metals [21,22], fecal indicator bacteria (FIB) [23,24,25], and other pathogens and pollutants. Although often considered a secondary benefit, LID has also recently been shown to increase recharge to local aquifers [26,27] beneath BMPs without an impermeable bottom layer and underdrain system. Newcomer et al. (2014) [27] demonstrated that recharge rates beneath some LID BMPs can be orders of magnitude larger than beneath other sources of urban recharge. The treatment of stormwater pollutants and enhancement of local groundwater recharge are benefits of using LID BMPs for stormwater management [28,29,30]. However, few studies have examined how the design features of LID BMPs influence the biogeochemical processes that occur in the vadose zone and the implications for the quality of recharge to urban aquifers. Because recharge is enhanced by some LID BMPs, they could be direct sources and preferential pathways for the transport of stormwater derived contaminants to reach the groundwater and (or) indirectly alter the biogeochemical conditions within the aquifer that subsequently mobilize contaminants beneath the water table.

Several biogeochemical processes are likely major controls on the quality of stormwater that infiltrates and eventually recharges urban aquifers. In particular, reduction/oxidation (redox) processes can: (i) affect the mobility and availability of organic micro-pollutants such as pharmaceutically-active compounds, pesticides, organic halogens, and other halogenated organic compounds; (ii) affect the fate and mobility of potentially toxic metals associated with naturally occurring aquifer material; (iii) contribute to the degradation or preservation of anthropogenic contaminants; and (iv) generate undesirable byproducts such as dissolved manganese (Mn^{2+}), ferrous iron (Fe^{2+}), hydrogen sulfide (H_2S), and methane (CH_4) [31,32,33,34]. Therefore, characterizing and quantifying the controls on redox conditions beneath LID BMPs is important to better understand and manage urban stormwater effects on groundwater quality.

A few previous studies have explored characteristics of LID that affect subsurface redox conditions. Datry et al. (2004) [35] reported that redox conditions in soils and groundwater are affected by the residence times of infiltrating water in the vadose zone beneath a stormwater infiltration basin. Similarly, von Rohr et al. (2014) [36] found that the hydraulic residence time, which is influenced by the infiltration rate, may affect the redox conditions during riverbank filtration. Greskowiak et al. (2005) [37] showed that the hydraulic regime immediately below an artificial recharge pond near Lake Tegel, Germany is characterized by cyclic changes between saturated and unsaturated conditions that influence the redox conditions. The saturation conditions in the soil, vadose zone, and groundwater influence microbial activity and, in turn, redox conditions because most redox processes are microbially catalyzed [32,33]. Additionally, microbial activity is temperature-dependent and thus a function of diel and seasonal cycles [32,33]. Datry et al. (2004) [35] reported that seasonal microbial respiration in an infiltration basin was stimulated by the increase in temperature, thereby leading to higher oxygen deficit in summer, whereas cold winter stormwater slightly re-oxygenated groundwater. During the seasons of cooler soil temperatures, often from November through March in the Northern Hemisphere, saturated soils take longer to become reducing [38].

Molecular oxygen (O_2) acts as the preferred electron acceptor through aerobic respiration because it is the most energetically favorable reaction [39]. If the supply of O_2 is consumed or absent, alternative electron acceptors are used by microbes. After O_2, the descending sequential order of acceptor preferences is NO_3^-, manganese (IV) (Mn^{4+}), ferric iron (Fe^{3+}), sulfate (SO_4^{2-}), and finally carbon dioxide (CO_2). This order of preferential electron acceptor utilization—$O_2 > NO_3^- > Mn^{4+} > Fe^{3+} > SO_4^{2-} > CO_2$—is referred to as the ecological succession of terminal electron-accepting processes [32,33]. For the purpose of this study, a threshold concentrations of dissolved oxygen (DO) will be used to identify general redox conditions and processes following a framework developed by McMahon and Chapelle (2008) [32]. Under this framework, oxic conditions are present when the observed DO concentration is ≥0.5 $mg{\cdot}L^{-1}$ (≥15.6 $\mu mol{\cdot}L^{-1}$ O_2) and thus O_2 reduction is the predominant redox process. Anoxic conditions are present when the observed DO concentration is <0.5 $mg{\cdot}L^{-1}$ (<15.6 $\mu mol{\cdot}L^{-1}$ O_2).

Similar to stormwater quality, the rate of O_2 reduction has a significant influence on groundwater quality and the susceptibility of aquifers to contamination because the fate and transport of many groundwater contaminants largely depends on the redox processes that occur along flow paths [32,39]. Therefore, quantifying the rate of O_2 reduction within LID features is an important indicator of the progression of redox reactions in the vadose zone and aquifer beneath the LID and is critically important to predict the risk of groundwater contamination. Under anoxic conditions, NO_3^- is the preferred electron acceptor and denitrification (NO_3^- reduction) is the predominant redox process when NO_3^- (as N) concentrations are >0.5 $mg{\cdot}L^{-1}$ (and DO < 0.5 $mg{\cdot}L^{-1}$, Mn^{2+} < 0.05 $mg{\cdot}L^{-1}$, and Fe^{2+} < 0.1 $mg{\cdot}L^{-1}$) [32]. After NO_3^- has been reduced, redox processes may continue along the ecological succession of terminal electron-accepting processes [32,33]. Understanding and predicting the onset of denitrification beneath LID BMPs is important because stormwater often contains elevated nutrients, including NO_3^-. Nitrate is also the most ubiquitous contaminant of groundwater worldwide, and denitrification can release the greenhouse gas nitrous oxide and pose well-known ecological and human-health risks [40,41,42].

Therefore, the overall goal of this study is to better characterize and quantify the redox conditions in a LID BMP, which can be used to better understand and manage urban stormwater effects on groundwater quality. Here we use stormwater chemistry to evaluate redox conditions in a recently installed (2009) LID BMP infiltration trench. Specifically, this study addresses two primary research questions: (1) What is the dynamic relationship between water saturation (%) and temperature (°C) and the magnitude and duration of fluctuations in DO concentrations and redox conditions within the infiltration trench? (2) What are the O_2 reduction rates in the infiltration trench? Addressing these research questions provides important insight into how to design LID BMPs toward optimizing the water saturation and temperature within a desired redox range (oxic versus anoxic) for infiltrating stormwater to help minimize the transport of redox sensitive pollutants and risk of groundwater contamination.

2. MATERIALS AND METHODS

2.1. Study Area

The study area is a gravel infiltration trench site at the San Francisco State University (SFSU) LID research network, within the city of San Francisco, CA, USA (Figure 1a). The infiltration trench was installed in 2009 and instrumented in 2011, and has a surface area of 11 m^2 (0.90 m × 12 m). The infiltration trench is located in the center of a vegetated depression that collects runoff from an impervious bike path, parking lot, surrounding roof tops, and vegetative landscape with an effective drainage area of 430 m^2 (Figure 1c). The gravel trench has a maximum storage capacity of approximately 3.0 m^3 assuming a gravel porosity of 0.35 [27]. The design of the infiltration trench follows the volume-based sizing method and a goal of 80% stormwater volume capture [43] for 1-h, 6-h, and 24-h precipitation events (design storms) with a two-year return interval.

The site of the infiltration trench is located above the Westside Basin aquifer (104 km^2), which is part of the regionally important California Coastal Basin aquifers (Figure 1b). The Westside Basin aquifer system has a shallow (<30 m below sea level), unconfined aquifer and two deeper confined aquifers, and is bounded in the north and south by Franciscan

Complex bedrock [44,45]. The aquifer sediments include coastal deposits of sand, silt, mud, gravel, and peat from the Merced and Colma Formations [45].

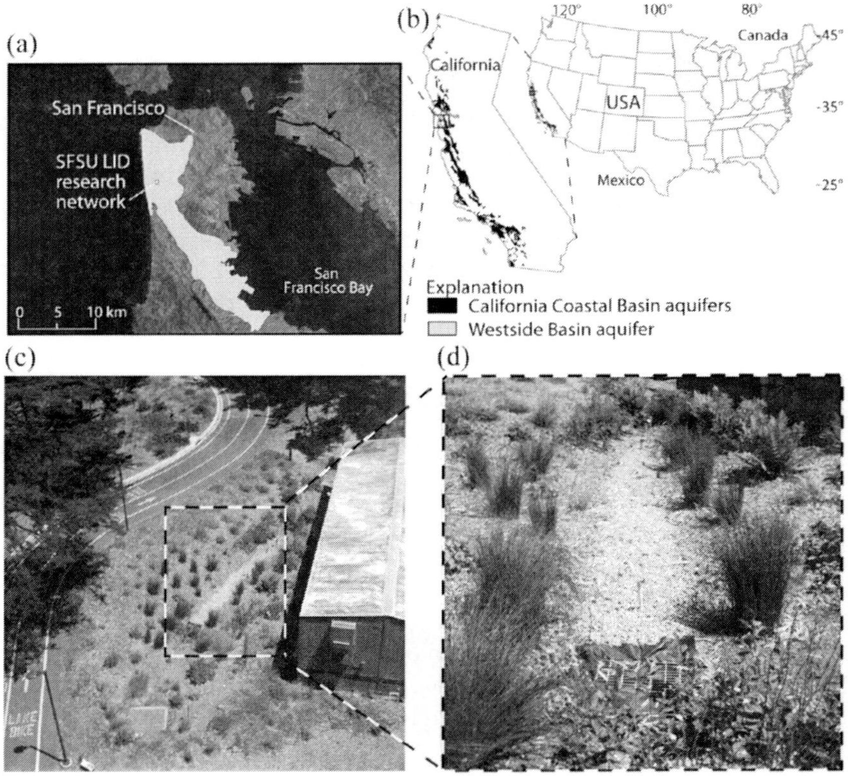

Figure 1. (**a**) Map showing the location of the San Francisco State University (SFSU) low impact development (LID) research network; and (**b**) the Westside Basin aquifer, which is part of the California Coastal Basin aquifers. The aquifer location data are modified from the California Department of Water Resources and the U.S. Geological Survey. Photographs of the infiltration trench from: (**c**) aerial view; and (**d**) zoomed-in view are modified from Newcomer et al. 2014 [27].

2.2. LID Data Collection

The infiltration trench site is instrumented with five Decagon 5TM soil temperature and moisture sensors, five Decagon MPS-1 matric potential sensors, and one Decagon G3 drain gauge (passive capillary lysimeter), which continuously record measurements at a two-minute interval to detect event-based changes in drainage beneath the gravel in the vadose zone (Figure 2). Additionally, one Decagon ECRN-100 rain gauge, one Solinst Inc. (Georgetown, ON, Canada) barometer, and three piezometers with Solinst Inc. pressure transducers (model 3001 LTC Levelogger junior) were installed at the infiltration trench site (Figure 2). The pressure transducers in each piezometer record the water level (cm), temperature (°C), and conductivity (µS/cm). A fourth piezometer was installed 2.13 m below land surface beneath the infiltration site to manually record water levels. Each piezometer was constructed of pvc pipe with a slotted screen. Soil texture and hydraulic properties of the vadose zone beneath the LID site are reported in [27].

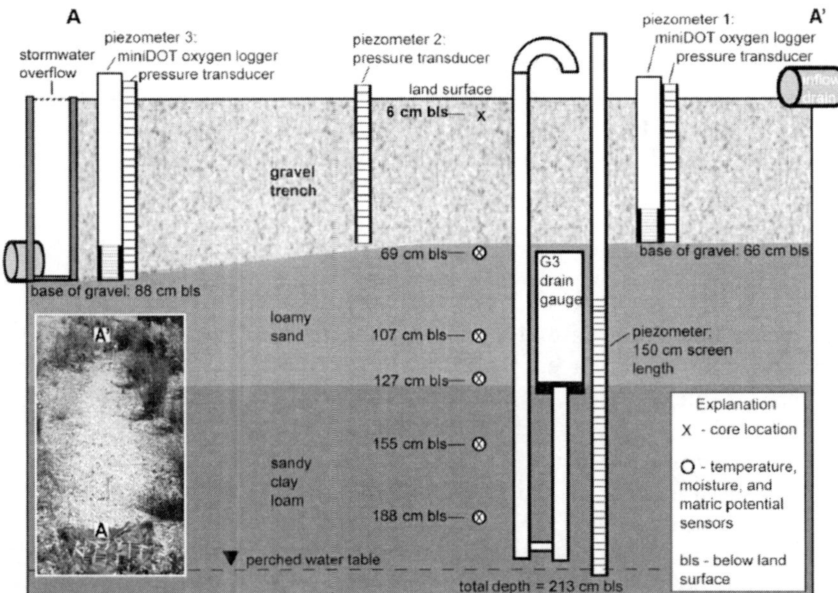

Figure 2. Schematic depiction (not to scale) of the instrumentation and core locations along the longitudinal cross-section A–A' (inset) of the infiltration trench site at San Francisco State University (SFSU) low impact development (LID) research network. Modified from Newcomer et al. 2014 [27].

MiniDOT oxygen loggers [46] were installed on 11 November 2013 to measure DO concentrations along the flow path in the infiltration trench. The miniDOT oxygen loggers were installed in 5 cm diameter piezometers adjacent to piezometers 1 and 3 (Figure 2). These loggers were programmed to make measurements at two-minute intervals, which is consistent with the sampling frequency of the other field instruments. For the DO sensors in the miniDOT oxygen loggers to be in contact with water, water levels in the 5 cm piezometers at locations 1 and 3 have to be at least 26 cm and 34 cm above the base of the gravel, respectively. The miniDOT oxygen loggers measure and record DO concentrations with an accuracy of ±10 μmol/L, DO range of 0%–150% saturation, and DO resolution of at least 0.05 μmol/L [46]. The miniDOT oxygen loggers also measure and record temperature with an accuracy of ±0.1 °C, range of 0–30 °C, and resolution of 0.01 °C [46].

2.3. Saturation of the Gravel Trench

Saturation of the gravel trench was evaluated using water levels in the three piezometers and hourly precipitation during the period of field observation (July 2012–June 2014). Here, saturation of gravel (%) refers to the ratio of observed water level in the gravel to the thickness of the gravel at each of the three piezometer locations (Figure 2). Saturation of gravel (%) at each piezometer location is calculated as:

$$\text{Saturation (\%)} = \frac{\text{Observed water level above base of gravel}}{\text{Gravel thickness}} \times 100 \tag{1}$$

where the observed water level above the base of the gravel (L) is the average water levels for each piezometer during an hourly time interval, and gravel thickness (L) is the thickness of the gravel at the piezometer location. The stormwater that collects in the gravel is assumed to infiltrate and eventually become recharge. Water loss due to evapotranspiration (ET) and evaporation is negligible because the gravel is not vegetated and the rapid infiltration into the gravel trench minimizes surface ponding and direct evaporation. When the saturation of gravel for the entire trench exceeds 100%, the excess stormwater overflows into the municipal stormwater system (Figure 2).

The duration of saturation of gravel (%) is defined as the length of time (reported as either hours or days) when water levels were elevated above the pre-rainfall event water level, and is analogous to the excess saturation of the gravel for a given storm. The duration of saturation of gravel (%) was determined using observed time series of water levels and precipitation.

2.4. Precipitation

Historical hourly and daily precipitation data from the San Francisco, Downtown station (NOAA Station ID: 047772) were evaluated for the entire data collection period at the LID BMP (July 2012–June 2014) [47]. Hourly precipitation time series were compared to the water level, temperature, and conductivity time series for the entire collection period and with DO concentrations between 22 February 2014 and 30 June 2014. The shorter collection period of DO concentrations reflects the fact that the miniDOT oxygen loggers were installed at a later time than the LTC Leveloggers. Hourly and daily precipitation data were used to develop hourly and daily runoff volumes for the entire data collection period.

2.5. Runoff

The infiltration trench collects stormwater runoff from an effective drainage area of 430 m^2 that includes an impervious bike path, parking lot, and surrounding roof tops. Daily and hourly runoff to the infiltration trench was calculated for the period 2012 to 2014 using the TR-55 SCS curve number method [48]. The stormwater runoff was calculated as:

$$R = \frac{(P - 0.2S)^2}{(P + 0.8S)} \tag{2}$$

$$S = \frac{1000}{CN} - 10 \tag{3}$$

where R is the runoff depth (L), P is the precipitation (L), S is the maximum retention (unit less), and CN is the composite curve number (unit less). A CN of 98, representing impermeable surfaces such as roofs and pavement was applied to the impermeable surfaces draining to the infiltration trench. Runoff volumes (L^3) were calculated by multiplying daily or hourly runoff depths by the effective drainage area of 430 m^2.

2.6. Rate of O_2 Reduction

The rate of O_2 reduction in the gravel infiltration trench was estimated by relating measured DO concentrations to the elapsed time since stormwater runoff into the trench, by modifying the approach described in [32,39]. Here, elapsed time is defined as the duration of time (minutes) from the beginning of a rain event to the time of the lowest measured DO concentrations before the recovery of DO levels to pre-event conditions.

Zero-order reaction rates were estimated from changes in concentration of DO as a function of elapsed time using:

$$-K_0 = \frac{\Delta(C - C_0)}{\Delta t} \tag{4}$$

where $-K_0$ is the rate (mg·L^{-1}·min^{-1}) for zero-order reaction kinetics, C is the concentration of DO at the time of interest (mg·L^{-1}), C_0 is the initial concentration of DO (mg·L^{-1}), and Δt is the elapsed time (min). Zero-order rates were determined by fitting a linear regression curve to a plot of DO concentration versus elapsed time.

2.7. LID Oxygen Demand (LIDOD) Rate

The LID oxygen demand (LIDOD) rate is first defined here as the rate of DO consumption by microbial respiration processes in the gravel infiltration trench. Using a modified method from Heckathorn and Gibs (2010) [49] and Wilson (2014) [50], LIDOD was calculated during three storm events that represent a range of O_2 reduction rates that were calculated during the period of field observation for DO. The measured DO concentrations for piezometers 1 and 3 were plotted as a function of elapsed time, resulting in a DO depletion curve. Elapsed time for each rain event was defined from the beginnings of the rain event to the time of lowest measured DO concentration before the recovery of DO levels to pre-event conditions. The slope of the linear segment of the oxygen-depletion curve was used to calculate the LIDOD rate as:

$$LIDOD_T = \frac{V_{water}}{V_{trench}} \times b \tag{5}$$

where $LIDOD_T$ is the LID oxygen demand rate (mg·m^{-3}·min^{-1}) at water temperature T; V_{water} is the volume of water in the gravel (m^3); V_{trench} is the maximum volume capacity of the infiltration trench (m^3); and b is the slope of the oxygen-depletion curve for each storm event (mg·m^{-3}·min^{-1}). The variable b is similar to $-K_0$, but expressed in different units. In order to standardize DO measurements among different rain events and temperatures, the measured LIDOD rates are adjusted to 20 °C using the standard van't Hoff equation [49]:

$$LIDOD_{20} = LIDOD_T / 1.065^{(T-20)} \tag{6}$$

where $LIDOD_{20}$ is the LID oxygen demand rate ($mg \cdot m^{-3} \cdot min^{-1}$) at 20 °C, and T is the water temperature (°C).

Because DO concentrations and water temperature were measured at a 2-min interval, the $LIDOD_T$ and the $LIDOD_{20}$ rate were similarly calculated at 2-min intervals for each rain event for piezometers 1 and 3. Volume of water in the gravel was calculated based on an average water level from piezometers 1 and 3. An average of the calculated $LIDOD_{20}$ rates for all rain events for piezometers 1 and 3 was calculated and this average represents the $LIDOD_{20}$ for the entire infiltration trench.

3. RESULTS AND DISCUSSION

3.1. Water Levels, Saturation of Gravel, Temperature, and Conductivity in Response to Stormwater Runoff

During the period of observations (July 2012–June 2014), all of the hourly and daily precipitation were below the 2–5-year design storm values (<15 mm (Figure 3) and <60 mm (shown in [51]), respectively). The maximum hourly precipitation was 12.7 mm (Figure 3). Precipitation and stormwater events throughout the period of observation resulted in sharp and substantial increases in water levels in all three piezometers (Figure 3). A short time-lag (less than an hour) was observed between the beginning of the precipitation events to the first increase in water levels above pre-event values. This fast response in water levels to precipitation events is a function of the high CN number (98) of the effective drainage area and the relatively high infiltration capacity in the gravel trench, which enables effective capture and storage of stormwater. The infiltration of the stormwater below the gravel trench is recorded as water levels decline in the piezometers. A previous study [27] has reported that the infiltration of stormwater below the infiltration trench can result in very high recharge rates (1750 to 3710 $mm \cdot year^{-1}$).

Time series of hourly precipitation and saturation of gravel (%) at each piezometer for events that occurred during the rainy season (November to March) of 2013 and 2014 are shown in Figure 4. During the rainy season, water levels were observed in the piezometers and saturation of gravel was always greater than 0%. Although the precipitation did not exceed the 2–5-year design storm, saturation of gravel exceeded 100% during several precipitation event (Figure 4), resulting in the ponding of stormwater at land surface. Ponding of stormwater at land surface was more frequently detected at the inflow to the LID near piezometer 1 than at locations near piezometers 2 or 3 (Figure 4).

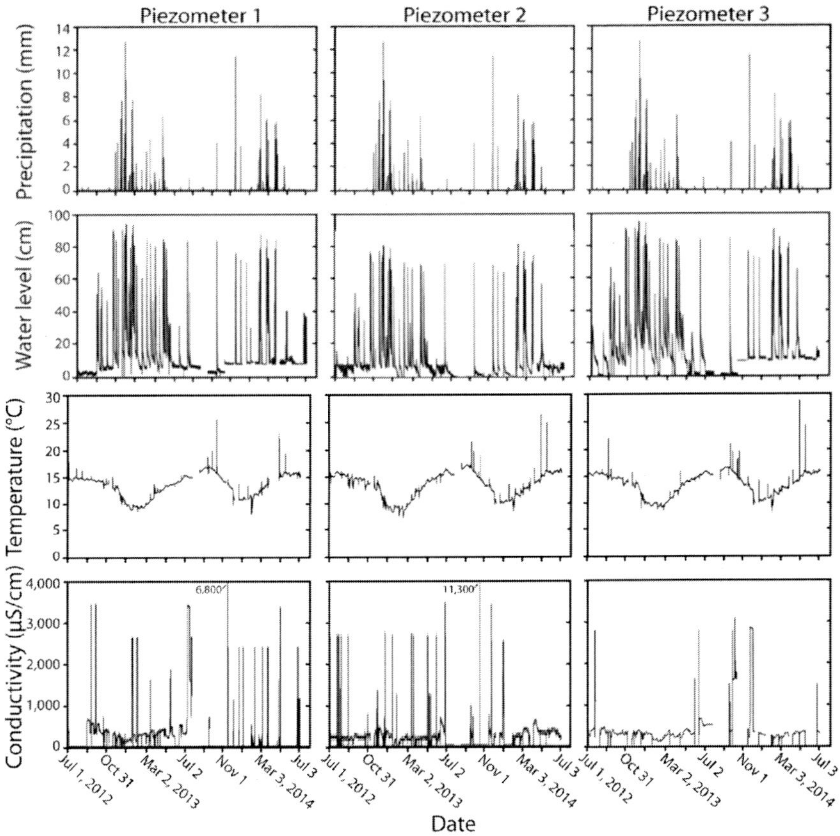

Figure 3. Hourly time series of precipitation (mm), water levels (cm), temperature (°C), and conductivity (µS/cm) in piezometers 1, 2, and 3 during the period of observation (July 2012 to July 2014).

The time series of the water levels and corresponding saturation of gravel (%) indicate spatial variation in the flow and storage of stormwater within the LID during the same rain events (Figure 4). Higher water levels and greater saturation of gravel (%) generally occur near piezometer 1 than piezometer 2 (Figure 4) as a result of piezometer 1 receiving more stormwater sooner due to proximity to the inlet pipe, causing temporary ponding as water infiltrates downward or laterally along the infiltration trench. However, water levels at piezometer 3 (located farthest from inlet pipe (Figure 2)) were occasionally observed to be higher than water levels in piezometers 1 and 2. Piezometer 3 is located near the cement stormwater overflow structure at the downstream end of the infiltration trench (Figure 2). The walls of the stormwater overflow structure acts as an impermeable surface, which prevents lateral water flow and occasionally results in greater saturation of gravel (%) near piezometer 3. However, no events were observed where the saturation of gravel exceeded 100% at all piezometer locations simultaneously (Figure 4), indicating that the maximum storage capacity of the LID trench was not exceeded during the period of observation.

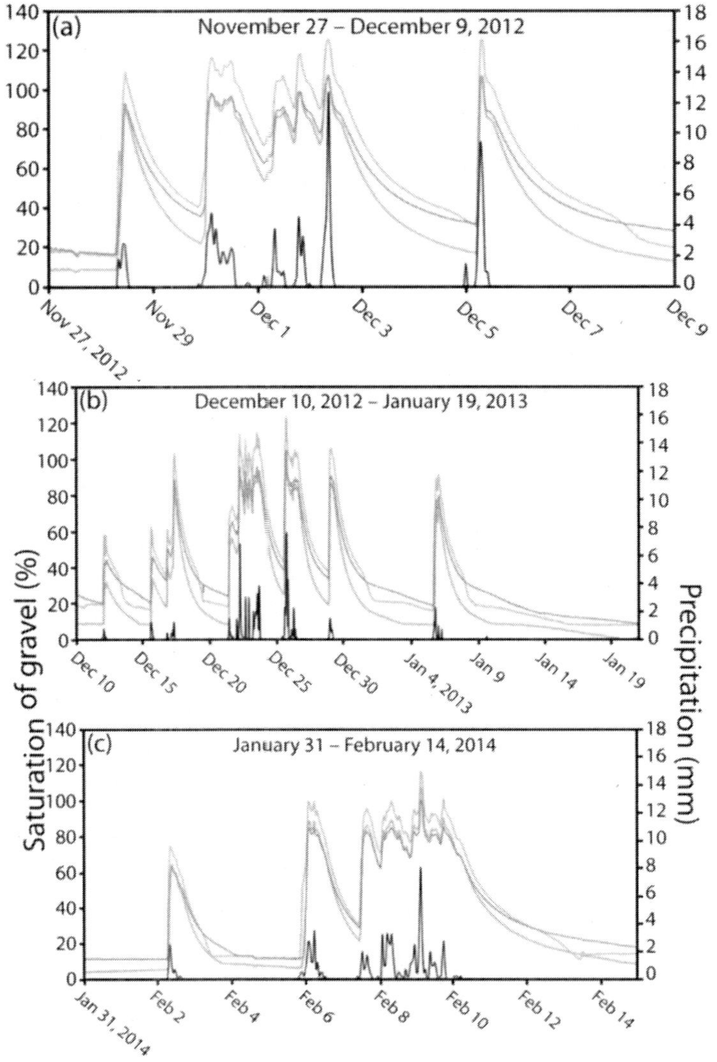

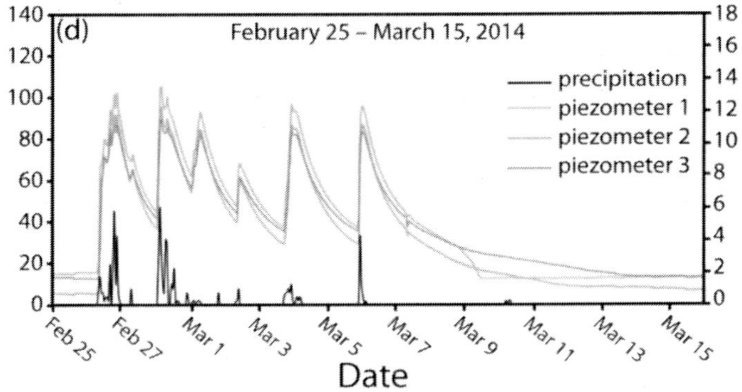

Figure 4. Hourly time series of precipitation (mm) and saturation of gravel (%) for piezometers 1, 2, and 3 during the rainy season of: (**a**,**b**) 2012–2013; and (**c**,**d**) 2014.

The duration of saturation of gravel is a function of the intensity and duration of the precipitation, but generally lasted for several days after each storm event (Figure 4). The maximum duration that the infiltration trench was saturated above pre-event water levels was about 12–14 days for all three wells (Figure 4). Between 21–29 December 2012 (Figure 4b) and 26 February–5 March 2014 (Figure 4d), successive precipitation events resulted in some of the longest periods of saturation above pre-event water levels. Lower intensity and single rain events generally resulted in shorter periods of saturation above pre-event water levels, as was the case for the storm event on 2 February 2014 that took 2–3 days for saturation levels to return to pre-event conditions for all three piezometers (Figure 4b).

Time series of water temperatures for the period of observation are shown in Figure 3. Measured water temperatures in the three piezometers average about 15 °C, but fluctuate depending on the temperature of stormwater runoff. Observed water temperatures show variation before and after precipitation events during the rainy seasons of 2013 and 2014. Mixing of existing water in the gravel with relatively colder or warmer stormwater decreased or increased water temperature in the gravel, respectively (Figure 3).

Time series of conductivity (μS/cm) for the period of observation are shown in Figure 3. In general, the conductivity values decreased immediately following rising water levels from the inflow of stormwater during precipitation events (Figure 3; reference [51] has a detailed time series plots of conductivity and water levels). This inverse relationship is attributed to the dilution of stored water in the gravel trench by the inflow of stormwater runoff and is consistent with previous studies. Datry et al. (2004) [35] found that inflow to a stormwater infiltration basin resulted in a sharp decrease in conductivity in the infiltration bed. Appleyard (1993) [52] demonstrated similar findings that stormwater recharge through infiltration basins receiving runoff from large impervious catchments resulted in a substantial reduction in total dissolved solids to depths of more than 10 m below the water table.

3.2. DO Dynamics in Response to Saturation of Gravel Magnitude and Duration

To illustrate the relation among precipitation, saturation of gravel (%), and DO concentrations, the time series for a period of the 2013–2014 rainy season (24 February to 5 April 2014) are shown in Figure 5. The rapid responses in water levels and saturation of gravel (%) to precipitation events and stormwater runoff have an important control of DO dynamics in the gravel infiltration trench (Figure 5). Due to the geometry of the miniDOT oxygen loggers inside the piezometers, the saturation of gravel (%) has to be ≥39% for the DO sensors to be in contact with stormwater that has collected in the gravel trench (Figure 5). When the saturation of gravel (%) is <39%, DO concentrations are highly oxic conditions (>8 mg·L^{-1}) and indicate that the miniDOT oxygen loggers are in direct contact with the atmosphere. After stormwater flows into the gravel trench and the saturation of gravel (%) exceeds 39%, we generally observed that DO concentrations decline rapidly, reaching anoxic conditions (<0.5 mg·L^{-1}) during some rain events (e.g., 1 March 2014 at piezometer 1 and 3) (Figure 5a,b). However, we also observed that DO concentrations increase with input of stormwater when DO concentrations in the gravel were anoxic or at lower concentrations than DO of the stormwater runoff (e.g., 6 March 2014 at piezometer 1 and 3) (Figure 5a,b). To confirm the DO concentrations of the stormwater runoff, we measured the DO of stormwater directly from the inflow drain (Figure 2) during the precipitation event on 26 February 2014, which had DO concentrations that ranged from 6.86 to 8.25 mg·L^{-1}. These DO concentrations are consistent with stormwater DO (6.00 to 9.90 mg·L^{-1}) reported from previous studies [35,53].

The time lag between the stormwater runoff, increased saturation of gravel (%), and corresponding changes to DO concentration was very short, often on the order of less than an hour to several days (Figure 5). For example, during the precipitation event on 26 February 2014 DO concentrations in piezometer 1 declined from 9.2 to 3.4 mg·L^{-1} within one day (Figure 5a). Although a subsequent smaller precipitation event on 27 February resulted in a slight increase in DO to about 4.0 mg·L^{-1} in piezometer 1, the DO continued a rapid decrease toward anoxic conditions (<0.5 mg·L^{-1}) by 28 February (Figure 5a). Subsequent precipitation events and stormwater runoff from 28 February to 2 March 2014 resulted in rapid DO fluctuations from anoxic to oxic (<4.0 mg·L^{-1}) conditions that closely matched the magnitude of fluctuations in saturation of gravel (%) (Figure 5a). Between 26 February and 2 March, anoxic conditions were observed for a relatively short period of about 15 h (Figure 5a). The DO returned to pre-event oxic conditions as the result of either the replenishment with oxic stormwater or infiltration-induced lowering of water levels in piezometer below 39% saturation of gravel (Figure 5a).

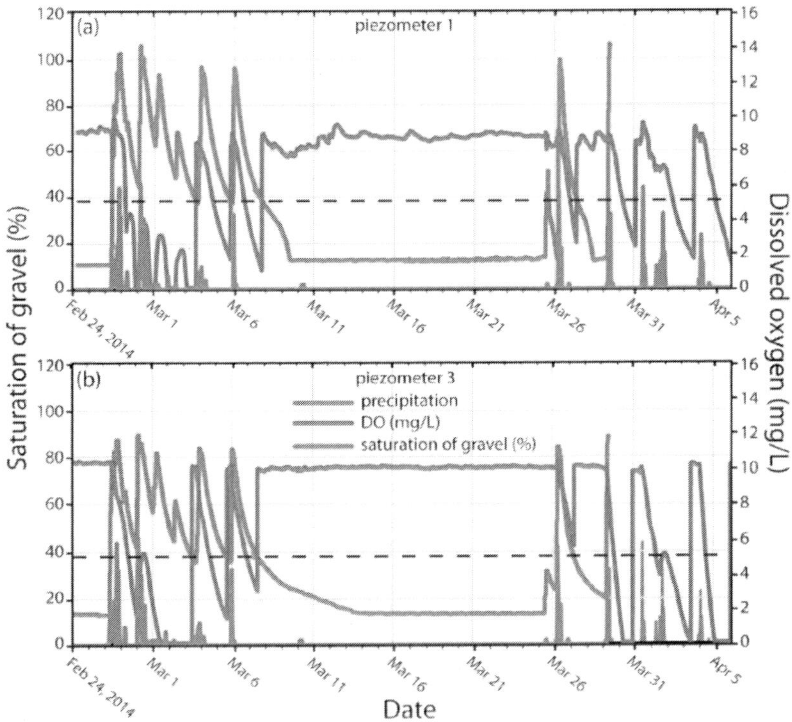

Figure 5. Hourly time series of saturation of gravel (%) (red line), dissolved oxygen (DO) concentration (mg·L^{-1}) (blue line), and precipitation (mm) (gray line) for: (**a**) piezometer 1; and (**b**) piezometer 3 for 24 February to 5 April 2014. The dashed line at 39% saturation of gravel is the depth of the DO sensor. Saturation of gravel data are missing for 29 March to 5 April 2014.

The DO fluctuations in piezometer 3 (Figure 5b) as a response to the magnitude of saturation of gravel (%) follow the same general pattern as those observed in piezometer 1. However, anoxic conditions were observed for almost 2 days in piezometer 3 during the same period of 26 February and 2 March (Figure 5b). These results indicate that even within a relatively short distance (<10 m), the duration of anoxic conditions can be substantially (about 3×) different between the two ends of the gravel infiltration trench (Figure 2). These findings indicate considerable spatial variability in the redox conditions and reactions that may occur within a LID BMP, even during the same precipitation and stormwater events.

The duration of saturation of gravel (days with >39% saturation of gravel) and corresponding DO concentrations in piezometers 1 and 3 for six precipitation events during the period of observation are summarized in Table 1 and Figure 6. The maximum values of DO shown in Table 1 refer to DO concentrations in the gravel at the time the precipitation event began, and minimum values refer to the DO concentrations at the end of saturation period when the water levels start recovering (increasing). Due to the geometry of the LTC Leveloggers and miniDOT oxygen loggers inside the piezometers, the reported values of duration of saturation of gravel (%) and DO concentrations (Table 1) underestimate the actual duration of time that the entire thickness of the gravel trench is saturated.

Table 1 and Figure 6 indicate an inverse relation between duration of saturation of the gravel (days) and DO concentration. Significant (α = 90%) and inverse Spearman correlations (N = 6) were observed between the duration of saturation of gravel (days) and average and minimum DO values for piezometer 1 (average rho = −0.77 (p = 0.072); minimum rho = −0.83 (p = 0.041)) and piezometer 3 (average rho = −0.71 (p = 0.110); minimum rho = −0.83 (p = 0.041)), respectively. These results indicate that a large percentage of the variation in DO concentrations can be explained by the corresponding duration of saturation of gravel (days). Due to the relatively low number of observations (N = 6) in Figure 6, the data point with duration of saturation >5 days has a strong influence on the linear regression. The duration of saturation >5 days is caused by the relatively longer period of precipitation (26 February to 3 March) than the other observations that typically had precipitation events lasting less than two days (Table 1). If the >5 days of saturation is treated as an outlier and not included in the analysis, we note less significant and inverse Spearman correlations (N = 5) for piezometer 1 (avg. rho = −0.60 (p = 0.284); min. rho = −0.70 (p = 0.188)) and piezometer 3 (avg. rho = −0.90 (p = 0.037); min. rho = −0.70 (p = 0.188)), respectively.

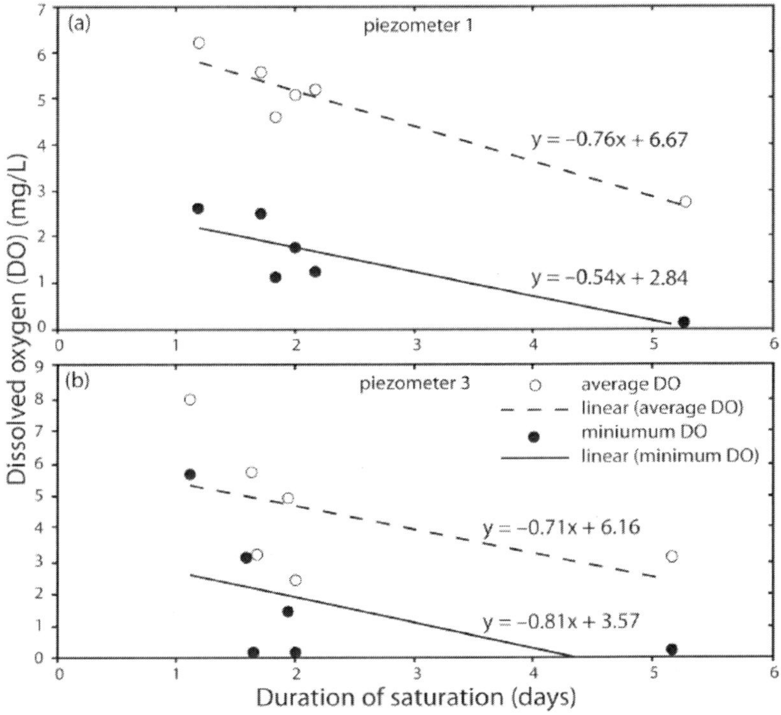

Figure 6. Average and minimum dissolved oxygen (DO) concentration ($mg{\cdot}L^{-1}$) at: (**a**) piezometer 1; and (**b**) piezometer 3 as a function of duration of saturation (days) for six precipitation events between February and June 2014 (Table 1). Note that several of the DO measurements are below the MiniDOT reporting level of 0.3 mg/L (10 μmol/L).

The less significant Spearman correlations between duration of saturation of gravel and DO for piezometer 3 than piezometer 1 is likely related to the previously mentioned spatial variation in flow geometry within the infiltration trench. Anoxic conditions in stormwater near piezometer 3 occurred more often and prevailed longer than piezometer 1 (Figure 5) and more rapidly during the first two days during saturation (Figure 6b), as compared to piezometer 1 (Figure 6a). The spatial variation in redox dynamics between piezometer 1 and piezometer 3 is a result of the proximity to the inflow drain where piezometer 1 (closest to inlet pipe) has a more immediate response to precipitation events due to receiving larger amount of oxic stormwater input sooner than piezometer 3, and the restricted lateral flow near piezometer 3 caused by the impermeable cement stormwater overflow structure.

Table 1. Durations of saturation of gravel and dissolved oxygen (DO) concentrations for piezometers 1 and 3 during precipitation events February to April 2014. The standard error (SE) of the average is shown in parentheses. Note that MiriDOT reporting level for DI is 0.3 mg/L (10 μmol/L).

piezometer 1	54	5.24	0.13	2.78 (0.05)	9.9	13.6
piezometer 3		5.16	0.10	3.12 (0.05)	10.41	12.9
3–4 March 2014						
piezometer 1	5.3	1.99	1.77	5.06 (0.06)	8.22	13.4
piezometer 3		1.93	1.51	4.92 (0.07)	10.15	13.0
5–6 March 2014						
piezometer 1	5.8	1.81	1.12	4.71 (0.06)	8.86	14.3
piezometer 3		1.63	3.09	5.76 (0.05)	9.79	14.2
	25–26 March 2014					
piezometer 1	13	1.17	2.85	6.3 (0.07)	9.26	14.3
piezometer 3		1.11	5.68	8.04 (0.05)	10.39	13.2
29 March 2014						
piezometer 1	12	1.69	2.47	5.57 (0.06)	8.87	15.2
piezometer 3		1.64	0.10	3.1 (0.1)	9.99	14.8
4 April 2014						
piezometer 1	6.4	2.14	1.23	5.16 (0.06)	8.96	13.8
piezometer 3		2	0.10	2.44 (0.08)	10.2	12.8

The duration of saturation of gravel is also a function of drainage rate of the gravel trench and the saturated hydraulic conductivity of the underlying loamy sand (see Table 1 in [27]). There are likely spatial heterogeneities in soil texture and corresponding hydraulic properties of the loamy sand and sandy clay loam beneath the gravel trench that may have an effect on the spatial variations in stormwater draining beneath the gravel trench. Additionally, infiltration rates beneath surface ponding are known to decrease with time, which leads to longer residence times of stormwater and reaction progress for DO consumption by microbial respiration in the gravel trench. Lower infiltration rates have been shown to result in more microbial oxygen consumption, leading to anoxic conditions in the vadose zone [36].

The observed DO dynamics in response to the hydraulic conditions and stormwater runoff exhibit a strong inverse relation between the saturation of gravel (%) and DO concentrations (Figure 5 and Figure 6). The rapid consumption of DO while the gravel is at >39% saturation is likely caused by microbial respiration in the gravel trench. Although not measured in this study, stormwater is often rich in organic matter [38,54], which we infer as an important electron donor for microbially mediated oxygen reduction in the gravel trench. Furthermore, observed water temperatures for the period of field observation of DO (February–June 2014) range between 7 °C and 23 °C (Figure 3), which are above the benchmark temperature referred to as biological zero (5 °C), below which microbial activity generally ceases [55].

3.3. DO Dynamics in Response to Water Temperature

To illustrate the relation between precipitation events, water temperature, and DO concentrations, time series for a period of the 2013–2014 rainy season (24 February to 5 April 2014) are shown in Figure 7. Measured water temperatures in the infiltration trench during this period ranged between 12 °C and 17 °C (Figure 7). As previously described in Figure 3, there was no consistent pattern of stormwater effect on water temperature in the gravel trench (Figure 7). Runoff of relatively cooler stormwater decreased the water temperature in the gravel trench and runoff of relatively warmer stormwater increased water temperatures in the infiltration trench (Figure 7).

Warmer water temperatures are associated with greater microbial activities in soils, and therefore higher rates of DO consumption [38]. Vaughan et al. (2009) [26] observed that during periods of cooler soil temperatures, saturated soils took longer to become reducing. However in this study, the observed variations in stormwater temperature have no apparent influence on the DO dynamics in the gravel trench (Figure 7).

For example, the infiltration of relatively warmer stormwater on 29 March 2014 (16 °C) (Figure 7a) did not result in lower DO concentrations than infiltration of relatively cooler stormwater on 3 March 2014 (13.93 °C) and 4 April 2014 (12.62 °C) (Figure 7a). Using a t-test, we tested if the DO concentrations were statistically different during relatively warm or cold stormwater events ($N = 6$) for the period February to June 2014. Here we categorize warm or cold stormwater events relative to the average water temperature in the infiltration trench. Results of the *t*-test indicate no statistical difference ($p < 0.0001$) in DO concentration during the relatively warm or cold stormwater events.

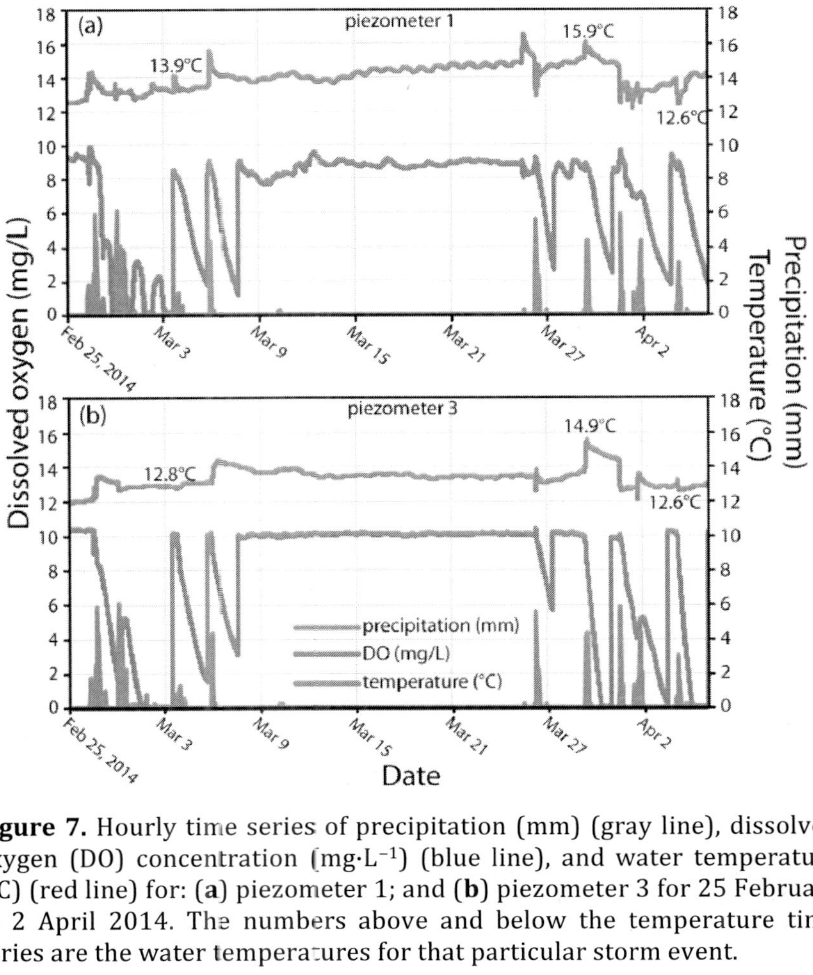

Figure 7. Hourly time series of precipitation (mm) (gray line), dissolved oxygen (DO) concentration ($mg{\cdot}L^{-1}$) (blue line), and water temperature (°C) (red line) for: (**a**) piezometer 1; and (**b**) piezometer 3 for 25 February to 2 April 2014. The numbers above and below the temperature time series are the water temperatures for that particular storm event.

3.4. Rates of O_2 Reduction and LID O_2 Demand (LIDOD)

Rates of O_2 reduction for stormwater in piezometers 1 and 3 were estimated for six precipitation events during the period February–April 2014 (Table 2). O_2 reduction plots show a linear relation with time from the beginning of the rain event to the lowest DO values during that event (plots shown in [51]), indicating a zero-order kinetics for the O_2 reduction process in the gravel trench. The linear fit for the O_2 reduction plots results in a relatively small range of O_2 reduction zero-order rates: 0.002–0.004 $mg{\cdot}L^{-1}{\cdot}min^{-1}$ (0.10–0.25 $mg{\cdot}L^{-1}{\cdot}h^{-1}$) for stormwater in piezometer 1 and 0.002–0.005 $mg{\cdot}L^{-1}{\cdot}min^{-1}$ (0.12–0.27 $mg{\cdot}L^{-1}{\cdot}h^{-1}$) for stormwater in piezometer 3 (Table 2). The O_2 reduction rate for the entire gravel trench was calculated as the average of all the zero-order rates from piezometers 1 and 3, resulting in a value of 0.003 $mg{\cdot}L^{-1}{\cdot}min^{-1}$ (0.179 $mg{\cdot}L^{-1}{\cdot}h^{-1}$).

The $LIDOD_T$ and $LIDOD_{20}$ (Table 2) were calculated using the measured DO concentrations for piezometers 1 and 3 plotted as a function of elapsed time (in minutes) from the beginning of the precipitation event until the lowest DO concentrations measured before recovery of DO levels to pre-event conditions (Figure 8). The slope of the best-fit regression line for each oxygen-depletion curve ($mg{\cdot}m^{-3}{\cdot}min^{-1}$) (b, Equation (5)) was calculated (Figure 8) and used to estimate the $LIDOD_T$ and $LIDOD_{20}$ for each event (Table 2). The range of $LIDOD_T$ is 1.48–2.49 $mg{\cdot}m^{-3}{\cdot}min^{-1}$ for stormwater in piezometer 1 and 1.52–1.83 $mg{\cdot}m^{-3}{\cdot}min^{-1}$ for stormwater in piezometer 3. The $LIDOD_T$ were converted to $LIDOD_{20}$, which ranged between 2.23 and 3.59 $mg{\cdot}m^{-3}{\cdot}min^{-1}$ for piezometer 1 and 2.80 and 2.81 $mg{\cdot}m^{-3}{\cdot}min^{-1}$ for piezometer 3. The $LIDOD_{20}$ for the entire gravel trench was calculated as the average of all the $LIDOD_{20}$ values from piezometers 1 and 3, resulting in a value of 2.71 $mg{\cdot}m^{-3}{\cdot}min^{-1}$.

The average O_2 reduction rate for the gravel trench (0.003 $mg{\cdot}L^{-1}{\cdot}min^{-1}$) is two to five orders of magnitude greater than the O_2 reduction rates for groundwater estimated in recent studies [39,56]. Tesoriero and Puckett (2011) [39] estimated a range of zero-order O_2 reduction rates of <3–140 $\mu mol{\cdot}L^{-1}{\cdot}year^{-1}$ (equivalent to $<1.826 \times 10^{-7}$–8.523×10^{-6} $mg{\cdot}L^{-1}{\cdot}min^{-1}$) in 12 shallow aquifers in the U.S., by relating dissolved O_2 concentrations to groundwater age. Similarly, McMahon et al. (2008) [56] estimated zero-order O_2 reduction rates in four Principal Aquifers of the U.S. that ranged between 0.03 and 6 $mg{\cdot}L^{-1}{\cdot}year^{-1}$ (equivalent to 5.707×10^{-8} and 1.1415×10^{-5} $mg{\cdot}L^{-1}{\cdot}min^{-1}$). The higher rates of O_2reduction reported here are attributed to the differences in the spatial and temporal scales of measurements between the LID infiltration trench and groundwater in aquifers.

Table 2. Zero-order rates of oxygen reduction, LIDODT, and LIDOD20 for piezometers 1 and 3 during precipitation events February to April 2014.

Precipitation Event	Total Rainfall (mm)	Average Temp. (°C)	Zero-Order Reaction Rates ($-K_0$)		$LIDOD_T$	$LIDOD_{20}$
			($mg \cdot L^{-1} \cdot min^{-1}$)	($mg \cdot L^{-1} \cdot h^{-1}$)	($mg \cdot m^{-3} \cdot min^{-1}$)	($mg \cdot m^{-3} \cdot min^{-1}$)
26 February 26–3 March 2014						
piezometer 1	54	–	0.003	0.15	–	–
piezometer 3			0.003	0.18	–	–
3–4 March 2014						
piezometer 1	5.3	13.0	0.003	0.19	1.48	2.23
piezometer 3			0.003	0.16	1.80	2.80
5–6 March 2014						
piezometer 1	5.8	14.2	0.004	0.25	1.84	2.63
piezometer 3			0.003	0.18	1.52	2.20
25–26 March 2014						
piezometer 1	13	12.6	0.003	0.17	2.49	3.59
piezometer 3			0.005	0.27	1.83	2.81
29 March 2014						
piezometer 1	12	–	0.002	0.10	–	–
piezometer 3			0.002	0.12	–	–
4 April 2014						
piezometer 1	6.4	–	0.003	0.16	–	–
piezometer 3			0.003	0.20	–	–

The dynamic flow system of the LID infiltration trench is in more direct contact with the atmosphere and receives more regular inflow of oxic recharge in the form of stormwater runoff as compared to groundwater in most aquifers that is deeper and more isolated from the atmosphere. Additionally, the stormwater runoff in the infiltration trench likely has greater organic matter content than recharge to deeper aquifers [54], and thus the more readily available electron donor promote more rapid microbial O_2 reduction in the infiltration trench. Similarly, aerobic respiration rates are expected to be high in aquifers contaminated with organic compounds because the availability of electron donors as the primary factor affecting O_2reduction rates [39].

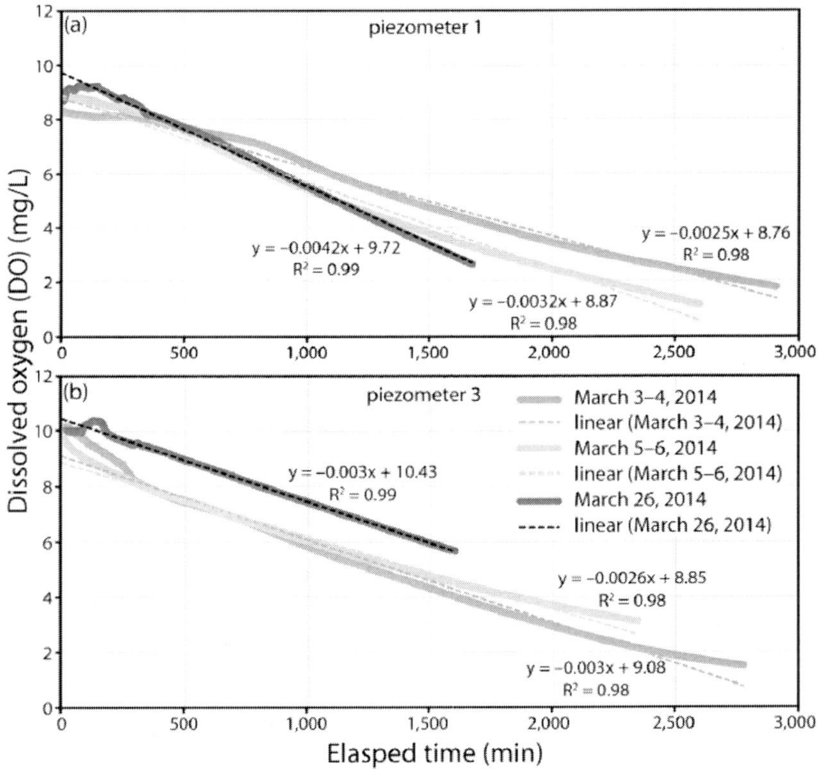

Figure 8. Dissolved oxygen (DO) depletion curves for: (**a**) piezometer 1; and (**b**) piezometer 3 for three precipitation events in March 2014. The slope of the linear fit for each DO depletion curve ($mg{\cdot}L^{-1}{\cdot}min^{-1}$) is used to calculate the average slope of DO depletion curve for the LID infiltration trench.

The importance of organic matter driving O_2 reduction rates in the LID infiltration trench is further supported by the observations of Tesoriero and Puckett (2011) [39] that concentrations of dissolved organic carbon (DOC) were positively correlated with groundwater age at sites with high O_2 reduction rates and negatively correlated at sites with lower rates.

The observation of anoxic conditions and rapid O_2 reduction rates in the infiltration trench has important implications for many stormwater pollutants, and particularly for denitrification of NO_3^- in stormwater runoff. Previous studies identified <0.5 mg·L^{-1} [32] as the O_2 concentration threshold required for the onset of denitrification in groundwater, but the threshold could be as high as <1 mg·L^{-1}[39] and <2 mg·L^{-1} in some aquifers [57,58]. The estimated rates for O_2 reduction in this study strongly indicate that redox conditions favorable for denitrification may be present in the infiltration trench. Based on O_2 reduction rate of 0.003 mg·L^{-1}·min^{-1} and a maximum initial DO concentration of 10 mg·L^{-1}, redox conditions in the infiltration trench may reach the lower end of O_2 concentration threshold (<0.5 mg·L^{-1}) required for the onset of denitrification in about 2–3 days (and <2 days to reach <2 mg·L^{-1}). The time to reach anoxic conditions in the infiltration trench varies depending on the initial oxic conditions present at the beginning of the precipitation event and the duration the infiltration trench is saturated. Future work here at the SFSU LID research network will evaluate denitrification rates and removal of NO_3^- in stormwater runoff during anoxic conditions observed in the LID infiltration trench, and whether the observed 15 h to two days of continuous anoxic conditions in the gravel trench is representative of wetter precipitation scenarios or the vadose zone sediments beneath the gravel trench. We suspect that anoxic conditions in the underlying vadose zone sediments persist longer than what is observed in the trench gravel.

4. CONCLUSIONS AND IMPLICATIONS FOR DESIGNING LID BMPS

We find that the magnitude and duration of fluctuations in DO concentrations in the infiltration trench were influenced by hydraulic conditions in the gravel with an apparent inverse relation to the duration of saturation of gravel (%). DO was quickly consumed and conditions reached anoxic levels within hours to days of stormwater events, and strongly indicates that microbial respiration is a limiting factor for DO in the gravel trench. Anoxic conditions remained for short periods of time (from a few hours to maximum of two days) and returned relatively quickly to oxic conditions after the end of each precipitation event by either rapid infiltration of stormwater and drying of the gravel or by re-oxygenation with oxic stormwater runoff. In contrast, the temperature of stormwater runoff was not found to exhibit a major influence on redox dynamics in the infiltration trench, which may be a function of the relatively constant Mediterranean climate of San Francisco. Statistical analysis of stormwater events with different temperatures shows no significant difference between the DO concentrations in the gravel trench during relatively warm and cold stormwater runoff. Follow-on studies that monitor changes in dissolved organic carbon (DOC) and nitrogen (NO_3^-, NH_4^+) and phosphorus (PO_4^{3-}) species will gain additional insight into important biogeochemical controls on DO consumption and have a greater ability to generalize our findings to other LID systems.

We also find that the average O_2 reduction rate for the infiltration trench is 0.003 $mg \cdot L^{-1} \cdot min^{-1}$, which is two to five orders of magnitude higher than the O_2 reduction rates for groundwater from previous studies. Higher rates of O_2 reduction reported here reflect the more dynamic flow system of the infiltration trench that receives more oxic water and organic carbon in stormwater runoff that drives faster microbial O_2 reduction. The estimated O_2 reduction rate and observed DO concentrations strongly indicates that anoxic conditions are temporally present and could possibly enable denitrification processes and removal of NO_3^-, which is the most ubiquitous contaminant of groundwater resources worldwide. To confirm our hypothesis that denitrification is a possible O_2 consumption pathway when anoxic conditions are reached, additional characterization of the pore-water quality, microbial communities in the biomass, and nitrification/denitrification or mineralization processes in the LID system is necessary and recommended for follow-on studies.

The findings from this study have important implications for the design of infiltration trenches and potentially other types of LID BMPs. The important hydraulic controls, such as the magnitude and duration of saturation of gravel on DO concentrations, O_2 reduction rates, and resulting redox conditions could be altered through the design of the infiltration trench. In general, the relative dimensions of the surface area and average depth of the gravel trench could be manipulated to achieve desired redox dynamics and conditions within and beneath the infiltration trench. For illustration purposes of these concepts, we assume an infiltration trench with similar stormwater storage capacity (3.0 m^3) and underlying soil texture as in this study, but consider two hypothetical design scenarios by altering the ratio of surface area to depth of the gravel trench. A similar stormwater storage capacity could be achieved by using a slightly lower ratio of gravel surface area to depth (smaller area and greater depth) or a slightly higher ratio of gravel surface area to depth (greater area and smaller depth). Therefore, the two hypothetical design scenarios would have the same saturation of gravel (%) under identical precipitation events, but different duration of saturation of gravel (minutes).

The lower ratio of area to depth design would create relatively higher water levels within the gravel and corresponding longer time to infiltrate into the underlying soil. This longer duration of saturation of gravel would enable longer time for microbial O_2 reduction, resulting in possibly more frequent and longer periods of anoxic conditions within and beneath the gravel trench. Such a hypothetical LID design that enhances the frequency and duration of anoxic conditions may be best used to promote denitrification of elevated NO_3^- concentrations in stormwater from areas of high fertilizer use on lawns, golf courses, or agricultural land-use settings. Future work should address more complex systems that require coupled oxic-anoxic zones to oxidize ammonia to nitrate and then promote denitrification and reduce NO_3^-. In contrast, the higher ratio of area to depth design would create relatively lower water levels within the gravel and corresponding shorter time to infiltrate into the underlying soil. This shorter duration of saturation of gravel would enable shorter time for microbial O_2 reduction, resulting in possibly less frequent and shorter periods of anoxic conditions within and beneath the gravel trench. Such a hypothetical LID design that diminishes the frequency and duration of anoxic conditions may be best used to reduce the mobilization of redox sensitive heavy metals in stormwater from parking lots, roads, and other urban or industrialized land-use settings.

Although we found that the temperature of stormwater runoff was not a significant control on redox dynamics in the infiltration trench studies here, the two hypothetical design scenarios could potentially enhance a temperature effect on redox dynamics. The lower ratio of area to depth design could potentially result in relatively cooler stormwater temperatures in the gravel trench, while the higher ratio of area to depth design could potentially result in relatively warmer stormwater temperatures because of the relative differences in the penetration of solar radiation through the gravel trench. Additional study is needed to evaluate how these and other design scenarios may affect the microbial respiration rates and redox conditions of the stormwater within and beneath the LID BMPs. Altering the design of LID BMPs to manipulate redox processes could be an important tool to help protect groundwater quality from the infiltration and recharge of contaminated stormwater runoff beneath LID BMP features.

ACKNOWLEDGMENTS

Funding for this research was provided by the Dawdy Hydrology Research Grant at San Francisco State University; G.A. Harris Research Instrumentation Grant from Decagon Devices, Inc.; and San Francisco State University ORSP-FOA 2010-2011. We thank Leonard Sklar and Bruce Manning (SFSU) for constructive comments on previous versions of this paper.

REFERENCES

1. Paul, M.J.; Meyer, J.L. Streams in the urban landscape. *Annu. Rev. Ecol. Evol. Syst.* **2001**, *32*, 333–365.
2. Pitt, R.; Field, R.; Lalor, M.; Brown, M. Urban stormwater toxic pollutants: Assessment, sources, and treatability. *Water Environ. Res.* **1995**, *67*, 260–275.
3. Elliott, A.; Trowsdale, S. A review of models for low impact urban stormwater drainage. *Environ. Model. Softw.* **2007**, *22*, 394–405.
4. Maniquiz, M.C.; Lee, S.-Y.; Kim, L.-H. Long-Term Monitoring of Infiltration Trench for Nonpoint Source Pollution Control. *Water Air Soil Pollut.* **2010**, *212*, 13–26.
5. Fletcher, T.D.; Andrieu, H.; Hamel, P. Understanding, management and modelling of urban hydrology and its consequences for receiving waters: A state of the art. *Adv. Water Resour.* **2013**, *51*, 261–279.
6. Eriksson, E.; Baun, A.; Scholes, L.; Ledin, A.; Ahlman, S.; Revitt, M.; Noutsopoulos, C.; Mikkelsen, P.S. Selected stormwater priority pollutants—A European perspective. *Sci. Total Environ.* **2007**, *383*, 41–51.
7. Joshi, U.M.; Balasubramanian, R. Characteristics and environmental mobility of trace elements in urban runoff. *Chemosphere* **2010**, *80*, 310–318.
8. Barbosa, A.E.; Fernandes, J.N.; David, L.M. Key issues for sustainable urban stormwater management. *Water Res.* **2012**, *46*, 6787–6798.
9. Gurdak, J.J.; Hanson, R.T.; Green, T.R. *Effects of Climate Variability and Change on Groundwater Resources of the United States*; U.S. Geological Survey: Lakewood, CO, USA, 2009; p. 4.
10. Kuss, A.J.M.; Gurdak, J.J. Groundwater level response in U.S. principal aquifers to ENSO, NAO, PDO, and AMO. *J. Hydrol.* **2014**, *519*, 1939–1952.
11. Barbu, I.A.; Ballestero, T.P.; Roseen, R.M. LID-SWM Practices as a Means of Resilience to Climate Change and Its Effects on Groundwater Recharge. In Proceedings of the World Environmental and Water Resources Congress, Great Rivers, Kansas City, MO, USA, 17–21 May 2009; pp. 1–10.
12. U.S. Environmental Protection Agency. *Low Impact Development (LID)—A Literature Review*; USEPA, Office of Water: Washington, DC, USA, 2000; p. 35.
13. Department of Environmental Resources. *Low-Impact Development Hydrologic Analysis*; Prince George's County, Department of Environmental Resources: Marlboro, MD, USA, 1999.
14. Facility for Advancing Water Biofiltration (FAWB). *Adoption Guidelines for Stormwater Biofiltration Systems*; FAWB, Monash University: Melbourne, Australia, 2009.
15. Dietz, M.E.; Clausen, J.C. Saturation to improve pollutant retention in a rain garden. *Environ. Sci. Technol.* **2006**, *40*, 1335–1340.
16. Prince George's County. *Design Manual for Use of Bioretention in Stormwater Management*; Prince George's County (MD) Government, Department of Environmental Protection, Watershed Protection Branch: Landover, MD, USA, 1993.

17. Kim, H.; Seagren, E.A.; Davis, A.P. Engineered bioretention for removal of nitrate from stormwater runoff. *Water Environ. Res.* **2003**, *75*, 355–367.
18. Zinger, Y.; Blecken, G.-T.; Fletcher, T.D.; Viklander, M.; Deletić, A. Optimising nitrogen removal in existing stormwater biofilters: Benefits and tradeoffs of a retrofitted saturated zone. *Ecol. Eng.* **2013**, *51*, 75–82.
19. Zhang, Z.; Rengel, Z.; Liaghati, T.; Antoniette, T.; Meney, K. Influence of plant species and submerged zone with carbon addition on nutrient removal in stormwater biofilter. *Ecol. Eng.* **2011**, *37*, 1833–1841.
20. Payne, E.G.I.; Fletcher, T.D.; Cook, P.L.M.; Deletic, A.; Hatt, B.E. Processes and drivers of nitrogen removal in stormwater biofiltration. *Crit. Rev. Environ. Sci. Technol.* **2013**, *44*, 796–846.
21. Blecken, G.-T.; Zinger, Y.; Deletić, A.; Fletcher, T.D.; Viklander, M. Impact of a submerged zone and a carbon source on heavy metal removal in stormwater biofilters. *Ecol. Eng.* **2009**, *35*, 769–778.
22. Davis, A.P.; Shokouhian, M.; Sharma, H.; Minami, C.; Winogradoff, D. Water quality improvement through bioretention: Lead, copper, and zinc removal. *Water Environ. Res.* **2003**, *75*, 73–82.
23. Chandrasena, G.I.; Pham, T.; Payne, E.G.; Deletic, A.; McCarthy, D.T. E. *E. coli* removal in laboratory scale stormwater biofilters: Influence of vegetation and submerged zone. *J. Hydrol.* **2014**, *519*, 814–822.
24. Rippy, M.A. Meeting the criteria: Linking biofilter design to fecal indicator bacteria removal. *Wiley Interdiscip. Rev. Water* **2015**, *2*, 577–592.
25. Li, Y.L.; Deletic, A.; Alcazar, L.; Bratieres, K.; Fletcher, T.D.; McCarthy, D.T. Removal of Clostridium perfringens, Escherichia coli and F-RNA coliphages by stormwater biofilters. *Ecol. Eng.* **2012**, *49*, 137–145.
26. Holman-Dodds, J.K.; Bradley, A.A.; Potter, K.W. Evaluation of hydrologic benefits of infiltration based urban storm water management. *J. Am. Water Resour. Assoc.* **2007**, *39*, 205–215.
27. Newcomer, M.E.; Gurdak, J.J.; Sklar, L.S.; Nanus, L. Urban recharge beneath low impact development and effects of climate variability and change. *Water Resour. Res.* **2014**, *50*, 1716–1734.
28. Dietz, M.E.; Clausen, J.C. A field evaluation of rain garden flow and pollutant treatment. *Water Air Soil Pollut.* **2005**, *167*, 123–138.
29. Lu, P.; Yuan, T. Low impact development design for urban stormwater management—A case study in USA. In Proceedings of the International Symposium on Water Resource and Environmental Protection (ISWREP), Xi'an, China, 20–22 May 2011; Volume 4, pp. 2741–2744.
30. U.S. Environmental Protection Agency. *Incorporating Low Impact Development into Municipal Stormwater Programs*; U.S. Environmental Protection Agency: Washington, DC, USA, 2009; p. 11.
31. Korom, S.F. Natural denitrification in the saturated zone: A review. *Water Resour. Res.* **1992**, *28*, 1657–1668.
32. McMahon, P.B.; Chapelle, F.H. Redox processes and water quality of selected Principal Aquifer systems. *Ground Water* **2008**, *46*, 259–271.
33. McMahon, P.B.; Cowdery, T.K.; Chapelle, F.H.; Jurgens, B.C. *Redox Conditions in Selected Principal Aquifers of the United States*; U.S. Geological Survey: Lakewood, CO, USA, 2009; p. 6.
34. Kumar, A.R.; Riyazuddin, P. Seasonal variation of redox species and redox potentials in shallow groundwater: A comparison of measured and calculated redox potentials. *J. Hydrol.* **2012**, *444–445*, 187–198.
35. Datry, T.; Malard, F.; Gibert, J. Dynamics of solutes and dissolved oxygen in shallow urban groundwater below a stormwater infiltration basin. *Sci. Total Environ.* **2004**, *329*, 215–229.
36. Von Rohr, M.R.; Hering, J.G.; Kohler, H.-P.E.; von Gunten, U. Column studies to assess the effects of climate variables on redox processes during riverbank filtration. *Water Res.* **2014**, *61*, 263–275.

37. Greskowiak, J.; Prommer, H.; Massmann, G.; Johnston, C.D.; Nützmann, G.; Pekdeger, A. The impact of variably saturated conditions on hydrogeochemical changes during artificial recharge of groundwater. *Appl. Geochem.* **2005**, *20*, 1409–1426.
38. Vaughan, K.L.; Rabenhorst, M.C.; Needelman, B.A. Saturation and temperature effects on the development of reducing conditions in soils. *Soil Sci. Soc. Am. J.* **2009**, *73*, 663.
39. Tesoriero, A.J.; Puckett, L.J. O_2 reduction and denitrification rates in shallow aquifers. *Water Resour. Res.* **2011**, *47*.
40. Fan, A.M.; Steinberg, V.E. Health implications of nitrate and nitrite in drinking water: An update on methemoglobinemia occurrence and reproductive and developmental toxicity. *Regul. Toxicol. Pharmacol.* **1996**, *23*, 35–43.
41. Spalding, R.F.; Exner, M.E. Occurrence of nitrate in groundwater—A Review. *J. Environ. Qual.* **1993**, *22*, 392–402.
42. Gurdak, J.J.; Qi, S.L. Vulnerability of recently recharged groundwater in principle aquifers of the United States to nitrate contamination. *Environ. Sci. Technol.* **2012**, *46*, 6004–6012.
43. California Stormwater Quality Association (CASQA). *Stormwater Best Management Practice Handbook: New Development and Redevelopment*; California Stormwater Quality Association: Menlo Park, CA, USA, 2003; p. 378.
44. Clifton, H.E.; Hunter, R.E.; Gardner, J.V. Analysis of eustatic, tectonic, and sedimentologic influences on transgressive and regressive cycles in the Upper Cenozoic Merced Formation, San Francisco, California. In *New Perspectives in Basin Analysis*; Kleinspehn, K.L., Paola, C., Eds.; Springer: New York, NY, USA, 1988; pp. 109–128.
45. Nzewi, O.; Gilman, J.A.; Bartow, G. *2009 Annual Groundwater Monitoring Report, Westside Basin*; San Francisco Public Utilities Commission: San Francisco, CA, USA, 2010.
46. Precision Measurement Engineering (PME). *miniDOT User's Manual*; Precision Measurement Engineering: Vista, CA, USA, 2014.
47. National Oceanic and Atmospheric Administration (NOAA). National Centers for Environmental Information. Available online: https://www.ncdc.noaa.gov/ (accessed on 13 May 2016).
48. Pazwash, H. *Urban Storm Water Management*; CRC Press, Taylor & Francis Group: Florence, KY, USA, 2011.
49. Heckathorn, H.A.; Gibs, J. *Sediment Oxygen Demand in the Saddle River and Salem River Watersheds, New Jersey, July–August 2008*; U.S. Geological Survey: Reston, VA, USA, 2010; p. 10.
50. Wilson, T.P. *In-Situ Sediment Oxygen Demand Rates in Hammonton Creek, Hammonton, New Jersey, and Crosswicks Creek, near New Egypt, New Jersey, August–October 2009*; U.S. Geological Survey: Reston, VA, USA, 2014; p. 18.
51. Danfoura, M.N. Effects of Low Impact Development Design on Redox Dynamics in the Vadose Zone and Urban Groundwater. Master's Thesis, San Francisco State University, San Francisco, CA, USA, 2015.
52. Appleyard, S.J. Impact of stormwater infiltration basins on groundwater quality, Perth metropolitan region, Western Australia. *Environ. Geol.* **1993**, *21*, 227–236.
53. Sansalone, J.J.; Hird, J.P.; Cartledge, F.K.; Tittlebaum, M.E. Event-Based Stormwater Quality and Quantity Loadings from Elevated Urban Infrastructure Affected by Transportation. *Water Environ. Res.* **2005**, *77*, 348–365.
54. Datry, T.; Malard, F.; Vitry, L.; Hervant, F.; Gibert, J. Solute dynamics in the bed sediments of a stormwater infiltration basin. *J. Hydrol.* **2003**, *273*, 217–233.
55. Brady, N.C.; Weil, R.R. *The Nature and Property of Soils*; Prentice Hall: Upper Saddle River, NJ, USA, 2001.
56. McMahon, P.B.; Böhlke, J.K.; Kauffman, L.J.; Kipp, K.L.; Landon, M.K.; Crandall, C.A.; Burow, K.R.; Brown, C.J. Source and transport controls on the movement of nitrate to public supply wells in selected principal aquifers of the United States. *Water Resour. Res.* **2008**, *44*.

57. Böhlke, J.K.; Wanty, R.; Tuttle, M.; Delin, G.; Landon, M. Denitrification in the recharge area and discharge area of a transient agricultural nitrate plume in a glacial outwash sand aquifer, Minnesota. *Water Resour. Res.* **2002**, *38*.
58. McMahon, P.B.; Böhlke, J.K.; Christenson, S.C. Geochemistry, radiocarbon ages, and paleorecharge conditions along a transect in the central High Plains aquifer, southwestern Kansas, USA. *Appl. Geochem.* **2004**, *19*, 1655–1686.

CHAPTER 14

Rain Drop Measurement Techniques: A Review

Gopinath Kathiravelu †, Terry Lucke *,† and Peter Nichols †

Stormwater Research Group, School of Science and Engineering, University of the Sunshine Coast, Sippy Downs, Queensland 4556, Australia

ABSTRACT

For over a century there have been many studies that describe the use of rain drop measurement techniques. Initial manual measurement methods evolved due to improved technology to include photographic and, more recently, automated disdrometer and laser measurement techniques. Despite these numerous studies, there have been few comparative reviews of the range of methodologies, and their relative performance. This review explores the raindrop measurement techniques available, and summarizes and classifies the techniques according to the method or principle involved. The requirements of a robust raindrop measurement technique are suggested, and these are reviewed against existing rain drop measurement techniques to provide a comparative guide to the use of the range of techniques available for any research study. This review revealed that while advances in technology have allowed many of the deficiencies of early techniques to be eliminated, challenges remain in relation to the precision of the measurement of the size, shape, and velocity of rain drops.

Keywords: raindrop measurement techniques; impact and optical disdrometers; laser precipitation monitor; pluviometer

1. INTRODUCTION

An appreciation of rain drop characteristics such as size, shape, velocity, kinetic energy, and drop size distribution is crucial for many scientific, commercial and industrial applications. Some examples of these include remote sensing, meteorology (weather prediction), telecommunications (signal distortion), and agriculture and horticulture (crop yield) radar meteorology, atmospheric physics, cloud photodetection, and measurement of tropospheric precipitation microstructure [1,2,3].

The characteristics of rain drops are also important for stormwater management purposes, particularly in relation to understanding how pollution wash off processes affect stormwater quality. For example, larger rain drops that possess more kinetic energy are known to result in higher pollution concentrations being washed off impervious surfaces and into downstream aquatic environments [4].

The objective of this review is to provide a summary of the development of rainfall measurements techniques, and to review and compare the different rain drop measurement techniques used in previous research studies. The scope of this review has been limited to include only those measurement techniques with the ability to measure rain drop size, shape, distribution, velocity, kinetic energy, and intensity. Different raindrop measurement techniques have been characterised according to the method used, and the relative merits of each method are discussed. In order to compare the merits of each technique, rain drop measurement methods included in this review have been broadly categorised into manual and automated techniques.

Manual rain drop measurement techniques include the stain method (measurement of stains on dyed absorbent paper), flour pellet method (measurement of rain drops that fall into finely sieved flour and produce dough pellets), and oil immersion method (measurement of rain drops in a vessel containing oil). Despite these manual methods being simple, they are time consuming, have limited measurement accuracy, and do not give real time data records. Also, these manual techniques cannot provide terminal velocity data, which is required to estimate the kinetic energy of rainfall [5]. Manual techniques are reviewed in the first part of this article.

Recent advances in technology and electronics have enabled an exploration of automated rain drop measurement techniques. These are reviewed in the second part of this article and include techniques such as the using devices to measure the displacement and mechanical energy caused by raindrops hitting a surface, optical imaging to measure the velocity, diameter, and shape of the raindrops using camera technology, acoustic techniques which measure the noise produced by rain drops hitting a diaphragm and optical scattering, whereby rain drop size, shape, velocity, and diameter are measured passing through a light or laser beam.

In the final section of this article, the essential characteristics required for an accurate rain drop measurement technique are suggested and explained. A summary and conclusions of the review are presented.

2. MANUAL RAIN DROP MEASUREMENT TECHNIQUES

Early studies (*ca.* 1900–1960) attempted to describe rain drop size and velocity using manual measurement techniques such as chemically treated paper, and sugar or soot coated nylon screen [6,7,8]. These very early, functional techniques were found to provide inaccurate results, and have been superseded due to technological advancement over the last century.

2.1. Stain Method

The stain method was one of the earliest accepted techniques to be developed and it is still in use today. First described by Lowe [9], this method involves the use of chemically treated paper to measure the size of raindrops. For a short period of time rain drops are allowed to land on a sheet of absorbent paper covered with a water-soluble dye. A variety of absorbent papers have previously been used including filter paper, blotting paper, blueprint paper, paper towelling, photographic paper, and adding machine tape. Upon impact, the embedded dye reacts with the rain drops and this leaves permanent marks on the paper. The marks are then carefully measured and counted to provide information about the rain drops. One of the limitations of this method is that during prolonged sampling, the rain drop stains can overlap, which can make it difficult to accurately measure and count individual drops.

Several iterations of this method over time improved measurement accuracy, and increased size range measurement capacity including developments of the method described by Marshall and Palmer [10], and Marshall *et al.* [11] who used dyed filter paper. Two filter papers were used simultaneously to increase the accuracy of rain drop measurement. Ink blotters dusted with potassium manganese used by Anderson [12] and known water densities used by Abudi *et al.* [13], incorporated weights of raindrops to infer size. Several studies [14,15,16,17,

18,19,20,21,22] used Whatman's No. 1 filter paper, which was identified as yielding the most accurate results.

Bowen and Davidson [23] trialled an improvement to the stain method by using a semi-automated technique which produced a continuous record of the drop size distribution. The improved method involved deflection of rain drops onto moving absorbent paper embedded with dye. The diameters of the stains were categorised into five different size classes, from which drop size distributions were calculated. A similar recording instrument in which paper tape was used to record rain drops was developed in conjunction with an equation which described drop size in relation to stain size and time lag between sampling and analysis [24,25,26,27]. Calculations resulted in a calibration chart which translated the stained area caused by rain drops into raindrop diameter [14,15]. Limitations of this methodology included the uncertainty of allowing for terminal velocity of the rain drop prior to measurement [15,28,29], and maintenance of paper temperatures, which were both found to influence stain sizes [30].

2.2. Flour Pellet Method

First developed in 1904 by Bentley [31], the flour pellet method (Figure 1) was used to study drop size distributions of rain events in Washington D.C., USA. A number of studies have since used slightly different versions of the flour pellet method to successfully analyse rainfall (Table 1).

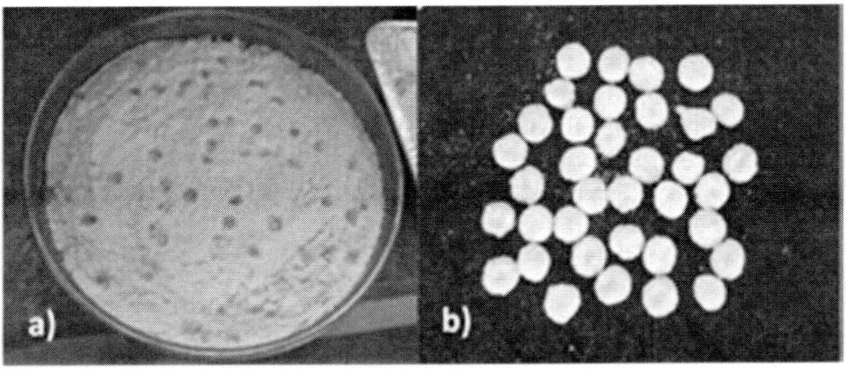

Figure 1. (**a**) Raindrop flour pellet samples collected in a pan filled with 2 cm depth of plain flour; (**b**) Flour pellets after oven drying.

Compaction of the flour over time was found to affect measured pellet size, and large sample numbers are usually required to account for the high variability in the number of rain drops observed during testing [32]. Test area should be restricted to the centre of the chosen collection tray to avoid splash effects. The test duration should also be brief (1–2 s) to avoid duplicate drop counts [4,33,34,35,36,37,38,39].

The chief technological advance to the flour pellet method was developed by Arnaez *et al.* [40], who used digital analysis of photographs to determine drop sizes. The main fields that still currently use the flour pellet method include soil erosion, and stormwater quality research studies [4,35,36,38,39,40,41,42,43,44,45,46].

2.3. Oil Immersion Method

An early manual rain drop measurement method first developed by Fuchs and Petrjanoff [50], the oil immersion method involves the collection of drops on a glass trough containing a fresh mixture of lightly viscose liquids, such as Vaseline® and light mineral oil which prevents evaporation and condensation [51,52,53,54,55,56,57,58]. Using a camera and microscope, this technique does not require calibration or special equipment [55].

Table 1. The research studies used the flour pellet method.

Research Study and Location	Purpose of Use	Method used
Laws & Parsons (1943) [47]	To measure drop sizes from natural storms	After sampling with raindrops, the formed pellets were dried in an oven. Pellets were sized with sieves and weighed. The size was calibrated by weighing dried pellets produced by drops of a known size.
Hudson (1963) [33]	To measure drop sizes from natural storms	A tray (0.05 m^2) of flour was exposed to simulated rainfall for a period of 1 s. The flour was then dried for 24 h at ambient temperature (28–30 °C) and the pellets formed were passed through a series of sieves (4.75, 3.35, 2.36, 1.18 and 0.85 mm). The pellets were then dried for 24 h at 105 °C, weighed and measured.
Kohl (1974) [32]	To verify the nozzle produced drop sizes in the rainfall simulation studies	Circular pans 21 cm in diameter and 2 cm deep were filled with flour and made level with a straight edge. After exposure to rain drops, the flour was dried (24 h at 38 °C). An 18.3 cm diameter sample was taken from the centre of the pan to avoid splash effects. The pellets were sieved (U.S. series 5 to 50 mesh) and weighed.

Carter *et al.* (1974) [48]	To study drop size distribution of natural rainfall	A circular pan (31 cm diameter) of flour (1.6 cm deep), was exposed in a rain for a short period of time. The pellets formed were first air- and later oven-dried and weighed. Raindrop diameter was estimated from the weight of the pellets.
Navas *et al.* (1990) [49]	To verify the nozzle produced drop sizes in the rainfall simulation studies	A 25.4 cm diameter plate containing an uncompacted, layer of flour (2.54 cm thick) is exposed to rainfall for 1–4 s. The small flour balls are dried for 24 h at 105 °C, and sieved (5000, 3000, 1000, 630, 500 and 250 μm) the fractions are weighed. Calibration of drops is required.
Ogunye and Boussabaine (2002) [35]	To verify the simulated drop sizes in the rainfall simulation studies	Exposure time is restricted to 1 s to minimise coalescence of the pellets in the flour. A large sample size is required to minimise the variability in counts of the rare large drops.
Arnaez *et al.* (2007) [40]	To verify the nozzle produced drop sizes in the rainfall simulation studies.	Rain drops formed small pellets in the flour that were photographed and analysed by computer.
Herngren (2005) [4]; Egodawatta (2007) [44]; Miguntanna (2009) [38]	To verify the nozzle produced drop sizes in the rainfall simulation studies.	A tray (diameter 240 mm) of uncompacted flour was exposed to simulated rainfall for a period of 2 s. Flour was dried for 12 h at 105 °C, and the pellets sieved (4.75 mm; 3.35 mm; 2.36 mm; 1.18 mm; 0.6 mm; and 0.5 mm).
Pérez-Latorre *et al.* (2010) [39]	To verify the nozzle produced drop sizes in the rainfall simulation studies.	A flour layer (1 cm depth) was placed over a surface of 50 cm × 50 cm and compacted using a ruler. The flour surface was covered to protect it from rainfall except when the cover was removed for 2 s during the simulation to collect drop samples. The diameter of pellets was measured using a calibre (±0.1 mm).
Asante (2011) [45]	To verify the nozzle produced drop sizes in the rainfall simulation studies.	A thin layer of cassava flour, and wheat flour were spread on separate trays and passed through a rain shower. The flour was dried and the pellets separated according to their size ranges using a nest of sieves. The size of raindrops was calculated from the size of pellets.
Parsakhoo *et al.* (2012) [46]	To verify the nozzle produced drop sizes in the rainfall simulation studies.	The drop impact on flour was estimated using a ruler.

The low viscosity and hydrophobic nature of the oil causes rain drops to form discreet spherical shapes, allowing drop counting and measurement by microscope [59] or via photograph [57,58,60]. Generally any low viscosity oil can be used [61], and several alternative liquids have been utilised in a range of studies, including Apiezon oil A, Shell 33, vacuum pump oil, paraffin oil and hydraulic fluid mixture, hydro carbon solvent, silicone oil, anisole mineral oil mixture, cold hexane, and grease (−20 °C) [57,58,59,60,62,63,64,65,66,67]. Courshee and Byass [59] found that the use of two oils of different densities improved drop shape measurement. Using a microscope or a photograph, they found it easier to identify the drops trapped at the liquid interface (two liquids) rather than one.

2.4. Photographic Method

The photographic method has been used extensively to measure rain drop size and velocity, and undergone many iterative improvements since its development by Mache in 1904 [68] (Table 2). Initially, Laws [69] measured drop sizes using a 9 cm × 12 cm still camera mounted behind a chopper-disc driven by a small synchronous motor (Figure 2).

Light infiltration problems have restricted some use of the photographic method to night time sampling [7,70,71]. Use of the Illinois camera resulted in drop count errors due to superimposition of multiple drops [70]. Digital pixilation also limited the accuracy of several photographic techniques [72].

Table 2. Range of photographic methods used in rainfall measurements studies.

Research Study	Methodology and other Comments
Abudi *et al.* (2012) [13]	A Motion-Scope® PCI-8oosc camera (Redlake Imaging Corp., San Diego, CA, USA) was used in conjunction with special software capture falling drops. Calibration of images resulted in drop velocity and size measurement.
De Jong (2010) [73]	A Canon Powershot® camera (Canon Inc., Tokyo, Japan) was with a Stopshot® module (Cognisys Inc., Traverse City, MI, USA) which triggered two successive flashes. The process was activated by an infrared sensor passed by a raindrop drop. Drop images were captured twice allowing velocity measurement (Figure 3).
Salvador *et al.* (2009) [72]	Low shutter speeds result in drops appearing as cylinders in a photograph. Drop diameter and velocity were calculated based on the selected shutter speed.
Sudheera and Panda (2000) [75]	High resolution photographs were digitised using a scanner. A digital single lens reflex (SLR) camera produced digital images converted by a CCD (charge couple device) camera connected to a MVP/AT computer system. Pixel aggregation was used to partition images to allow drop size and count measurement.
Cruvinel *et al.*, & Cruvinel *et al.* (1996, 1999) [58,61]	A Sony® TR50BR handycam video (Sony, Minato, Tokyo) and a MATROX® PIP-640B (Matrox, QC, Canada) were used in conjunction with oil immersion to calculate drop sizes.
Eigel and Moore & Kincaid *et al.* (1983, 1996) [57,76]	Drops were photographed using a 35 mm Fujichrome® 100 (Fujifilm, Tokyo, Japan) and illuminated with a circular fluorescent light. Slides projected on a screen resulting in a 30:1 magnification, supporting small drop measurement (0.1 mm diameter).
Mueller (1966) [77], Jones (1959) [78], Jones and Dean (1953) [79], Jones (1956) [80]	An Illinois camera was used to capture raindrops in an area of 1 m^3 of air every 10 s. This involved two synchronised cameras at perpendicular angles. The three-dimensional image of the shape of the raindrops was then calculated. The accuracy of this method was limited to >0.5 mm in drop size [81].
Laws (1941) [69]	A still camera was used mounted behind a chopper-disk driven by a small synchronous motor (Figure 2). A collimating lens resulted in accurate drop size measurement. Dark field illumination and the chopper-disk made it possible to obtain multiple images of a drop on a single film.

In addition, photographic techniques are subject to environmental influences such as wind which may cause drop drift and measurement errors [73]. The time consuming nature of some experimental photographic techniques were found to limit their practical use [74].

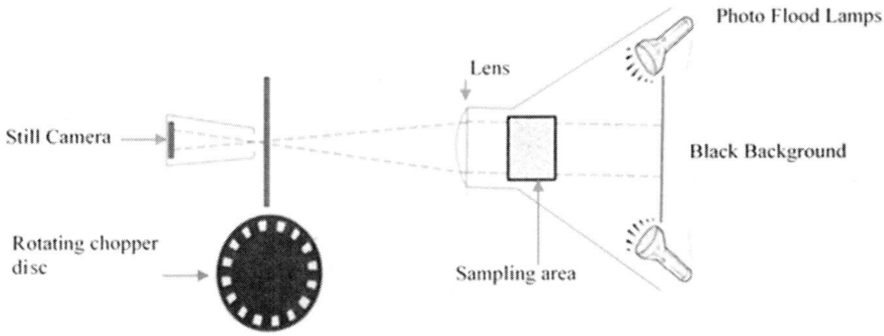

Figure 2. Schematic still camera setup developed to measure the velocity of falling drops [68].

Figure 3. Vaisala Rain Cap disdrometer (Vaisala, Vantaa, Finland)[82].

3. AUTOMATED RAIN DROP MEASUREMENT TECHNIQUES

3.1. Impact Disdrometers

The kinetic energy of rain drops is critical to soil erosion and stormwater pollutant wash off studies because it is indicative of the potential of drops to displace particles normally bound to a surface, causing to soil particles to enter surface water flows. The combination of drop size distribution and drop velocity can provide an estimation of kinetic energy, however there have been several previous attempts to take measurements directly [83,84,85]. This has been done using either acoustic or displacement methods.

3.1.1. Acoustic Disdrometers

Acoustic disdrometers involve the generation and recording of an electric signal via a piezoelectric sensor when drops fall on a specialized diaphragm. Based on the relationship between kinetic energy and drop size calculations [53,69], this electrical signal is converted to kinetic energy via the measured acoustic energy [73,83,84,85,86,87,88, 89,90, 91,92,93,94,95].

Modifications to the sensors used in acoustic disdrometers by Nystuen *et al.* [96] enabled use in marine environments, however difficulties remained during high rainfall intensity measurement. Jayawardena and Rezaur [83] also successfully modified the acoustic disdrometers, and improved drop size distribution, rain intensity and kinetic energy measurement accuracy. Other commercial devices have been successfully developed by Salmi and Ikonen [97], Salmi and Elomaa [84], Winder and Paulson [86], Bagree [98] and Vaisala [82] (Figure 3).

Limitations to accuracy in drop size estimation arise using acoustic disdrometers due to the difficulty in obtaining a uniform acoustic response over the entire diaphragm. Difficulties in the accurate measurement of smaller drop sizes also remain because of insensitive diaphragms, and splash effects. In addition, higher intensity storms are not able to be measured due to background noise which decreases measurement accuracy.

3.1.2. Displacement Disdrometers

Energy generated by drops falling on the top surface of a displacement disdrometer is translated via magnetic induction, and converted via electrical pulse to estimate the size of a rain drop (Figure 4).

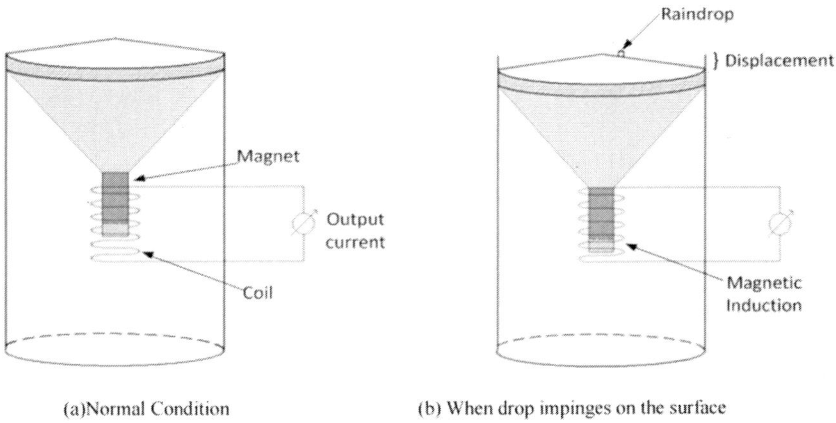

Figure 4. Schematic of the principle of operation of displacement disdrometers [98].

In addition to magnetic induction, several mechanisms have previously been trialled to accurately measure drop size including elastic springs [98], bonded strain gauges [99], and pressure transducers [84,100,101,102,103]. Arguably the most widely used displacement disdrometer is the Joss-Waldvogel Disdrometer [85] (Figure 5) which has been commercially available for past 45 years.

This unit has undergone several iterations to improve the composition of the cone which is the principle measurement component. Successful modifications have included the addition of a digital converter [104,105,106] (Table 3).

Although this disdrometer may have provided advantages such as measurement over a wide range of drop sizes, and the ability to continuously sample over longer durations, limitations remain including accurate drop counting, and accurate measurement of velocity, kinetic energy, intensity, and drop shape.

3.2. Optical Disdrometers

Optical technologies (optical imaging or optical scattering) are non-intrusive rain drop measurement techniques. These methods do not influence drop behaviour during measurement, and have successfully resolved drop break up, and drop splatter problems experienced by other measurement methods [126,127].

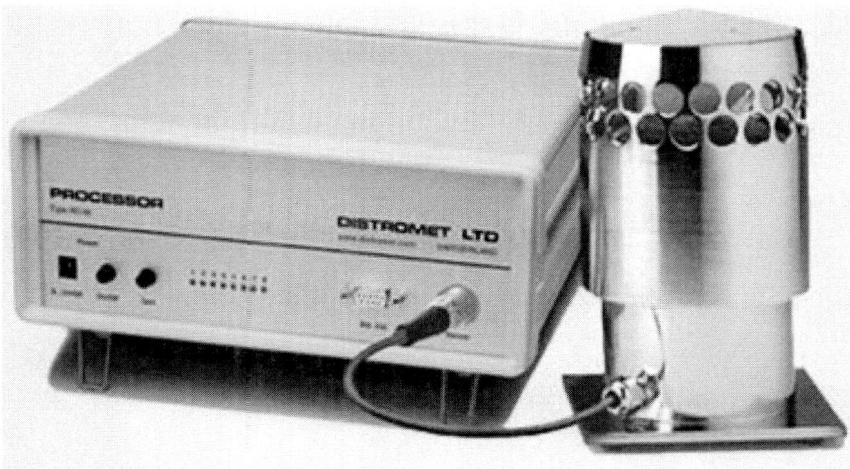

Figure 5. Joss-Waldvogel impact disdrometer (Distromet Ltd., Basel, Switzerland) [106].

Table 3. Capability summary of a range of optical disdrometers.

Device Name	Study	Rainfall Intensity	Drop Size	Fall Speed	Kinetic Energy	Sampling Area (Thickness of Light Beam)
Thies Clima® Laser Precipitation Monitor (LPM) (Adolf Thies GmbH & Co. KG, Göttingen, Germany)	Bloemink & Lanzinger (2005) [107]; Clima (2007) [108]; Upton & Brawn (2008) [109]; Anderson (2009) [110]; de Moraes Frasson (2011) [111]	<250 mm/h	<8.5 mm	<11 m/s	Not Measurable	45.6 cm² (22.5 cm × 2 cm)
OTT Parsivel® disdrometer (OTT Hydromet, Loveland, Colorado, USA)	Krajewski *et al.* (2006) [112]; Thurai *et al.* (2009) [113]; Friedrich *et al.* (2013) [114]	<1200 mm/h	0.2–5 mm	0.2–20 m/s	<30 KJ	54 cm² (18 cm × 3 cm)
Particulate Measurement System (PMS) 2DG spectrometer (Particle Measuring Systems, Airport Blvd Boulder, Colorado, USA)	Hawke (2003) [115]	Not Measurable	0.15–9.6 mm (in 64, 0.15 mm size categories)	<25 m/s	Measurable	100 mm²
Paired-pulse optical disdrometer (P-POD)	Grossklaus *et al.* (1998) [116]	Not Measurable	0.35–6.4 mm	Measurable	Not Measurable	Cylindrical volume with 120 mm length and 22 mm diameter

Particle Measuring System GBPP-100S	Solomon *et al.* (1991) [117]	Measurable	0.2–13 mm in 0.2 mm increments	Measurable	Not Measurable	13 × 500 mm^2
Paired pulse optical disdrometer (P-POD)	Illingworth and Stevens (1987) [118]	Not Measurable	0.72–3.62 mm in 0.21 mm steps, <0.72 and >3.62 mm also detectable	Measurable	Not Measurable	Measurable
VIDIAZ spectro Pluvio meter	Donnadieu (1980) [119]	Not Measurable	>0.6 mm	Measurable	Not Measurable	80 cm^2
Optical spectro pluviometer (OSP)	Picca &Trouilhet (1964) [120], Donnadieu *et al.* (1969) [121], Klaus (1977) [122], Hauser *et al.* (1984) [123], Salles & Poesen (1999) [124]; Salles *et al.* (1999) [125]	<35 mm/h underestimates intensity by 12%. >35 mm/h, underestimates intensity by 38%.	0.3–4.7 mm (±6%) (Larger drops are detected but without quantification of their diameter)	0.2–10 m/s	Not Measurable	Not reported

3.2.1. Optical Imaging

Recent imaging techniques developed have involved two motion cameras (2DVD) to show raindrop microstructure, including front and side drop contours, fall velocity, drop cant and horizontal velocity. General rainfall parameters such as rain intensity and drop size distributions have also been accurately measured [128]. Two motion cameras record images of drops which have been used to accurately measure drop velocity, diameter, and shape (including oblateness, Figure 6). Measurement errors arising from drop drift caused by the tall unit design have led to design modifications, including the development of an indoor model [129], and one specifically designed for outdoor use [127].

Liu *et al.* [130] developed a video system capable of accurate drop shape and velocity measurement (Figure 7). The set up consists of optical and processing units, and a unique imaging unit comprised of a planar array charge-coupled device (CCD) sensor. The shape, size, and velocity of drops can be accurately measured by a single CCD sensor.

3.2.2. Optical Scattering

Optical scattering techniques involve the generation of a horizontal light beam which travels to a receiver where electrical measurements are taken. Drops that pass through the light beam cause the light to scatter. The attenuation of the light caused by each drop is converted to an electrical pulse by the receiver which is then successfully converted to accurate drop velocity measurement [108] (Figure 8).

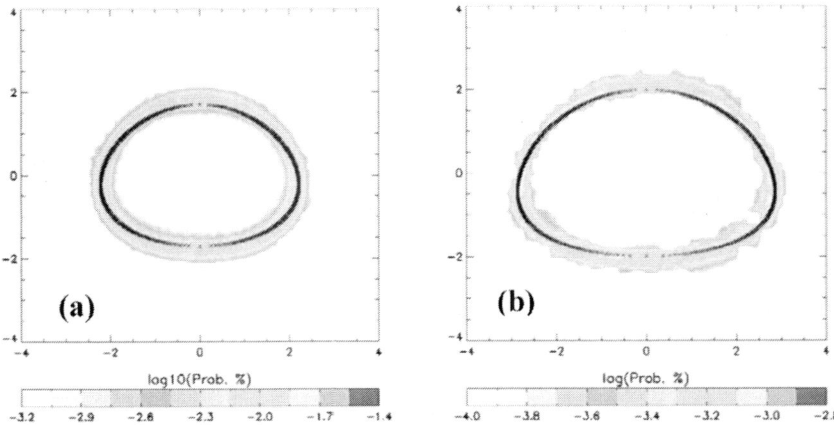

Figure 6. Drop shapes in terms of probability of (**a**) 4 mm and (**b**) 5 mm obtained from 2DVD [113].

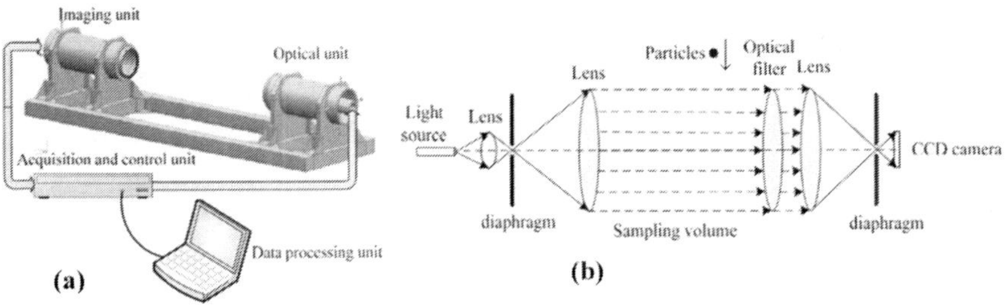

Figure 7. (**a**) Video setup; (**b**) Schematic of imaging process [129].

Since the mid-20th century, optical disdrometers have been used to successfully count and size individual rain drops [114,119,120, 121,122,123,131,132,133,134] (Table 4). Performance evaluations have suggested that optical disdrometers may be limited to measuring larger drop sizes and that the rainfall intensity measurements were inaccurate [9,124]. Although optical disdrometers have also been found to be sensitive to wind effects [116], a modified version [118] included a paired pulse and was successfully used in windy conditions (wind speeds up to 20 m/s). Several models are also capable of successfully differentiating between solids and liquids, enabling use in the snow [135,136].

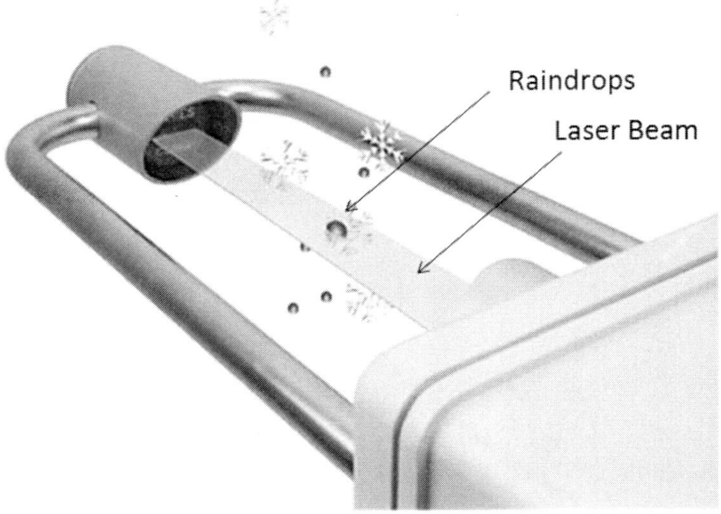

Figure 8. Schematic of optical disdrometer [108].

4. CHARACTERISTICS OF A ROBUST RAINFALL DROPLET MEASUREMENT TECHNIQUE

For a realistic prediction and monitoring of droplet characteristics, a robust rainfall measurement instrument must be able to:

- Measure both larger (up to 10 mm) and smaller (down to 0.3 mm) drop sizes precisely;
- Count the drop sizes accurately;
- Measure the fall velocity precisely;
- Measure the rainfall intensities across all expected ranges;
- Sample continuously; and
- Tolerate wind effects while retaining drop measurement precision.

The possibility of achieving these target characteristics using the full range of available techniques is discussed below.

Table 4. Summary of the Characteristics of Rain Droplet Measurement Techniques.

	Stain Method	FPM	Oil Immersion Technique	Photography Technique	JWD RD 80 & RD 69 Disdrometer	VR—WXT520 Disdrometer	2 Dimensional Video Disdrometer	OTT Parsivel Disdrometer	Laser Optical Disdrometer
Principle	Manual	Manual	Manual	Optical Technology	Impact Displacement Technology	Impact Acoustic Technology	Optical Technology	Optical Laser Technology	Optical Laser Technology
Measurability of larger drops	2.0 mm	5 mm	2.1 mm	Not reported	5.0–5.5 mm	5.0 mm	Yes Range not reported	5.0–5.5 mm	8.5 mm
Measurability of smaller drops	0.3 mm	0.75 mm	Not reported	Not reported	1.0 mm	0.8 mm	Yes Range not reported	0.2 mm	0.125 mm
Measurability of counting the number of droplets	Yes	Yes	Yes	No	No	No	Yes	Yes	Yes
Measurability of the rain fall velocity	No	No	No	Yes	No	No	Yes	20 m/s	11 m/s
Measurability of the rain kinetic energy	No	No	No	No	No	No	No	Yes up to 30 kJ	No
Measurability of the rain intensity	No	No	No	No	No	No	Yes	Yes	Yes
Ability to account the oblateness	No	No	No	No	No	No	Yes	No	No
Ability to sampling continuously for longer durations	No	No	No	No	Yes	Yes	Yes	Yes	Yes
Resilience to the wind effects	No	No	No	No	No	No	No	No	No
* Resolution					127 classes	8 classes		1014 (32 size × 32 velocity)	430 classes (23 × 20)
Temporal resolution					1 min	1 min		10 s to 60 min	1 min

* The resolution is defined as the number of classes into which the drops can be classified.

4.1. Precise Measurement of Larger Rain Drops

Because larger rain drops (>6 mm) are correlated with larger pollution wash off from urban areas [110], their accurate measurement is essential to stormwater quality studies that utilise rainfall simulation. Accurate manual rain drop measurement is limited to a maximum of approximately 2 mm in diameter due to splashing effects distorting results [7,17,137]. Large size drops are often overestimated due to the drop size growth over time on absorbent paper during use of the stain method [138]. The flour pellet method is limited to measurement of drops larger than about 0.5 mm due to sieve size limitations [5]. The measurement of larger drops using the oil immersion method is limited to 2.1 mm diameter due to drop splatter and amalgamation of drops [7,53,59].

Automated measurement techniques such as the impact disdrometers are also limited to the measurement of drops less than 5.5 mm because of a reliance on calculations using the relationship between velocity-diameter which plateaus beyond this diameter range using current formulae [53,113,139]. Although the two dimensional video and laser particle methods claim to have the capacity to precisely measure drop sizes as large as 10 mm, peer reviewed studies to confirm this are yet to be published.

4.2. Precise Measurement of Smaller Rain Drops

The accurate measurement of small raindrops is challenging using both manual and automated techniques. Manual raindrop measurement techniques are restricted to precise measurement of drops of greater than 0.3 mm diameter [15,33,44,59,61,66,140]. Automated measurement techniques are also restricted to measurement of drops greater than 1 mm in size [141].

Several studies reported difficulties in relation to impact disdrometers and the measurement of smaller raindrop diameters [85,134,142,143]. Intense rainfall causes splashing and vibrations is also reported to distort the measurement of smaller drop sizes [142,144,145]. Measurement accuracy difficulties caused by laser technology recovery time (dead time error), and noise distortion that affect the measurement of smaller drop sizes are known to restrict precise measurement to 1 mm effects [145,146,147,148].

Acoustic disdrometers have limitations arising from the duration of the decaying waveform which when measured leads to distorted results [94,149]. Optical disdrometers claim to measure smaller drops with more precision, however, the reliability of these measurements remains unreported.

4.3. Accurate Measurement of the Number of Rain drops

Accurate measurement of the number of drops is critical for the generation of a drop size distribution for any given rainfall event. This is the most widely used characteristic used to describe rainfall [150,151]. Manual measurement techniques have a good capacity to accurately count drop numbers over short durations. Automated disdrometers such as laser precipitation monitors have the capacity to accurately sample over the longer durations required by rainfall sampling studies. However, limitations on the measurement accuracy of drops below 0.2 mm remain due to drop splatter [152] and background noise [112]. Optical disdrometers claim to accurately count drop numbers, however, the reliability of these measurements has also not been verified or reported.

4.4. Precise Measurement of Rain Drop Velocity

Manual techniques are not capable of measuring drop velocity with an adequate level of precision [153]. Although limited to static measurements, and defined by frame capture rate per second, photographic techniques are capable of precise drop velocity measurement. Video technique measurement (2DVD), laser precipitation monitors, and optical spectral pluviometers can also provide precise, continuous drop velocity measurements.

4.5. Ability to Measure a Wide Range of Rainfall Intensities

Because of the particular features of each technique, none of the manual rain drop measurement techniques, nor the impact disdrometer methods are capable of measuring rainfall intensity reliably. Although overestimation of higher rainfall intensities (>20 mm/h) is common, optical laser and video (2DVD) measurement of rainfall intensity are generally accepted as more accurate [154]. Due to the limitations regarding accurate measurement of high rainfall intensities, it is recommended that optical laser techniques are used in combination with a conventional pluviometer to enable measurements to be verified and to ensure accurate rain intensity measurement.

4.6. Precise Measurement of Rain Drop Shape (Oblateness)

Initially spherical due to surface tension forces, with increasing size, fall velocity and drag forces, rain drops tend to flatten out at the base, and sometimes develop a concave shape (Figure 6). The degree of oblateness may affect the kinetic energy of the drop, and thus the potential wash-off process caused by drop impact. Efforts to precisely measure the oblateness of larger drops have been limited, and it has been suggested that as yet, it may not be accurately described [155].

4.7. Capacity to Accurately Sample Rainfall over a Long Duration

Restricted sampling durations are synonymous with manual raindrop measurement techniques [156,157]. Automated disdrometers (laser, optical, acoustic, and impact) are known to measure rain drops in real time with virtually no time duration limitation. However, as discussed above, the accuracy of these devices can be limited.

4.8. Capacity to Perform Precise Rain Drop Measurement during Adverse wind Conditions

Because sampling time is usually quite brief, all of the manual measurement techniques are known to be accurate (within their individual output limitations) during windy conditions. Air movement and wind noise around automated samplers (disdrometers, video, and acoustic) are known to influence rain drop trajectory and sound filtering, which have been shown to lead to inaccuracies in drop size measurements in previous studies [86,152,158]. Wind effects were reduced to an acceptable level in one previous study by tilting an acoustic disdrometer parallel to wind direction [114]. However, this does not offer a reliable or permanent solution.

5. SUMMARY AND CONCLUSION

Every rainfall measurement technique has strengths and weaknesses and these generally result in some limitation in the accuracy of rain drop characteristic measurements. The precise requirements of any proposed study will determine the most appropriate method or combination of methods that should be used to produce the most suitable and accurate raindrop measurements. This is particularly the case for stormwater management purposes, particularly in relation to understanding how different raindrop characteristics affect pollution wash off processes and how this influences stormwater runoff quality from urban areas.

The main findings of this review have been:

- The use of manual rain drop measurement techniques have been successfully used in studies involving drop size measurements. However, these methods are generally not suitable for the measurement of smaller and larger drop sizes outside the normal range (0.3–6 mm), they are not capable of precise drop counts, they are not suitable for continuous rainfall monitoring studies, and they are less effective during intense and windy storm conditions. In addition, manual rain drop measurement techniques cannot be used to measure or report drop velocity.
- Automated (impact and optical) disdrometers are generally able to sample continuously over long durations. However, inaccuracies in drop size and velocity measurements are likely during heavy rain. It is recommended that optical disdrometers should be used in combination with a conventional rain gauge to enable validation of results and ensure precise rain intensity measurement.
- The common limitations of all the rain drop measurement techniques includes their inability to precisely measure both the smaller, and the larger drop sizes outside the normal size range, their inaccuracy during high intensity rainfall events, and their reduced measurement precision during windy conditions. With improvements in technology occurring on a nearly daily basis, it is anticipated that the accuracy and precision of automated rainfall measurement techniques will significantly improve in the near future. This will enable more precise measurements to be undertaken and result in a much better understanding of real rainfall characteristics.

REFERENCES

1. Schönhuber, M.; Urban, H.; Baptista, J.P.; Randeu, W.L.; Riedler, W. Measurements of precipitation characteristics by a new distrometer. In Proceedings of Atmospheric Physics and Dynamics in the Analysis and Prognosis of Precipitation Fields, Rome, Italy, 15–16 November 1994; pp. 15–18.
2. Testud, J.; Oury, S.; Black, R.A.; Amayenc, P.; Dou, X. The Concept of "Normalized" Distribution to Describe Raindrop Spectra: A Tool for Cloud Physics and Cloud Remote Sensing. *J. Appl. Meteorol.* **2001**, *40*, 1118–1140.
3. Krajewski, W.F.; Ciach, G.J.; Habib, E. An analysis of small-scale rainfall variability in different climatic regimes. *Hydrol. Sci. J.* **2003**, *48*, 151–162.

4. Herngren, L.F. Build-up and Wash-off Process Kinetics of PAHs and Heavy Metals on Paved Surfaces Using Simulated Rainfall. Ph.D. Thesis, Queensland University of Technology, Brisbane, Australia, 2005.
5. Hudson, N.W. Instrument for Studies of the Erosive Power of Rainfall, Erosion and Sediment Transport Measurement. In Proceedings of the Florence Symposium, 22-26 June 1981; International Association of Hydrological Sciences (IAHS): Ottawa, ON, Canada; pp. 383–393.
6. Wiesner, J. Beitrage zur Kenntnis des troischen regens (About Contributions to the knowledge of the tropical rain). *Atmos. Electr.* **1895**, *104*, 1397–1434.
7. Pearson, J.E.; Martin, G.E. An evaluation of raindrop sizing and counting techniques. *Sci. Rep.* **1957**, *1*, 1–17.
8. Blanchard, D.C. *Raindrop Size Distribution and Associated Phenomena in Hawaiian Rains*; No. R52; Woods Hole Oceanographic Institution: Woods Hole, MA, USA, 1952.
9. Lowe, E.J. Rain drops. *Q. J. R. Meteorol. Soc.* **1892**, *18*, 242–245.
10. Marshall, J.S.; Palmer, W.M.K. The distribution of raindrops with size. *J. Meteorol.* **1948**, *5*, 165–166.
11. Marshall, J.S.; Langille, R.C.; Palmer, W.M.K. Measurement of rainfall by radar. *J. Meteorol.* **1947**, *4*, 186–192.
12. Anderson, L.J. Drop-size distribution measurements in orographic rain. *Bull. Am. Met. Soc.* **1948**, *29*, 362–366.
13. Abudi, I.; Carmi, G.; Berliner, P. Rainfall simulator for field runoff studies. *J. Hydrol.* **2012**, *454*, 76–81.
14. Best, A.C. The size distribution of raindrops. *Q. J. R. Meteorol. Soc.* **1950**, *76*, 16–36.
15. Hall, M.J. Use of the stain method in determining the drop-size distributions of coarse liquid sprays. *Am. Soc. Agric. Eng. Trans.* **1970**, *13*, 33–37.
16. Shiotsuki, Y. On the flat size distribution of drops from convective rainclouds. *J. Meteorol. Soc. Jpn.* **1974**, *52*, 42–60.
17. Shiotsuki, Y. An estimation of drop-size distribution in the severe rainfall. *J. Meteorol. Soc. Jpn.* **1976**, *54*, 259–263.
18. Brandt, C.J. Transformation of the Kinetic Energy of Rainfall with Variable Tree Canopy. Ph.D. Thesis, University of London, London, UK, 1986.
19. Brandt, C.J. The transformation of rainfall energy by a tropical rain forest canopy in relation to soil erosion. *J. Biogeogr.* **1988**, *15*, 41–48.
20. Brandt, C.J. Simulation of the size distribution and erosivity of raindrops and throughfall drops. *Earth Surf. Process. Landf.* **1990**, *15*, 687–698.
21. Cerdà, A. Rainfall drop size distribution in the Western Mediterranean basin, València, Spain. *Catena* **1997**, *30*, 169–182.
22. Lu, J.Y.; Su, C.C.; Lu, T.F.; Maa, M.M. Number and volume raindrop size distributions in Taiwan. *Hydrol. Process.***2008**, *22*, 2148–2158.
23. Bowen, E.G.; Davidson, K.A. A raindrop spectrograph. *Q. J. R. Meteorol. Soc.* **1951**, *77*, 445–449.
24. Blanchard, D.C. A simple recording technique for determining raindrop size and time of occurrence of rain showers. *EOS Trans. Am. Geophys. Union* **1953**, *34*, 534–538.
25. Imai, I.; Fujiwara, M.; Ichimura, I.; Yoshihara, Z. On the radar reflectivity and the drop-size distribution of rain. *J. Meteorol. Tokyo* **1955**, *7*, 422–433.
26. Spencer, A.T.; Blanchard, D.C. *An Automatic Raindrop Sampler*; No. TR-11; Woods Hole Oceanographic Institution: Woods Hole, MA, USA, 1956.
27. Gillespie, T. The spreading of low vapor pressure liquids in paper. *J. Colloid Sci.* **1958**, *13*, 32–50.
28. Becker, A. Zur messung der Tropfengroessen bei Regenfaellen nach der Absorption methode. (About measurement of drop sizes in rainfall according to the filter paper method). *Meteorol. Z.* **1907**, *24*, 247–261.
29. Neuberger, H. Notes on measurement of rain-drop sizes. *Bull. Am. Meteorol. Soc.* **1942**, *23*, 274–276.

30. Quinn, N.W.T. Properties of Transformed Simulated Rainfall under a Corn Canopy at Different Growth Stages and Row Widths. Masters Thesis, Iowa State University, Ames, IA, USA, 1981.
31. Bentley, W.A. Studies of raindrops and raindrop phenomena. *Mon. Weather Rev.* **1904**, *32*, 450–456.
32. Kohl, R.A. Drop size distributions from medium-sized agricultural sprinklers. *Am. Soc. Agr. Eng. Trans.* **1974**, *17*, 690–693.
33. Hudson, N.W. Raindrop size distribution in high intensity storms. *Rhodesian J. Agric. Res.* **1963**, *1*, 6–11.
34. Regmi, T.P.; Thompson, A.L. Rainfall simulator for laboratory studies. *Appl. Eng. Agric.* **2000**, *16*, 641–652.
35. Ogunye, F.O.; Boussabaine, H. Development of a rainfall test rig as an aid in soil block weathering assessment. *Constr. Build. Mater.* **2002**, *16*, 173–180.
36. Júnior, S.S.; Siqueira, E.Q. Development and Calibration of a Rainfall Simulator for Urban Hydrology Research. In Proceedings of 12th International Conference on Urban Drainage, Porto Alegre, Brazil, 11–16 September 2011; pp. 226–231.
37. Lascelles, B.; Favis-Mortlock, D.T ; Parsons, A.J.; Guerra, A.J. Spatial and temporal variation in two rainfall simulators: Implications for spatially explicit rainfall simulation experiments. *Earth Surf. Process. Landf.* **2000**, *25*, 709–721.
38. Miguntanna, N.P. Nutrients build-up and wash-off processes in urban land uses. Ph.D. Thesis, Queensland University of Technology, Brisbane, Australia, 2009.
39. Pérez-Latorre, F.J.; de Castro, L.; Delgado, A. A comparison of two variable intensity rainfall simulators for runoff studies. *Soil Tillage Res.* **2010**, *107*, 11–16.
40. Arnaez, J.; Lasanta, T.; Ruiz-Flaño, P.; Ortigosa, L. Factors affecting runoff and erosion under simulated rainfall in Mediterranean vineyards. *Soil Tillage Res.* **2007**, *93*, 324–334.
41. Simone, A.; Vignali, V.; Bragalli, C.; Maglionico, M. Surface run-off: A rainfall simulator for wash-off modelling and road safety auditing under different rainfall intensities. In Proceedings of the second International SIIV Congress, Firenze, Italy, 27–29 October 2004.
42. Parsons, A.J.; Stone, P.M. Effects of intra-storm variations in rainfall intensity on interrill runoff and erosion. *Catena***2006**, *67*, 68–78.
43. Clarke, M.A.; Walsh, R.P. A portable rainfall simulator for field assessment of splash and slopewash in remote locations. *Earth Surf. Process. Landf.* **2007**, *32*, 2052–2069.
44. Egodawatta, P.K. Translation of Small-plot Scale Pollutant Build-up and Wash-off Measurements to Urban Catchment Scale. Ph.D. Thesis, Queensland University of Technology, Brisbane, Australia, 2007.
45. Asante, E.A. Effect of Mulch Type, Mulch Rate and Slope on Soil Loss, Runoff and Infiltration Under Simulated Rainfall for Two Agricultural Soils in Ghana. Ph.D. Thesis, University of Science and Technology, Kumasi, Ghana, 2011.
46. Parsakhoo, A.; Lotfalian, M.; Kavian, A.; Hoseini, S.A.; Demir, M. Calibration of a portable single nozzle rainfall simulator for soil erodibility study in hyrcanian forests. *Afr. J. Agric. Res.* **2012**, *7*, 3957–3963.
47. Laws, J.O.; Parsons, D.A. The relation of raindrop size to intensity. *Am. Geophysical Union Trans.* **1943**, *26*, 452–460.
48. Carter, C.E.; Greer, J.D.; Braud, H.J.; Floyd, J.M. Raindrop characteristics in South Central United States. *Am. Soc. Agric. Eng. Trans.* **1974**, *17*, 1033–1037.
49. Navas, A.; Alberto, F.; Machín, J.; Galán, A. Design and operation of a rainfall simulator for field studies of runoff and soil erosion. *Soil Technol.* **1990**, *3*, 385–397.
50. Fuchs, N.; Petrjanoff, I. Microscopic examination of fog, cloud and rain droplets. *Nature* **1937**, *139*, 111–112.
51. May, K.R. The cascade impactor: An instrument for sampling coarse aerosols. *J. Sci. Instrum.* **1945**, *22*, 187–195.
52. Lane, W.R. A Microburette for producing small liquid drops of known size. *J.Sci. Instrum.* **1947**, *24*, 98–101.
53. Gunn, R.; Kinzer, G.D. The terminal velocity of fall for water droplets in stagnant air. *J. Meteorol.* **1949**, *6*, 243–248.

54. Tate, R.W. Immersion sampling of spray droplets. *AIChE J.* **1961**, *7*, 574–577.
55. Nawaby, A.S. A Method of Direct Measurement of Spray Droplets in an Oil Bath. *J. Agric. Res.* **1970**, *15*, 182–184.
56. McCool, D.K.; Williams, J.D.; Morse, J.R. Raindrop characteristics in the Pacific Northwest. In Proceedings of American Society of Agricultural and Biological Engineers (ASABE) Annual International Meeting, Reno, NV, USA, 11–14 October 2009; Volume 97441, pp. 21–24.
57. Eigel, J.D.; Moore, I.D. A simplified technique for measuring raindrop size and distribution. *Amer. Soc. Agr. Eng. Trans.* **1983**, *24*, 1079–1083.
58. Cruvinel, P.E.; Minatel, E.R.; Mucheroni, M.L.; Vieira, S.R.; Crestana, S. An Automatic Method Based on Image Processing for Measurements of Drop Size Distribution from Agricultural Sprinklers. In Proceedings of Simpósio Brasileiro de Computação Gráfica e Processamento de Imagens (SIBGRAPI), Caxambú- Minas Gerais, Brazil, 29 October–1 November 1996; pp. 39–46.
59. Courshee, R.J.; Byass, J.B. *A Study of the Methods of Measuring Small Spray Drops*; National Institute of Agricultural Engineering: Bedford, UK, 1953.
60. Hacker, P.T. An Oil-stream Photomicrographic Aeroscope for Obtaining Cloud Liquid-water Content and Droplet Size Distributions in Flight. National Advisory Committee for Aeronautics: Lewis Flight Propulsion Laboratory, Cleveland, OH, USA, 1956; Technical Note 3592. pp. 1–36. Available online: http://digital.library.unt.edu/ark:/ 67531/ metadc53544/m2/1/high_res_d/19810068735.pdf (accessed on 10 January 2016).
61. Cruvinel, P.E.; Vieira, S.R.; Crestana, S.; Minatel, E.R.; Mucheroni, M.L.; Neto, A.T. Image processing in automated measurements of raindrop size and distribution. *Comput. Electron. Agric.* **1999**, *23*, 205–217.
62. Golitzine, N. *Method for Measuring the Size of Water Droplets in Clouds, Fogs, and Sprays*; Note 6; National Aeronautical Establishment: Ottowa, ON, Canada, 1951; pp. 1–52.
63. Bigg, F.J.; Abel, G.C. *Note on Sampling and Photographing Cloud Droplets in Flight*; RAE TN-ME-156; Royal Aircraft Establishment: Farnborough, UK; September; 1953; pp. 46–57.
64. Brown, E.N.; Willett, J.H. A three-slide cloud droplet sampler. *Bull. Am. Meteorol. Soc.* **1955**, *36*, 123–127.
65. Blanquies, J.; Scharff, M.; Hallock, B. The Design and Construction of a Rainfall Simulator, International Erosion Control Association (IECA). In Proceedings of 34th Annual Conference and Expo, Las Vegas, NV, USA, 24–28 February 2003; p. 10.
66. Houghton, H.G.; Radford, W.H. *On the Measurement of Drop Size and Liquid Water Content in Fogs and Clouds*; Physical Oceanography and Meteorology: Cambridge, MA, USA, 1938; Volume 6, pp. 5–31.
67. Baumgardner, D.; Jonsson, H.; Dawson, W.; O'Connor, D.; Newton, R. The cloud, aerosol and precipitation spectrometer: A new instrument for cloud investigations. *Atmos. Res.* **2001**, *59–60*, 251–264.
68. Mache, H. Ueber die Geschwindigkeit un Grosse der Regentrpfen. *Meteorol. Z.* **1904**, *39*, 278.
69. Laws, J.O. Measurements of the fall-velocity of water-drops and raindrops. *Am. Geophys. Union Trans.* **1941**, *22*, 709–721.
70. Mason, B.J.; Ramanadham, R. A photoelectric raindrop spectrometer. *Q. J. R. Meteorol. Soc.* **1953**, *79*, 490–495.
71. Mikirov, A.E. A photoelectric method of investigating the distribution of particle size precipitation. *Izv. Akad. Nauk. SSSR* **1957**, *1*, 104.
72. Salvador, R.; Bautista-Capetillo, C.; Burguete, J.; Zapata, N.; Serreta, A.; Playán, E. A photographic method for drop characterization in agricultural sprinklers. *Irrig. Sci.* **2009**, *27*, 307–317.
73. De Jong, S. Low Cost Disdrometer. Masters Thesis, Delf University of Technology, Delft, The Netherlands, 2010.
74. Sadeghi, S.H.; Abdollahi, Z.; Darvishan, A.K. Experimental comparison of some techniques for estimating natural raindrop size distribution on the south coast of the Caspian Sea, Iran. *Hydrol. Sci. J.* **2013**, *58*, 1374–1382.

75. Sudheer, K.P.; Panda, R.K. Digital image processing for determining drop sizes from irrigation spray nozzles. *Agric. W Manag.* **2000**, *45*, 159–167.
76. Kincaid, D.C.; Solomon, K.H.; Oliphant, J.C. Drop size distributions for irrigation sprinklers. *Amer. Soc. Agr. Eng. Trans.* **1996**, *39*, 839–845.
77. Mueller, E.A. Radar cross sections from drop size spectra. Ph.D. Thesis, University of Illinois at Urbana-Champaign, Champaign, IL, USA, 1966.
78. Jones, D.M. The shape of raindrops. *J. Meteorol.* **1959**, *16*, 504–510.
79. Jones, D.M.; Dean, L.A. *A Raindrop Camera*; Illinois State Water Survey: Champaign, IL, USA, 1953.
80. Jones, D.M. *Rainfall Drop Size-distribution and Radar Reflectivity*; Defense Technical Information Center, Signal Corps Engineering Laboratories: Fort Monmouth, NJ, USA, 1956.
81. Fujiwara, M. Raindrop-size distribution from individual storms. *J. Atmos. Sci.* **1965**, *22*, 585–591.
82. Vaisala. Vaisala Weather Transmitter WXT510 with the Vaisala RAINCAP® Sensor. 2012, 1. Available online: www.vaisala.com (accessed on 15 June 2015).
83. Jayawardena, A.W.; Rezaur, R.B. Measuring drop size distribution and kinetic energy of rainfall using a force transducer. *Hydrol. Process.* **2000**, *14*, 37–49.
84. Salmi, A.; Elomaa, L. Measurements of the Terminal Velocity and Shape of Falling Raindrops at Vaisala Rain Laboratory. In Proceedings of Eighth European Conference on Application of Meteorology (ECAM), San Lorenzo de El Escorial, Spain, 1–5 October 2007; pp. 55–61.
85. Kinnell, P.I.A. Some observations on the Joss-Waldvogel rainfall disdrometer. *J. Appl. Meteorol.* **1976**, *15*, 499–502.
86. Winder, P.; Paulson, K.S. The measurement of rain kinetic energy and rain intensity using an acoustic disdrometer. *Meas. Sci. Technol.* **2012**, *23*, 015801.
87. Schindelhauer, F. *Versuch einer Registrierung der Tropfenzahl bei Regenfallen (An Attempt to Record the Number of Drops in Rain)*; Scientific Report No. 1; Meteorologische Zeitschrift: Brunswick, Germany, 1925; pp. 25–27.
88. Perez, A.A. *The Analysis and Preliminary Synthesis of a Drop Size Distribution Measuring System*; Massachusetts Institute of Technology: Cambridge, MA, USA, 1949.
89. Cooper, B.F. A Balloon-borne Instrument for Telemetering Raindrop-size Distribution and Rainwater Content of Cloud. *Aust. J. Appl. Sci.* **1951**, *2*, 43–55.
90. Maulard, J. Measure du Nombre de Gouttes de Pluie (Measurement of the number of drops of rain). *J. Sci. Meteorol.***1951**, *3*, 69–73.
91. Katz, I. A momentum disdrometer for measuring raindrop size from aircraft. *Bull. Am. Meteorol. Soc.* **1952**, *33*, 365–368.
92. Smulowicz, B. Analysis of the Impactometer: An Instrument for Measuring the Distribution of Raindrop Sizes Encountered in Flight. Ph.D. Thesis, Massachusetts Institute of Technology, Cambridge, MA, USA, 1952.
93. Adderley, E.E. The growth of raindrops in cloud. *Q. J. R. Meteorol. Soc.* **1953**, *79*, 380–388.
94. Kinnell, P.I.A. The acoustic measurement of water-drop impacts. *J. Appl. Meteorol.* **1972**, *11*, 691–694.
95. Kimble, P.D. Measuring the Momentum of Throughfall Drops and Raindrops. Ph.D. Thesis, Western Kentucky University, Bowling Green, KY, USA, 1996.
96. Nystuen, J.A. Relative performance of automatic rain gauges under different rainfall conditions. *J. Atmos. Ocean. Technol.* **1999**, *16*, 1025–1043.
97. Salmi, A.; Ikonen, J. New piezoelectric Vaisala RAINCAPÒ precipitation sensor. In Proceedings of 19th Conference of Hydrology. Meteorolgical Society, San Diego, CA, USA, 9–13 January 2005; pp. 2–6.
98. Bagree, R. Characterization and Design of a Readout Circuit for a Piezoelectric-based Acoustic Disdrometer. Ph.D. Thesis, Delft University of Technology, Delft, The Netherlands, 2012.
99. Neal, J.H.; Baver, L.D. Measuring the impact of raindrops. *Agron. J.* **1937**, *29*, 708–709.

100. Webster, D.K. *An Investigation into the Use of Microprocessors for the Measurement of Rainfall Intensity*; National College of Agricultural Engineering: Silsoe, Bedfordshire, UK, 1980.
101. Bass, H.G. *Introduction to Engineering Measurements*; McGraw-Hill Education: Columbus, OH, USA, 1971.
102. Henry, Z.A.; Zoerb, G.C.; Birth, G.S. *Instrumentation and Measurement for Environmental Sciences*; SP-0375; American Society of Agricultural Engineers: St. Joseph, MI, USA, 1991; pp. 5–91.
103. Joss, J.; Waldvogel, A. Ein spektrograph für niederschlagstropfen mit automatischer auswertung. *Pure Appl. Geophys.* **1967**, *68*, 240–246.
104. Kinnell, P.I.A. Some observations on Joss-Waldvogel rainfall disdrometer-reply. *J. Appl. Meteorol.* **1977**, *16*, 113–114.
105. Joss, J.; Waldvogel, A. Some observations on Joss–Waldvogel rainfall disdrometer—Reply. *J. Appl. Meteorol.* **1977**, *16*, 112–113.
106. Distromet Ltd. Distromet. 2011. Available online: www.distromet.com (accessed on 15 June 2015).
107. Bloemink, H.I.; Lanzinger, E. Precipitation type from the Thies disdrometer. In *Instruments and Observing Methods Report 82 (TD1265),* In Proceedings of World Meterological Organization Technical Conference on Instruments and Methods of Observation (TECO-2005), Bucharest, Romania, 4–7 May 2005.
108. Clima, T. *Laser Precipitation Monitor Instruction for Use: 5.4110.X0.X00*; Software Version 1.04; Adolf Thies GmbH & Co. KG: Göttingen, Germany, 2007.
109. Upton, G.; Brawn, D. An investigation of factors affecting the accuracy of Thies disdrometers. In Proceedings of World Meterological Organization Technical Conference on Instruments and Methods of Observation (TECO-2008), St. Petersburg, Russian Federation, 27–29 November 2008; pp. 27–29.
110. Anderson, D. Preliminary evaluation of OTT and Thies Clima present weather sensors. Program operations and standards Instrument Test Report 691. Bureau of Meteorology, Australian Government: Melbourne, Australia, 2009.
111. De Moraes Frasson, R.P.; da Cunha, L.K.; Krajewski, W.F. Assessment of the Thies optical disdrometer performance. *Atmos. Res.* **2011**, *101*, 237–255.
112. Krajewski, W.F.; Kruger, A.; Caracciolo, C.; Golé, P.; Barthes, L.; Creutin, J.D.; Vinson, J.P. DEVEX-disdrometer evaluation experiment: Basic results and implications for hydrologic studies. *Adv. W. Res.* **2006**, *29*, 311–325.
113. Thurai, M.; Bringi, V.N.; Petersen, W.A. Rain microstructure retrievals using 2-D video disdrometer and C-band polarimetric radar. *Adv. Geosci.* **2009**, *20*, 13–18.
114. Friedrich, K.; Higgins, S.; Masters, F.J.; Lopez, C.R. Articulating and stationary PARSIVEL disdrometer measurements in conditions with strong winds and heavy rainfall. *J. Atmos. Ocean. Technol.* **2013**, *30*, 2063–2080.
115. Hawke, R.M. Using a spectrometer to assess simulated rainfall: Research in action. *S. Afr. J. Science* **2003**, *99*, 311.
116. Grossklaus, M.; Uhlig, K.; Hasse, L. An optical disdrometer for use in high wind speeds. *J. Atmos. Ocean. Technol.* **1998**, *15*, 1051–1059.
117. Solomon, K.H.; Zoldoske, D.E.; Oliphant, J.C. Laser Optical measurement of Sprinkler Drop Sizes. In Proceedings of Automated Agriculture for the 21st Century, St. Joseph, MI, USA, 16–17 December 1991; pp. 87–96.
118. Illingworth, A.J.; Stevens, C.J. An optical disdrometer for the measurement of raindrop size spectra in windy conditions. *J. Atmos. Ocean. Technol.* **1987**, *4*, 411–421.
119. Donnadieu, G. Comparison of results obtained with the VIDIAZ spectropluviometer and the Joss-Waldvogel rainfall disdrometer in a "rain of a thundery type". *J. Appl. Meteorol.* **1980**, *19*, 593–597.
120. Picca, B.; Trouilhet, G. Un pluviogranulometre photoelectrique (A photoelectric raindrop-size spectrometer). *J. Atmos. Res.* **1964**, *7*, 184–188.
121. Donnadieu, G.; Dubosclard, G.; Godard, S. Un pluviometre ohotoelctrique pour la determination simultanee des spectres dimensionnel at de vitesse de chute des gouttes de pluie. *J. Res. Atmos.* **1969**, *4*, 37–46.

122. Klaus, V. Study Spectropluviomètre Photoelectric Providing Real-time Integrated Parameters. Ph.D. Thesis, University of Paris, Paris, France, 1977.
123. Hauser, D.; Amayenc P.; Nutten, B.; Waldteufel, P. A new optical instrument for simultaneous measurement of raindrop diameter and fall speed distributions. *J. Atmos. Ocean. Technol.* **1984**, *1*, 256–269.
124. Salles, C.; Poesen, J. Performance of an optical spectro pluviometer in measuring basic rain erosivity characteristics. *J. Hydrol.* **1999**, *218*, 142–156.
125. Salles, C.; Poesen, J.; Borselli, L. Measurement of Simulated Drop Size Distribution with an Optical Spectro Pluviometer: Sample Size Considerations. *Earth Surf. Process. Landf.* **1999**, *24*, 545–556.
126. Lanzinger, E.; Theel, M.; Windolph, H. Rainfall amount and intensity measured by the Thies laser precipitation monitor. In *Instruments and Observing Method Report No. 94, 1–97*, Proceedings of TECO-2006-WMO Technical Conference on Meteorological and Environmental Instruments and Methods of Observation, 4–6 December 2006; World Meteorological Organization: Geneva, Switherland, 2006.
127. Schönhuber, M.; Lammer, G.; Randeu, W.L. The 2D-video-distrometer. In *Precipitation: Advances in Measurement, Estimation and Prediction*; Springer: Berlin, Germany, 2008; pp. 3–31.
128. Schönhuber, M.; Lammer, G.; Randeu, W.L. One decade of imaging precipitation measurement by 2D-video-distrometer. *Adv. Geosci.* **2007**, *10*, 85–90.
129. Nešpor, V.; Krajewski, W.F.; Kruger, A. Wind-induced error of raindrop size distribution measurement using a two-dimensional video disdrometer. *J. Atmos. Ocean. Technol.* **2000**, *17*, 1483–1492.
130. Liu, X.C.; Gao, T.C.; Liu, L. A video precipitation sensor for imaging and velocimetry of hydrometeors. *Atmos. Meas. Tech.* **2014**, *7*, 2037–2046.
131. Gucker, F.T., Jr. Determination of concentration and size of particulate matter by light scattering and sonic techniques. In Proceedings of First National Air Pollution Symposium, 10–11 November 1949; Stanford Research Institute: Los Angeles, CA, USA; pp. 14–25.
132. Knollenberg, R.G. The optical array: An alternative to scattering or extinction for airborne particle size determination. *J. Appl. Meteorol.* **1970**, *9*, 86–103.
133. Knollenberg, R.G. Three new instruments for cloud physics measurements: The 2-D spectrometer, the forward scattering Spectrometer probe and the active scattering aerosol spectrometer. In Proceedings of International Conference on Cloud Physics, Boulder, Colo., American Meteorological Society, Boston, MA, USA, 26–30 July 1976; pp. 544–561.
134. Löffler-Mang, M.; Joss, J. An optical disdrometer for measuring size and velocity of hydrometeors. *J. Atmos. Ocean. Technol.* **2000**, *17*, 130–139.
135. Battaglia, A.; Rustemeier, E.; Tokay, A.; Blahak, U.; Simmer, C. PARSIVEL snow observations: A critical assessment. *J. Atmos. Ocean. Technol.* **2010**, *27*, 333–344.
136. Jaffrain, J.; Berne, A. Experimental quantification of the sampling uncertainty associated with measurements from PARSIVEL disdrometers. *J. Hydrometeorol.* **2011**, *12*, 352–370.
137. Dingle, A.N.; Schulte, H.F., Jr. A research instrument for the study of raindrop-size spectra. *J. Appl. Meteorol.* **1962**, *1*, 48–59.
138. Mason, B.J.; Andrews, J.B. Drop-size distributions from various types of rain. *Q. J. R. Meteorol. Soc.* **1960**, *86*, 346–353.
139. Atlas, D.; Srivastava, R.C.; Sckhon, R.S. Doppler radar characteristics of precipitation at vertical incidence. *Rev. Geophys. Sp. Phys.* **1973**, *11*, 1–35.
140. Kobayaski, T. Measurement of Rain-drop Size by Means of Photographic Paper Treated with $COCl_2$. *J. Meteor. Soc. Jpn.* **1995**, *33*, 217.
141. Roy, S.S.; Datta, R.K.; Bhatia, R.C.; Sharma, A.K. Drop size distributions of tropical rain over south India. *Geofizika* **2005**, *22*, 105–131.
142. List, R.A. Linear radar reflectivity-rainrate relationship for steady tropical rain. *J. Atmos. Sci.* **1988**, *45*, 3564–3572.
143. Tokay, A.; Kruger, A.; Krajewski, W.F. Comparison of drop size distribution measurements by impact and optical disdrometers. *J. Appl. Meteorol.* **2001**, *40*, 2083–2097.

144. Nystuen, J.A.; Proni, J.R.; Lauter, C.A.; Bufkin, J.; Rivero, U.; Boland, M.; Wilkerson, J.C. *APL Disdrometer Evaluation*; National Oceanic and Atmospheric Administration/ Environmental Research Laboratories Technical Memorandum ERL AMOL-83 (PB95-181681): Miami, FL, USA, 1994.
145. Harikumar, R.; Sampath, S.; Kumar, V.S. An empirical model for the variation of rain drop size distribution with rain rate at a few locations in southern India. *Adv. Sp. Res.* **2009**, *43*, 837–844.
146. Sheppard, B.E.; Joe, P.I. Comparison of raindrop size distribution measurements by a Joss-Waldvogel disdrometer, a PMS 2DG spectrometer, and a POSS Doppler radar. *J. Atmos. Ocean. Technol.* **1994**, *11*, 874–887.
147. Tokay, A.; Bashor, P.G.; Habib, E.; Kasparis, T. Raindrop size distribution measurements in tropical cyclones. *Mon. Weather Rev.* **2008**, *136*, 1669–1685.
148. Das, S.; Maitra, A.; Shukla, A.K. Rain attenuation modeling in the 10-100 GHz frequency using drop size distributions for different climatic zones in tropical India. *Prog. Electromagn. Res. B* **2010**, *25*, 211–224.
149. De Wulf, F.D.; Gabriels, D.; Boodt, M.D. A device for analyzing the energy load of rainstorms. *Assess. Eros.* **1980**, 165–167.
150. Campos, E.F.; Zawadzki, I.; Petitdidier, M.; Fernandez, W. Measurement of raindrop size distributions in tropical rain at Costa Rica. *J. Hydrol.* **2006**, *328*, 98–109.
151. Konwar, M.; Sarma, D.K.; Das, J.; Sharma, S. Shape of the rain drop size distributions and classification of rain type at Gadanki. *Indian J. Radio Sp. Phys.* **2006**, *35*, 360–367.
152. Kruger, A.; Krajewski, W.F. Two-dimensional video disdrometer: A description. *J. Atmos. Ocean. Technol.* **2002**, *19*, 602–617.
153. Montero-Martínez, G.; Kostinski, A.B.; Shaw, R.A.; García-García, F. Do all raindrops fall at terminal speed? *Geophysical Res. Lett.* **2009**, *36*, 1–4.
154. do Khac, K.; Zanghi, F.; Tabary, P. Radar-disdrometer comparison. In Proceedings of Third European Conference on Radar Meteorology and Hydrology (ERAD), Visby, Sweden, 6–10 September 2004; pp. 272–277.
155. Beard, K.V.; Bringi, V.N.; Thurai, M. A new understanding of raindrop shape. *Atmos. Res.* **2010**, *97*, 396–415.
156. Erpul, G.; Gabriels, D.; Janssens, D. Assessing the drop size distribution of simulated rainfall in a wind tunnel. *Soil Tillage Res.* **1998**, *45*, 455–463.
157. Salles, C.; Poesen, J.; Sempere-Torres, D. Kinetic energy of rain and its functional relationship with intensity. *J. Hydrol.* **2002**, *257*, 256–270.
158. Habib, E.; Krajewski, W.F. An example of computational approach used for aerodynamic design of a rain disdrometer. *J. Hydraul. Res.* **2001**, *39*, 425–428.

CHAPTER 15

Low-Cost Stormwater Filtration System to Improve Urban Water Quality: The Case of Addis Ababa, Ethiopia

Dagnachew Adugna[1,2*], Larissa Larsen[2]

[1]*Ethiopian institute of Architecture, Building Construction and City Development, Addis Ababa University, Addis Ababa, Ethiopia*

[2]*Taubman College of Architecture and Urban Planning, University of Michigan, Ann Arbor, Michigan, USA*

ABSTRACT

Urbanization in developing countries often negatively impacts water resources by polluting surface waters. Addis Ababa, Ethiopia, is currently experiencing rapid urbanization accompanied by significant water shortages, unmanaged stormwater, and increasing river water pollution. To supplement the need for non-potable water and address stormwater runoff pollution, we constructed a low cost stormwater filtration system. The filtration system is comprised of a sedimentation area followed by three gravel grain sizes arranged horizontally from coarse to medium to fine filter media. We compared the quality of pretreatment water with post-treatment water by measuring physicochemical parameters, heavy metals and nutrients. We found that the filtration system reduced turbidity by 87%, TSS by 80%, Cu by 87% and Zn by 90%. Further, it positively increased the concentration of DO by 42%. However, the filtration system did not remove nitrates and nitrites. Implementing this system at outfalls in the rapidly expanding condominium housing areas can increase residents' supply of non-potable water and reduce the amount of polluted stormwater entering nearby streams and rivers.

Keywords: Low Cost, Stormwater, Filtration, Addis Ababa, Water Quality, Performance Test

1. INTRODUCTION

Urbanization in developing countries often negatively impacts water resources by polluting surface waters [1] [2] and decreasing infiltration to recharge aquifers [3] [4]. Simultaneously, water demand increases with population growth and per capita consumption rises with increasing standards of living. Addis Ababa, which is the capital of Ethiopia and the site of many international organizations, is currently experiencing rapid urbanization [5] [6]. Since 2007 when the population was 2.7 million [7] , it has expanded to an estimated 4 million. For new residents, access to water as well as housing is two significant problems.

In an effort to address Addis Ababa's housing shortage, the city administration is constructing an extensive number of multi-story, multi-family condominium projects. According to the Addis Ababa Housing Construction and Development Bureau (2014), these multistory condominiums, are located on more than 134 sites throughout the city and are projected to house 1.3 million people upon completion [6]. One of the problems linked with these new developments is a water shortage due to the imbalance between the available water supply sources (such as limited water supply sources and insufficient water infrastructure) and the growing population (the annual rate of population growth is 2.8%) [8].

The natural distribution of rainfall is another challenge to the provision of water for Addis Ababa residents. While the city receives an average annual rainfall of 1057 mm [9] , 75% falls during a four-month rainy season. Due to the erratic nature of the rainfall, improving water supplies would require huge reservoirs to harvest rain water for later use. These are unaffordable and practically impossible. Little attention is given to stormwater management as a resource and most stormwater is directly discharged into rivers reducing river water quality [10]. The rivers are highly polluted year round and many city residents consider the rivers as sewer lines [11]. Thus, providing a low cost stormwater filtration system could provide water for non-potable uses at the new condominium developments and reduce the amount of polluted stormwater reaching the rivers. Moreover, the filtration system addressed residents' stated concerns that standing water impeded children's play in the common areas.

Sustainable stormwater filtration systems are generally categorized either as coarse or fine filtration systems. Coarse media stormwater filtration is efficient for larger pollutant removal, but it is not effective in removing dissolved pollutants [12]. In addition, coarse media filtration is more likely to clog due to the migration of fine sediments and compaction of the media [13]. Conversely, fine filter media may provide for better removal of pollutants through mechanisms that include fine filtration, sorption, ionic adhesion, and precipitation [14] [15]. Recently sustainable stormwater management systems are increasingly used in both the United States and Australia [16] , although there is currently limited data evaluating their water quality impacts [17] [18].

In addition, sand filtration systems are also categorized as either slow or rapid sand filters. In general, the sand filtration systems work vertically at various depths and they are generally found to be more effective at greater depths. Slow sand filters have a grain size of 0.15 to 0.35 mm and are used as a secondary treatment to reduce pathogen concentrations. Slow sand filtration is used to produce drinking water—the size of the filtration system is determined based on the required flow rate [19]. Slow sand filter is a biological process which reduces bacteria from 90% - 99% without chemicals [19]. However, slow sand filtration requires low turbidity levels to work efficiently and filters water at a slow, constant rate so that the biological processes can occur [19]. In Gaza, a vertical slow sand filtration system was tested for fecal coliform and suspended solid to investigate the removal efficiency of the system at different filter depth. Using a 203 mm (or 8 inch) cylindrical PVC-U pipe, the slow sand filter removed fecal coliform at a depth of 150 cm [20].

Rapid sand filters use sand that varies from 0.4 to 12.0 mm in grain size and they are used for primary treatment [21]. Rapid sand filtration alone is not as effective for producing drinking water as slow sand filtration due to the fact that the biological systems are relatively less effective [19]. According to Zaman et al. [19] , rapid sand filtration is suitable for large urban areas where land is scarce. Conversely slow filtration is suitable to areas where land scarcity is not a problem as it requires a larger surface area to filter the same amount of water.

Gravel is the other commonly used filter media to treat polluted water by the formation of biofilm. The biofilm is a gelatinous layer consisting of bacteria, algae, protozoa and other aquatic insect larvae that develops after 10 to 20 days of operation and degrades some of the organic components. Further, the filter materials remove pollutants by adsorption, sedimentation, physical and mechanical straining [22]. Moreover, in a vermifiltration system, gravel ranging in grain size from 10 to 40 mm in combination with sawdust was used to filter grey wastewater. The average removal efficiency of the system for TSS was found to be 98% [23].

To correct some of the identified drawbacks of the coarse media filtration and the sand filtration systems, the present study combined three gravel grain sizes (coarse, medium and fine gravels). To maintain separation of the three gravel areas, we added layers of wire mesh. Importantly to reduce the possibility of clogging, a sedimentation pond was created nearest the inlet to trap sediments before it entered the filtration system.

This study focuses on a horizontal filtration system where water passes through the three grain sizes-from the coarse to the fine gravel. We selected this horizontal arrangement, as opposed to a vertical system, so the water could flow smoothly at shallow depths without mechanical pumping. We call this a low cost stormwater filtration system because it relies on locally available materials, is relatively low-cost, is simple to construct and requires no pumping. This system is appropriate for the context of Addis Ababa because it would help reduce water shortages in the condominium housing areas and reduce the contribution of polluted stormwater to the rivers. Municipalities or community groups could implement this low-cost stormwater filtration system at outfalls along with installing new drains and retrofitting the existing outfalls.

The specific objectives of this study are to:

- Characterize pre treatment stormwater quality at the condominium site,
- Develop low-cost stormwater filtration system and estimate the filtration capacity of the system,
- Test the performance of the low-cost stormwater filtration system in reducing pollutants.

2. MATERIALS AND METHODOLOGY

2.1. Study Area

The study area is located at the "Biruh tesfa" compound in Jemo No. 1 condominium in a southwestern part of Addis Ababa. There are 1264 people who live in "Biruh tesfa". The "Biruh tesfa" community gave permission for the construction and testing of this filtration system (Figure 1). In addition, there was a commitment from the community representatives to connect the filtration system with storage tanks to use the filtered water for non-drinking purposes (e.g. car washing, urban agriculture within the "Biruh tesfa" compound).

2.2. Materials and Methods

Stormwater at the "Biruh tesfa" compound was collected at a manhole where stormwater from a rooftop, local streets, and parking lots combined. The manhole is connected to our filtration system by underground PVC pipe. Based on the volume of the stormwater from the rooftop, local streets, and parking lots, we selected a 110 mm PVC pipe stormwater (Figure 2).

On site, a 2.22 meter by 1.22 meter rectangular brick masonry box was constructed in the ground at a depth of 0.60 meter (Figure 3). The box for the filtration system was made based on the volume of stormwater which can be generated from the rooftop, local street and parking lots. The filtration bed was made from concrete on top of compacted stones. The walls were plastered with mortar to prevent leakage and/or entrance of water from outside via the brick joints. To prevent the entrance of rainwater from the top, after the three filters media were in place, geo-synthetic material was laid on the top. The top of the filtration system was then covered with soil except for the area over the sedimentation pond that was covered by a reinforced concrete slab to permit maintenance. Within the rectangular masonry box, there are four different compartments. The sedimentation area at the inlet of the system traps is created by a wood baffle. This area permits sediments and some soluble pollutants to settle before water then enter into the filtration system.

After the sedimentation area, the next three compartments are filled with coarse (30 - 40 mm), medium (20 - 30 mm) and fine (10 - 20 mm) gravel respectively. Small holes were uniformly drilled at a depth of 25 centimeters from the top on the common wall between the sedimentation area and the filtration system (Figure 4). The length of the filter with the finest gravel was 50% of the total length of the filtration system and the medium was about 33%. The remaining 17% of the total length of the filtration system was the coarse gravel. The area with the finest gravel is largest due to its smaller hydraulic conductivity and relative importance in filtering the stormwater.

2.3. Stormwater Samples Collection for Performance Test

To evaluate the filtration system's performance, stormwater samples were collected before and after the filtration system. Pre treatment samples were taken at the manhole. Post treatment samples were taken after the stormwater passed through the filtration system. For reliability, we collected three sets of samples within a rainy season (between July and August), more than 10 days after construction (to let the biofilms develop).

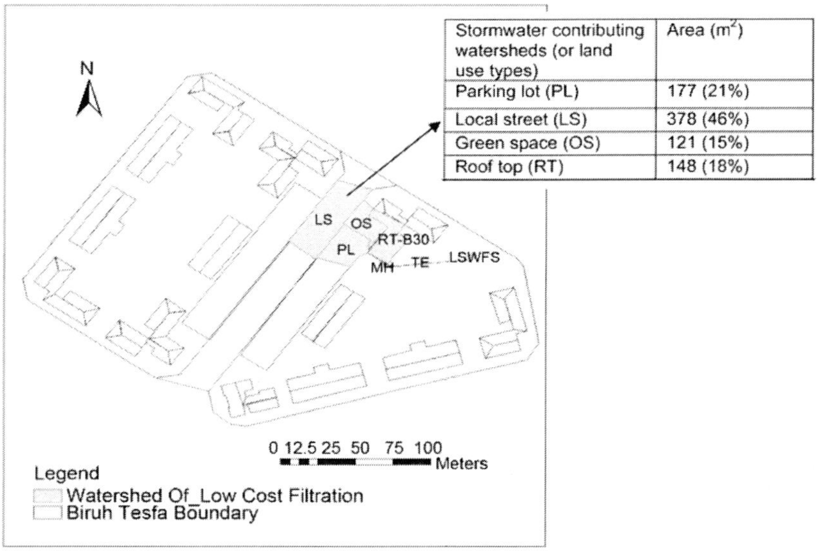

Figure 1. The condominium housing area, filtration system, and distribution of area land uses.

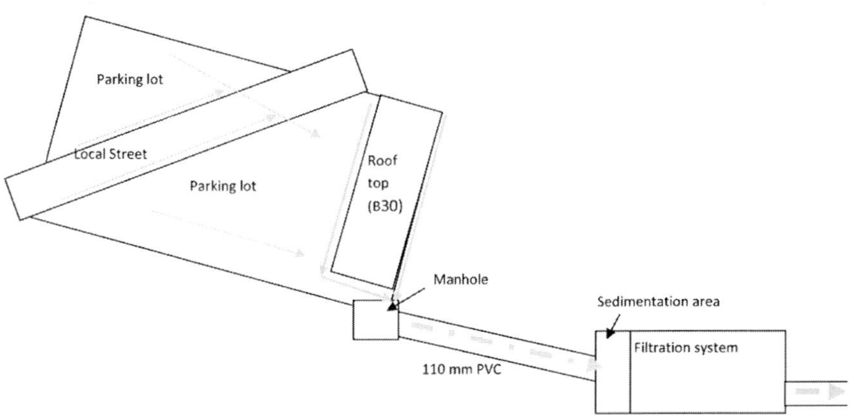

Figure 2. Schematic presentation of the filtration system with the contributing sub-watersheds that drain into the system (B30 stands for Block No. 30 apartment).

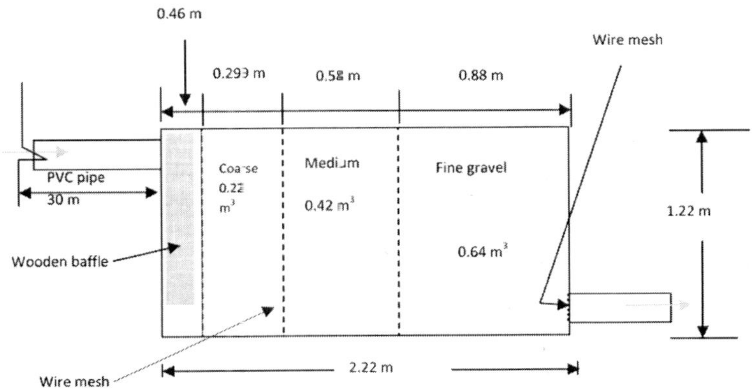

Figure 3. The schematic presentation of the elements of the low-cost stormwater filtration system with their dimensions and volumetric capacity.

The pre and post-treatment water samples were tested for levels of physicochemical parameters (pH, EC, Turbidity and DO), nutrients (NO_2-N, NO_3-N, NH_3-N and PO_4-P) and heavy metals (Cu, Cr, Pb and Zn).

2.4. Stormwater Samples Transportation

The stormwater samples were collected with a plastic container (20 liters in volume) pre and post treatment in the low-cost stormwater filtration system immediately after rain fall and measured in-situ for pH, DO (mg/l), EC (μs/cm) and Turbidity (NTU). The other stormwater samples for NO_2-N, NO_3-N, NH_3-N, PO_4-P, Cu, Pb, Zn, and Cr analysis were collected at same time at the same sampling point with plastic bottles rinsed with distilled water (1000 ml in volume) and then transported within two hours for laboratory analysis.

2.5. Stormwater Samples Measurement

The physicochemical parameters (DO, pH, EC and Turbidity) were measured using standard methods (APHA, 1998). The DO probe (LDO 10105, Model USPAT 6912050) and the Conductivity probe (CDC 40105) were mounted on the multi-meter (Hach-Model-HQ40d multi-parameter digital meter) which measures DO and pH, and EC respectively. The turbidity was measured using the portable turbidity meter (OAKTON T-100 Model).

The concentration of NO_2-N, NO_3-N, NH_3-N and PO_4-P were analyzed at the lab using spectrophotometer (JENWAY 6405 UV/Vis) following standard methods (APHA, 1995).

Figure 4. The low-cost stormwater filtration system from excavation to filling the system with filtration media (top) to making the system ready for performance test (bottom).

The heavy metals including Cr, Cu, Mn, and Zn were analyzed using Bicin Chonimate, Diphenyl carbohydride, Periodate oxidation and Zincon (HACH, Atomic Absorption Spectrometer) methods respectively.

2.6. Softwares Used to Delineate the Land Uses Draining in to the Filtration System

ArcGIS was used to delineate the contributing sub-watersheds (or land uses) that drain stormwater into the filtration system and to calculate the area of each contributing sub-watershed following ESRI's standard method. Furthermore, Microsoft excel was employed to analyze the in-situ and laboratory data of stormwater including the means and percentage removal efficiency of the filtration system.

3. RESULTS

3.1. The Low-Cost Stormwater Filtration System and Its Filtering Capacity

Based on Figure 2, the volume of the filtration system and volume of stormwater diverted from upstream contributing land uses in to the filtration system was computed as follows:

Volume of stormwater diverted from the rooftop (Rt), Local street (Ls), Parking lots (Pl) and green spaces (Gs) in to the filtration system:

= stormwater from (rooftops + local streets + parking lots + green spaces);

= Area of contributing land use × Runoff coefficient × Daily rainfall;

= (A of Rt × C × R) + (A of Ls × R) + (A of Pl × C × R) + (A of Gs × C × R);

= (148 m^2 × 0.85 × 31.4 mm/day) + (378 m^2 × 0.5 × 31.4 mm/day) + (177 m^2 × 0.4 × 31.4 mm/day) + (121 m^2 × 0.2 × 31.4 mm/day);

= 12,870 liters per day or 536 liters per hour or 8.9 liters per minute.

According to the volume of stormwater which can be generated from the contributing land uses, the volume and capacity of the filtration system was estimated.

Volume of filtration system = Volume of the (course + medium + fine) filtering media:

= (0.229 m × 1.22 m × 0.6 m) + (0.58 m × 1.22 m × 0.60 m) + (0.86 m × 1.22 m × 0.60 m);

= 1.28 m^3.

The PVC pipe transports 1.14 m^3 ($\pi \times r^2 \times 30$ m) of stormwater once per its entire length (30 m) showing that with a free board of nearly 10% the filtration system (1.28 m^3 capacity) has the ability to hold all the stormwater that passes through the PVC pipe. Further, stormwater from the common points of the contributing watersheds (i.e. at the manhole) to the filtration system were reached in 70 seconds. Thus, 10.42 liters {(12.8 m^3 × 70 sec) ÷ (24 hr × 3600 sec/hr)} of stormwater could be filtered (or treated) per 70 seconds revealing that 0.15 liters (10.42 liters ÷ 70 sec) of stormwater could be filtered per second by the filtration system.

In the future, we hope that water storage tanks will be connected to the outlet of the filtration system and local residents can use the filtered water for activities such as car washing and urban agriculture.

3.2. Pre Treatment Stormwater Quality

The concentration of stormwater pollutants and physicochemical parameters found in-situ and with laboratory measurements is reported in Table 1.

3.3. Performance of the Low-Cost Stormwater Filtration System

The performance of the filtration system was tested for physicochemical parameters, heavy metals, total suspended solids and nutrients (Tables 2-4). The results of the present study showed that, with the exception of the nutrients, quality of the stormwater had notably improved.

Physical observation (as presented in Figure 5) showed that the filtration system in clarity after filtration.

Table 1. The pre-treatment concentration of stormwater pollutants at "Biruh tesfa" compound.

Stormwater pollutants and physicochemical parameters		Replications			Mean value
		1	2	3	
Physicochemical parameters	DO (mg/l)	5.15	2.56	5.26	4.32
	Turbidity (NTU)	777	675.3	454	635.43
	EC (μs/cm)	44.9	41.7	63.4	50.00
	pH	8.58	7.96	8.13	8.22
Heavy metals	Cu (mg/l)	0.12	0.08	0.03	0.08
	Pb (mg/l)	BDL	BDL	BDL	-
	Zn (mg/l)	2.30	3.72	5.06	3.69
	Cr (mg/l)	BDL	BDL	BDL	-
Nutrients	NO_2-N (mg/l)	0.538	4.15	0.167	1.62
	NO_3-N (mg/l)	0.179	0.094	0.393	0.074
	NH_3-N (mg/l)	0.422	0.523	0.074	0.114
	PO_4-P (mg/l)	0.374	0.196	0.0178	0.065

Table 2. Physicochemical water quality parameters.

	Replication No.						Mean		% mean removal
	1		2		3				
	Before	After	Before	After	Before	After	Before	After	
Turbidity (NTU)	777	123.33	675.3	454	454	79.10	635.43	84.04	−87%
EC (μs/cm)	44.9	37.87	41.7	35.60	63.4	54.90	50.00	42.79	−14%
pH	8.58	7.88	7.96	7.88	8.13	8.00	8.22	7.92	4%
DO (mg/l)	5.15	6.2	2.56	4.7	5.26	7.5	4.32	6.15	+42%
Total suspended solids (mg/l)	2777.10	92.80	1040	415	1535	580	1784.033	362.60	−80%

Table 3. Heavy metals.

	Replication No.						Mean		% mean removal
	1		2		3				
	Before	After	Before	After	Before	After	before	After	
Cu (mg/l)	0.12	0.02	0.08	0.00	0.03	0.01	0.0767	0.01	−87%
Zn (mg/l)	2.30	0.20	3.72	0.21	5.06	0.73	3.69	0.38	−90%
Pb (mg/l)	Below detection limit (BDL)	BDL	BDL	BDL	BDL	BDL	BDL	BDL	-
Cr (mg/l)	BDL	BDL	BDL	BDL	BDL	BDL	BDL	BDL	-

4. DISCUSSION

The present study was limited to stormwater filtration as the entire condominium site where all the sanitation systems (e.g. toilets, showers, kitchens) are connected into a separate sewer system.

Table 4. Nutrients.

	Replication No.						Mean		% mean removal
	1		2		3				
	Before	After	Before	After	Before	After	before	After	
NH_3^+ -N (mg/l)	0.422	0.069	0.523	0.18	0.074	0.378	0.340	0.209	−38%
PO_4-P (mg/l)	0.374	0.168	0.196	0.223	0.0178	0.045	0.196	0.145	−26%
NO_2-N (mg/l)	0.538	3.086	4.154	0.374	1.675	19.22	2.122	7.560	+356
NO_3-N (mg/l)	0.179	3.299	0.094	0.607	0.383	0.393	0.219	1.433	+654

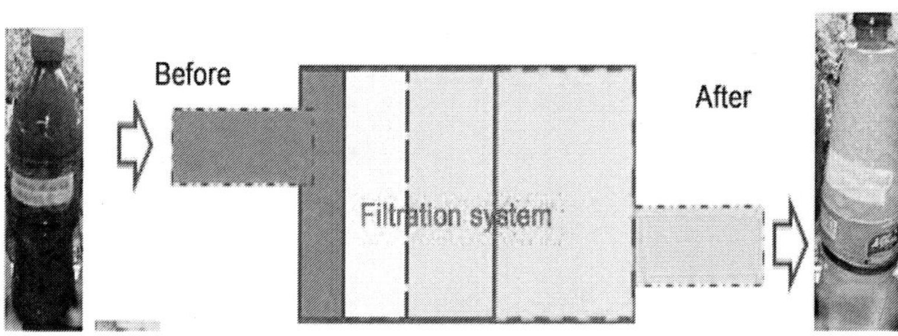

Figure 5. A physical comparison of the performance of the filtration system before and after filtration shows the improvement in the water quality.

4.1. Characterization of Stormwater Pollutants and Physicochemical Parameters

In Addis Ababa, Ethiopia, the absence of stormwater water quality data has hindered research. Therefore, the pretreatment stormwater pollutant data may serve as a baseline for further investigations. The data showed that stormwater pollutants common in urban areas [24] [25] [26] were also found at the condominium housing site.

4.2. Building a Low-Cost Stormwater Filtration System

As in many urban areas, stormwater management is a problem in Addis Ababa, Ethiopia [27] [28]. This stormwater filtration system was built to investigate a low-cost, on-site approach that provide residents with additional water and reduce stormwater pollutants from entering the river. In the short term (before city-wide structures are put in place), low-cost stormwater filtration system such as this one can be installed at outfalls in existing or newly constructed drains. This system costs approximately $1400 USD.

4.3. Performance of the Low-Cost Stormwater Filtration System

Based on a comparison of pre and post-treatment water samples, the filtration system removed turbidity by 87%, decreased conductivity by 14% and increased the desirable dissolved oxygen by 42%. The filtration system also removed the detected heavy metals of Cu and Zn by more than 85%. This is promising when compared with the heavy metal reduction achieved by other stormwater treatment technologies such as wetlands, infiltration trenches, retention ponds [29] [30]. The performance test provided evidence for the effectiveness of the system at outfalls (refer Figure 6) in areas of the city where no pathogens are not expected due to the presence of sewer systems (e.g. condominiums, industries, universities). This implies that if Addis Ababa is going to install the present technology at outfalls the rivers water quality will be significantly improved.

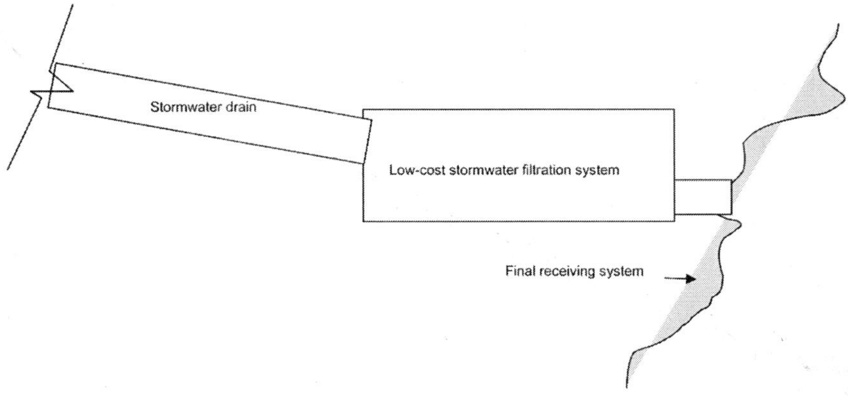

Figure 6. Installing the low-cost stormwater filtration system as outfall.

Such promising performances may be due to adsorption [31] [32] and the formation of biofilms (a thin layer of cells of microorganisms such as bacteria, and fungus) on the surfaces of the gravel (Headley & Tanner, 2012 and Wong et al., 1999). In addition, the sedimentation area combined with adsorption and biofilms might have increased the performance of the filtration system by settling the dissolved pollutants before they get in to the filtration system. The findings of Jellison et al. (2000) showed that biofilm can be developed 10 - 20 days after construction-which is consistent with the present study as samples were taken from the filtration system after 10 days.

The present study cannot be compared with sand filtration systems conducted by various researchers (e.g. Zaman et al., 2014; Nassar & Hajjaj, 2013 and Rowe & Abdel-Majid, 1995) as the grain size of sand filtration system varied significantly. In addition, our system involved three gravel grain sizes in a horizontal arrangement. Conversely, studies conducted by Rowe & Abdel-Majid (1995), Nassar & Hajjaj (2013) and Zamana et al. (2014) used different the filter media (i.e. the sands) and were vertically arranged.

5. CONCLUSION

In Addis Ababa, Ethiopia, stormwater runoff carries pollutants that impair river water quality. Thus, to supplement residents' need for non-potable water supplies and to improve river water quality, we constructed a low cost stormwater filtration system and tested its effectiveness in July through August 2016 (during the Ethiopian rainy season) by comparing water samples based on levels of physicochemical parameters, heavy metals and nutrients. The stormwater was collected from the rooftops, a local street, and parking lots. Our system used three gravel grain sizes (coarse, medium and fine) arranged horizontally from coarse to medium to fine. Based on water sample testing, our filtration system removed turbidity, total suspended solids, and heavy metals by more than 80%. Further, it increased the concentration of dissolved oxygen by 42%. However, our filtration system was not effective at lowering nitrates and nitrites levels. It is our hope that the filtered stormwater will eventually be connected to an underground storage tank at the "Biruh tesfa" condominium housing site. We will continue to monitor the water quality and we hope this low-cost filtration system will serve as a community asset that produces water quality improvements and assists residents in their daily tasks.

REFERENCES

1. Ouyang, T., Zhu, Z. and Kuang, Y. (2006) Assessing Impact of Urbanization on River Water Quality in the Pearl River Delta Economic Zone, China. Environmental Monitoring and Assessment, 120, 313-325. https://doi.org/10.1007/s10661-005-9064-x
2. Alemayehu, T. (2001) The Impact of Uncontrolled Waste Disposal on Surface Water Quality in Addis Ababa, Ethiopia. Ethiopian Journal of Science, 24, 93-104.
3. Pitt, R., Chen, S.E., Clark, S.E., Swenson, J. and Ong, C.K. (2008) Compaction's Impacts on Urban Storm-Water Infiltration. Journal of Irrigation and Drainage Engineering, 134, 652-658. https://doi.org/10.1061/(ASCE)0733-9437(2008)134:5(652)
4. Holman-Dodds, J.K., Bradley, A.A. and Potter, K.W. (2007) Evaluation of Hydrologic Benefits of Infiltration Based Urban Storm Water Management. Journal of the American Water Resources Association, 39, 205-215. https://doi.org/10.1111/j.1752-1688.2003.tb01572.x
5. Mazhindu, E., Gumbo, T. and Gondo, T. (2010) Living with Environmental Health Risks—The Case of Addis Ababa. Ecohydrology & Hydrobiology, 10, 281-286. https://doi.org/10.2478/v10104-011-0026-3
6. Haregewoin, Y.M. (2007) Integrated Housing Development Programs for Urban Poverty Alleviation and Sustainable Urbanization: The Case of Addis Ababa. ENHR International Conference on Sustainable Urban Areas, Institute for Housing and Urban Development Studies (IHS), Rotterdam.
7. CSA (2008) Summary and Statistical Report of the 2007 Population and Housing Census. Addis Ababa, Ethiopia.
8. AAWSA (2011) Business Plan from 2011-2020, Final Report. Addis Ababa, Ethiopia.
9. National Metrology Agency (2016) Rainfall data of Addis Ababa from 2000 to 2015. Addis Ababa, Ethiopia.
10. Belete, D.A. (2011) Road and Urban Storm Water Drainage Network Integration in Addis Ababa: Addis Ketema Sub-City. Journal of Engineering and Technology Research, 3, 217-225.
11. Prabu, P.C. (2009) Impact of Heavy Metal Contamination of Akaki River of Ethiopia on Soil and Metal Toxicity on Cultivated Vegetable Crops. Electronic Journal of Environmental, Agricultural & Food Chemistry, 8, 818-827.
12. Birch, G.F., Fazeli, M.S. and Matthai, C. (2005) Efficiency of an Infiltration Basin in Removing Contaminants from Urban Stormwater. Environmental Monitoring and Assessment, 101, 23-38.
13. Siriwardene, N.R., Deletic, A. and Fletcher, T.D. (2007) Clogging of Stormwater Gravel Infiltration Systems and Filters: Insights from a Laboratory Study. Water Research, 41, 1433-1440. https://doi.org/10.1016/j.watres.2006.12.040
14. Genc-Fuhrman, H., Mikkelsen, P.S. and Ledin, A. (2007) Simultaneous Removal of As, Cd, Cr, Cu, Ni and Zn from Stormwater: Experimental Comparison of 11 Different Sorbents. Water Research, 41, 591-602. https://doi.org/10.1016/j.watres.2006.10.024
15. Davis, A.P. (2005) Green Engineering Principles Promote Low-Impact Development. Environmental Science & Technology, 39, 338A-344A. https://doi.org/10.1021/es053327e
16. Wong, T.H. (2006) Australian Runoff Quality: A Guide to Water Sensitive Urban Design.
17. Henderson, C., Greenway, M. and Phillips, I. (2007) Removal of Dissolved Nitrogen, Phosphorus and Carbon from Stormwater by Biofiltration Mesocosms. Water Science and Technology, 55, 183-191. https://doi.org/10.2166/wst.2007.108
18. Davis, A.P., Shokouhian, M., Sharma, H. and Minami, C. (2001) Laboratory Study of Biological Retention for Urban Stormwater Management. Water Environment Research, 73, 5-14. https://doi.org/10.2175/106143001X138624
19. Zaman, S., Yeasmin, S., Inatsu, Y., Ananchaipattana, C. and Bari, M.L. (2014) Low-Cost Sustainable Technologies for the Production of Clean Drinking Water—A Review. Journal of Environmental Protection, 5, 42. https://doi.org/10.4236/jep.2014.51006

20. Nassar, A.M. and Hajjaj, K. (2013) Purification of Stormwater Using Sand Filter. Journal of Water Resource and Protection, 5, 1007. https://doi.org/10.4236/jwarp.2013.511105
21. Rowe, D.R. and Abdel-Magid, I.M. (1995) Handbook of Wastewater Reclamation and Reuse. CRC Press, Boca Raton, Florida.
22. Collins, M.R., Eighmy, T.T., Fenstermacher Jr., J.M. and Spanos, S.K. (1992) Removing Natural Organic Matter by Conventional Slow Sand Filtration. Journal (American Water Works Association), 80-90.
23. Adugna, A.T., Andrianisa, H.A., Konate, Y., Ndiaye, A. and Maiga, A.H. (2015) Performance Comparison of Sand and Fine Sawdust Vermifilters in Treating Concentrated Grey Water for Urban Poor. Environmental Technology, 36, 2763-2769. https://doi.org/10.1080/09593330.2015.1046951
24. Barbosa, A.E., Fernandes, J.N. and David, L.M. (2012) Key Issues for Sustainable Urban Stormwater Management. Water Research, 46, 6787-6798. https://doi.org/ 10.1016/j.watres.2012.05.029
25. Flint, K.R. and Davis, A.P. (2007) Pollutant Mass Flushing Characterization of Highway Stormwater Runoff from an Ultra-Urban Area. Journal of Environmental Engineering, 133, 616-626. https://doi.org/10.1061/(ASCE)0733-9372(2007)133:6(616)
26. Gobel, P., Dierkes, C. and Coldewey, W.G. (2007) Storm Water Runoff Concentration Matrix for Urban Areas. Journal of Contaminant Hydrology, 91, 26-42. https://doi.org/10.1016/j.jconhyd.2006.08.008
27. Kim, G., Yur, J. and Kim, J. (2007) Diffuse Pollution Loading from Urban Stormwater Runoff in Daejeon City, Korea. Journal of Environmental Management, 85, 9-16. https://doi.org/10.1016/j.jenvman.2006.07.009
28. Kao, C.M., Wang, J.Y., Lee, H.Y. and Wen, C.K. (2001) Application of a Constructed Wetland for Non-Point Source Pollution Control. Water Science and Technology, 44, 585-590.
29. Scholes, L., Revitt, D.M. and Ellis, J.B. (2008) A Systematic Approach for the Comparative Assessment of Stormwater Pollutant Removal Potentials. Journal of Environmental Management, 88, 467-478. https://doi.org/10.1016/j.jenvman.2007.03.003
30. Birch, G.F., Matthai, C., Fazeli, M.S. and Suh, J.Y. (2004) Efficiency of a Constructed Wetland in removing Contaminants from Stormwater. Wetlands, 24, 459-466. https://doi.org/10.1672/0277-5212(2004)024[0459:EOACWI]2.0.CO;2
31. Aziz, H.A., Adlan, M.N. and Ariffin, K.S. (2008) Heavy Metals Removal from Water in Malaysia. Bioresource Technology, 99, 1578-1583. https://doi.org/10.1016/ j.biortech.2007.04.007
32. Vymazal, J. (2002) The Use of Sub-Surface Constructed Wetlands for Wastewater Treatment in the Czech Republic: 10 Years Experience. Ecological Engineering, 18, 633-646. https://doi.org/10.1016/S0925-8574(02)00025-3

INDEX

M

O

P

R

S

U

W